Neuroculture

Neuroculture
On the implications of brain science

Edmund T. Rolls

Oxford Centre for Computational Neuroscience
Oxford
England

OXFORD
UNIVERSITY PRESS

OXFORD

UNIVERSITY PRESS

Great Clarendon Street, Oxford OX2 6DP

Oxford University Press is a department of the University of Oxford.
It furthers the University's objective of excellence in research, scholarship,
and education by publishing worldwide in

Oxford New York

Athens Auckland Bangkok Bogotá Buenos Aires Cape-Town
Chennai Dar-es-Salaam Delhi Florence Hong-Kong Istanbul Karachi
Kolkata Kuala-Lumpur Madrid Melbourne Mexico-City Mumbai Nairobi
Paris São-Paulo Shanghai Singapore Taipei Tokyo Toronto Warsaw
with associated companies in Berlin Ibadan

Oxford is a registered trade mark of Oxford University Press
in the UK and in certain other countries

Published in the United States
by Oxford University Press Inc., New York

British Library Cataloguing in Publication Data
Data available

Library of Congress Cataloging in Publication Data
Data available

Prelims typeset in by Glyph International, Bangalore, India
Printed in Great Britain
on acid-free paper by
CPI Antony Rowe, Chippenham, Wiltshire

ISBN 978–0–19–969547–8

10 9 8 7 6 5 4 3 2 1

Whilst every effort has been made to ensure that the contents of this book are as complete,
accurate and up-to-date as possible at the date of writing, Oxford University Press is not
able to give any guarantee or assurance that such is the case. Readers are urged to take
appropriately qualified medical advice in all cases. The information in this book
is intended to be useful to the general reader, but should not be used as a means of
self-diagnosis or for the prescription of medication.

Preface

Understanding how our brains work, and how evolution has shaped them, has interesting implications for understanding many aspects of human behaviour. To help understand ourselves this book describes the implications of our modern understanding of brain function for many different areas. They include emotion; social behaviour; rationality vs emotion; the philosophy of the relation between the mind and the brain, and of consciousness; aesthetics; ethics, economics; psychiatry; religion; and politics.

It is argued that a new type of understanding of these issues can emerge from the developments in understanding the brain and behaviour that are emerging from modern neuroscience, and how our brains have been shaped by evolution. This new understanding may have important implications for understanding the forces at work in areas such as emotional and social behaviour, economics, aesthetics, ethics, and politics.

The author brings a unique perspective to these issues, which can be grouped together as **neuroculture**, because of his approach to understanding the brain and biological mechanisms of emotion (*Emotion Explained*, 2005, Oxford University Press) which he now argues has implications for our understanding of aesthetics, sociality, ethics, and economics; and because of his approach to how the brain works at the mechanistic, i.e. neurobiological and computational, level (*Memory, Attention, and Decision-Making: A Unifying Computational Neuroscience Approach*, 2008, Oxford University Press; *The Noisy Brain: Stochastic Dynamics as a Principle of Brain Function*, 2010 with G.Deco, Oxford University Press) which is important for understanding human choices, decision-making, economics, psychiatry, normal aging, and relations between mental and physical events.

By combining rigorous neuroscientific approaches both to the evolutionary adaptive value of emotion and how this has shaped our brains, and to brain computation (i.e. how the brain works), the author brings a rational, fundamental, and new approach to these issues in the new area of **Neuroculture**. Each area covered is prefaced by the term 'neuro', to emphasize that it is the implications of our understanding of brain function that is being investigated in each chapter. The areas covered include Neuroaffect (Chapter 3); Neurosociality (Chapter 4); Neuroreason (Chapter 5); Neurophilosophy (Chapter 6); Neuroaesthetics (Chapter 7); Neuroeconomics (Chapter 8); Neuroethics (Chapter

9); Neuropsychiatry (Chapter 10); Neuroreligion (Chapter 11); and Neuropolitics (Chapter 12). Chapter 2, Neuroscience, introduces how we understand how the brain works, how it computes, at the level of the operations of brain cells, and of networks of brain cells, focusing on areas that help to provide a foundation for understanding these fascinating questions of emotion, rationality, decision-making, and memories that shape our lives, feelings, and behaviour. Some of these areas are developing rapidly, for example neuroethics, but the author seeks to bring an original approach to these areas, by building on his expertise in the brain mechanisms and evolutionary bases of emotion, and in computational neuroscience, to develop an understanding of why our brains operate as they do, and what the implications are.

The approach taken in this book is based on an understanding of the actual mechanisms by which the brain functions, as well as the evolutionary pressures that have led to the design of our brains. The approach thus goes beyond evolutionary psychology (or sociobiology), which considers behavioural adaptations shaped by evolutionary pressure, but has not taken the brain mechanisms involved into account in a direct way. Understanding the brain mechanisms involved helps us to understand reward processes and thereby to address aesthetics, emotion, decision-making, ethics, free will, and economics. Understanding the brain mechanisms involved in detecting causality helps one to address religion. Understanding the brain processing at the mechanistic level also provides a way to address how alterations in brain systems may lead to psychiatric and behavioural dysfunctions, which may then, in the light of our understanding of how they are implemented in the brain, become more easily treatable. The approach based on what the brain computes, and how it computes, is thus important in the approach taken in this book to a wide range of issues. The approach is fresh, for it seeks to understand many of the ways in which we behave by understanding some of the difficult computational problems faced by the brain, and the types of solution that the brain has found to these computational problems.

The overall aims of the book are developed further, and the plan of the book is described, in Chapter 1, Section 1.1.

The material in this text is the copyright of Edmund T. Rolls. Part of the material described in the book reflects work performed in collaboration with many colleagues, whose tremendous contributions are warmly appreciated. The contributions of many will be evident from the references cited in the text. I dedicate this book to them; to many scientific colleagues including Colin Blakemore, Marian Dawkins, and Larry Weiskrantz whose integrity and support have been outstanding; and to colleagues in other areas who have inspired me, including the sculptor, artist, and musician Penny Wheatley, and the philosopher David Rosenthal. Much of the work described would not have been possible without financial support from a number of sources, particularly the Medical Research Council of the UK, the Human Frontier Science Program, the Wellcome Trust, and the James S. McDonnell Foundation. The book was typeset by the author using LaTeX and WinEdt.

The cover shows part of the picture *The Birth of Venus* painted by Sandro Botticelli in c. 1486. The metaphor is that Botticelli was thinking about the origins of love, and beauty. I, in this book, am considering the scientific foundations of love, beauty, aesthetics, ethics, and many related aspects of our being in terms of how different evolutionary forces have shaped our brains, our emotions, and our rationality.

Updates to some of the publications cited in this book are available at **http://www.oxcns.org**.

I dedicate this work to the overlapping group: my family, friends, and colleagues – *in salutem praesentium, in memoriam absentium.*

Contents

1 Introduction

1.1 The framework

Understanding how our brains work, and how evolution has shaped them, has interesting implications for understanding many aspects of human behaviour. To help understand ourselves this book describes the implications of our modern understanding of brain function for many different areas, including the following areas. Each area is prefaced by the term 'neuro', to emphasize that it is the implications of our understanding of brain function that is being investigated in each chapter.

The chapters can be read to some extent independently, although many of the topics are illuminated by an understanding of how evolution has shaped our reward and emotion systems, which is described in Chapter 3. That chapter describes some of the reasons why we have emotion, and why some things appeal to us, and others do not. This book draws out the implications of these advances for our understanding of many topics such as aesthetics, ethics, economics, and religion. I bring a particular approach to such issues developed in this book, because I have worked on why and how the brain implements emotion and the importance of rewards in this, which I have developed more fully in *Emotion Explained* (Rolls 2005a).

The other expertise that helps to mark out the ideas developed in this book is in computational neuroscience: how the brain operates as a computer, but in ways very different to digital computers. I have developed these ideas in a number of books, sometimes written with colleagues with expertise in theoretical physics, including *Neural Networks and Brain Function* (Rolls and Treves 1998), *Computational Neuroscience of Vision* (Rolls and Deco 2002), *Memory, Attention and Decision-Making* (Rolls 2008c), and *The Noisy Brain* (Rolls and Deco 2010). Understanding how the brain operates computationally to implement for example short-term memory, long-term memory, and decision-making provides quite deep insight into what we are good at, and what we are less good at. For the latter, heuristics (short-cuts or 'rules of thumb') are used. Understanding how our brains solve problems in these different ways has many implications for understanding behaviour, including how we behave economically the way we do, how we are creative, and how the brain processes that promote creativity can sometimes if overdeveloped lead to states such as schizophrenia. Some of the foundations for a proper understanding of these types of processing are considered in a summary of some of the neuroscience foundations for understanding how the brain works, which are described in a neuroscience background chapter, Chapter 2. These foundations are important to help show the type of evidence that supports some of the points made in the remainder of the book, but a reader could skip to any of the other chapters and still follow the arguments.

The areas then that are approached in this brain-based way include the following:

Neuroaffect (Chapter 3): Why do we have emotions? Why are emotions so important in our behaviour? What role do genes play in emotion? Is emotion a Darwinian adaption to a major problem faced by genes? Emotions are adaptive, but are they rational? Are we conscious of all the emotional influences on our behaviour? How and why are women's and men's emotions different? Why do we as individuals place different values on different goals? What motivates us? Why do some people become obese? What makes people attractive? What is the role of hormones in such human processes as love and trust? These issues are considered in Chapter 3.

Neurosociality (Chapter 4): What are the fundamental bases of social behaviour? How is social behaviour adaptive, and is it adaptive in different ways in women and men? What are the roles of altruism towards our kin, and reciprocal altruism? Is behaviour selfish, or can it be truly altruistic?

Neuroreason: Chapter 5 describes the power of linguistic processing to enable us to plan ahead for the long term, and to defer or redefine our emotional goals. How do we choose between our emotional and rational, that is reasoning, systems? What are the mechanisms of decision-making, and why do we not always choose the same thing? Do our brains operate deterministically, or probabilistically?

Neurophilosophy (Chapter 6): What is the relation between the mind and the brain? Do mental, mind, events cause brain events? Do brain events cause mental effects? What can we learn from the relation between software and hardware in a computer about mind–brain interactions and how causality operates between brain events and mental events? How can we approach the 'hard' problem of consciousness: why some mental processing feels like something, and other mental processing does not? What type of processing is occurring when it does feel like something? Is consciousness an epiphenomenon, or is it useful? Are we conscious of the action at the time it starts, or later? Do we confabulate, that is invent 'reasonable', reasons for our behaviour? Do we have free will? How is the world represented in our brains?

Neuroaesthetics (Chapter 7): What are the foundations of what we appreciate in art? Is art – visual art, literature, music – related to fundamental adaptive capacities that help survival and thus reproduction, or is art a useless ornament, like a peacock's tail, shaped by sexual selection?

Neuroeconomics (Chapter 8): To what extent does our understanding of how our brains use heuristics to solve difficult problems involving costs and benefits provide a new way of thinking about economic choice made by individuals? Is the classical model of humans as selfish rational logical decision-makers satisfactory given the implications of modern neuroscience? What promotes trust between individuals? How stable is reciprocal altruism when the players are not perfectly matched? Is forgiveness related to its adaptive value in reinstating 'tit-for-tat' (reciprocally advantageous) interactions? What shapes our ideas about fairness? How does our behaviour and the choices we make change when the value we expect from an interaction is not met by the outcome? Could differences between individuals in how this is implemented in the brain be related to the susceptibility of some to gamble, and, if so, what are the implications?

Neuroethics (Chapter 9): Are there biological foundations to ethics? What role is played by adaptations for kin altruism and reciprocal altruism in what may be acceptable ethically? Is it likely that concepts of justice and laws should in practice be not too inconsistent with what we think is fair, where fairness has biological foundations? What is the relation between rights, justice, ethics, and the social contract?

Neuropsychiatry (Chapter 10): A mechanistic (brain computation based) approach with some randomness in the brain's computations leads to the concept of the stability of cognitive processes such as short-term memory, attention, and decision-making. This concept of the stability of cortical computational processes has fundamental implications for understanding and treating some psychiatric disorders such as schizophrenia, and obsessive-compulsive disorder. It also has implications for understanding some of the cognitive and memory changes in normal aging. Some psychiatric states can be considered as the ends of a spectrum of natural variation between individuals, where the variation is related to how evolution depends on variation.

Neuroreligion (Chapter 11): Why are religions so common in human society? What is the role of rationality in religion? Why might humans believe in or hope for life after death?

Neuropolitics (Chapter 12): Can an understanding of the forces of conflict between the emotional and the rational (reasoning) systems within humans lead to useful strategies for solving political problems? The rational approach might be to understand better how humans operate, and to build societies that utilize humans' rationality well.

It is argued that a new type of understanding of these issues can emerge from the developments in understanding the brain and behaviour that are emerging from modern neuroscience, and how our brains have been shaped by evolution. This new understanding may have important implications for understanding the forces at work in areas such as emotional and social behaviour, economics, aesthetics, ethics, and politics, and religion, and this is the subject of this book.

Before moving on from this introduction to this book, I wish to make a few preliminary points in the remainder of this chapter, to avoid misunderstanding.

1.2   Genetic influences on behaviour by adaptations of the brain

Part of the thrust of this book is that genes adapt our minds to make certain stimuli, events, and interactions rewarding or punishing. For example, we have genes that code for sweet taste, and activation of this system when we are hungry produces reward and pleasure. But this is an influence on behaviour, which is to the advantage of the genes that specify such rewards and punishers, but is not a determinant of behaviour. Genes rarely determine behaviour in any complete way, and the evidence does not lead us towards genetic determinism (Dawkins 1982). Instead, it is a matter of empirical test to what extent genes influence our behaviour, and indeed any genes promoting a goal for action must not do so too vigorously, for animals must choose many different rewards from time to time (e.g. food, water, warmth, sex) in order for their genes to be fit, to be passed into the next generation. This is an argument that genes for particular goals must not influence behaviour too strongly, but must operate in a partly cooperative way to let other genes that contribute to fitness influence behaviour too.

An important point is that we have only in the order of 30,000 genes, and this is insufficient to determine the connections of the brain. With in the order of 10^{11} (100,000,000,000) neurons in the brain, and in the order of 10^4 (10,000) synaptic connections onto each neuron, there are in the order of 10^{15} (1,000,000,000,000,000) connections to be specified. 30,000 genes cannot specify this number of connections. (Each connection between a pair of neurons, and the strength of the connection, would require, if the brain was determined in this way by genes, at least 1–2 genes. So the number of genes would be far too small for the genes to precisely determine brain connectivity, and thus behaviour.) Add to this that only a small proportion of the 30,000 genes are involved in specifying brain connectivity. (I think it likely to be less than 15%.) Thus, much of the connectivity of the brain must be specified by self-organizing processes including learning from the environment. (These self-organizing processes including competitive learning are considered in more detail in *Memory, Attention, and Decision–Making: A Unifying Computational Neuroscience Approach* (Rolls 2008c). The genes can only specify some of the general rules of brain wiring, such as the classes of neurons in a given brain region that should have some connections to each other (Rolls and Stringer 2000).

Considering how the brain might be specified by genes can be taken further. In this approach, I specified some rules that appeared to be being specified by genes, based on a comparison of the architecture of different cortical areas (Rolls and Stringer 2000). One such rule was for a given class of neuron, say hippocampal CA3 neurons, approximately what number of connections might be received from say other CA3 neurons, and what type of learning was implemented at those connections? Such a specification might take one or two genes. This type of specification rule has now received strong support, in that it has now been shown that it is possible by altering a single gene to abolish the CA3 to CA3 connectivity implemented with NMDA receptors, which implement a particular type of learning at synaptic connections between neurons (Nakazawa, Quirk, Chitwood, Watanabe, Yeckel, Sun, Kato, Carr, Johnston, Wilson and Tonegawa 2002).

We went on to allow a genetic algorithm, which is a simulation of the mathematics of evolution by gene recombination, mutation, and selection, to use our proposed set of specification rules to design the architecture of neuronal networks simulated on a computer (Rolls and Stringer 2000). The ability of the neural networks made by the genes of each individual in each generation to solve particular prototypical brain computations (pattern association, autoassociation, and competitive learning, see Chapter 2) was then tested, and the individuals that performed well at a problem were then allowed to breed the next generation of individuals (in the computer). We found that the system worked well, and that networks to solve different computational problems could be built using these specification rules (Rolls and Stringer 2000). This was an indication that the hypotheses about the genetic building blocks of brains were along the right lines.

Very interestingly, the number of genes required to build a brain with this approach is in the right range. In particular, the rules that I proposed might require in the order of 20–50 genes to specify a cortical area (such as the hippocampal CA3 region, or the primary visual cortex). With say 100 cortical areas, this could amount to 2,000–5,000 genes. This is probably about the right number, given that we have 30,000 genes or fewer, and the majority are needed for specifying things other than brains! Indeed, 1,000 genes are used just to specify odour receptors, of which there are 1,000 (Buck and Axel 1991, Zhang and Firestein 2002). This is the only approach I know that provides a realistic estimate of the number of genes that might be required to specify a brain, and moreover suggests how the brain might be specified by genes.

If we return to the main thrust of the argument, we are led firmly to the conclusion that every detail of brain structure, and in turn its function which relies on its connectivity, could not be determined by genes alone. The genes provide only quite general building instructions, leaving very much for self-organization during development based on the inputs being received, and on learning from the environment (Bateson and Gluckman 2011). (Given these points, it is fascinating to be able to start to unravel how a computational process as complex and massive as recognizing an object when it is seen from different views, in different sizes, and in different positions on the retina, can be understood in terms of quite simple rules for how cortical connectivity is specified genetically, and the influences of experience on how the cortical connections develop, as described in *Memory, Attention, and Decision-Making* (Rolls 2008c).)

Another argument against genetic determinism is that humans, and to some extent probably some related animals, have developed an ability to reason, and to plan ahead, which uses as one foundation a powerful short-term memory that can encode several items, and the order in which the items occur (Chapter 5). It is argued that this reasoning system can enable us to defer or not select (typically gene-specified) immediate goals, and to choose instead long-term goals selected by reasoning. Such reasoning might even lead to a failure to reproduce, as in some individuals who decide to devote themselves single-mindedly to say academic scholarship, medical research, or good deeds. So the reasoning system is a further adaptation of the human mind that makes humans' behaviour even less influenced by genes that specify goals for action, that is reinforcers. When I use the term gene-specified rewards, I thus do not refer at all to a deterministic or a 'hard-wired' system where genes completely determine behaviour, but instead to a system in which gene-defined reinforcers can have partial influences on behaviour.

The presence of 'noise' (randomness related to the neuronal firing times) in the brain as an influence on decision-making (Section 2.12, Rolls and Deco (2010)) is a further way in which one does not argue for genetic determinism at all, but for much weaker influences of genes, the impact of which can be empirically established. And it is important to establish this, for it turns out that we are each differently sensitive to different rewards and punishers. (This arises because we all have somewhat different genes, as gene variation coupled with selection is part of the basis for evolution.) This individual variation in reinforcers turns out to be an important part of the basis for personality (Section

3.6), and is important to appreciate and understand, as it has implications for the treatment of a number of problems, including obesity, addiction, gambling, and anxiety.

We may also emphasize that the genes that shape our behaviour through specifying our goals have evolved over many generations in the past, and that it takes many generations for genes to change. This means that the genes that for example may attract women to male power, wealth, and resources, may still operate, and influence our behaviour, even though it is now possible for many women to act economically independently of men, and for the women to themselves provide for the care of their young. Thus the genes that shape our behaviour may no longer be being selected for. Indeed, it is interesting to think about how the current changes in society may, over further generations, gradually lead to somewhat different genetic emphases on goals for action.

The intentional stance is adopted in much writing about sociobiology and evolutionary psychology, and is sometimes used here, but should not be taken literally. It is used just as a shorthand. An example is that it might be said that genes are selfish (Dawkins 1976). But this does not mean at all that genes think about whether to be selfish, and then take the decision. Instead it is just shorthand for a statement along the lines 'genes produce behaviour that operates in the context of natural selection to maximize the number of copies of the gene in the next generations'. Much of the behaviour produced is implicit or unconscious, and when the intentional stance is used as a descriptive tool, it should not be taken to mean that there is necessarily any explicit or conscious processing involved in the behavioural outcome.

2 Neuroscience

2.1 Introduction

The views I reach on neuroculture in this book are based on an understanding of some of the fundamental ways in which our brains work. In this chapter I describe some of the neuroscience background that is fundamental to the points made. The description is at a level intended to be approachable by a non-specialist in brain function. The description provides a foundation for further study, with many of the topics uncovered more fully in some of my previous books (Rolls and Deco 2002, Rolls 2005a, Rolls 2008c, Rolls and Deco 2010). In addition, there are a number of good books on neuroscience that summarize many parts of neuroscience (Kandel, Schwartz, Hudspeth, Siegelbaum and Jessell 2012, Gazzaniga 2009).

An approach we can now take to understanding brain function can be described as a mechanistic computational approach (Rolls and Deco 2010). This helps us to understand our behaviour in terms of the actual mechanisms that operate in our brains. Understanding at this level of the operations of brain cells (neurons), and of networks of neurons, helps us to see how the brain actually works, and is very important in understanding how to treat the system when it dysfunctions, in for example psychiatric states.

This mechanistic approach may be contrasted with a phenomenological approach, in which a process is modelled in a less brain-like way to account for the behaviour, but that approach is less good at integrating all the levels of detail that are necessary to understand the real working of the system and its dynamics. The mechanistic approach in contrast allows integration of understanding from the level of synaptic effects and ion channels in neurons and how they are affected by neurotransmitters, through the spiking activity of neurons, to computations performed in networks, and thus to effects that can be captured in functional neuroimaging, and importantly that can account for behaviour in terms of what are sometimes 'emergent' properties of the mechanistic system, as described in this book and elsewhere (Rolls 2008c, Rolls and Deco 2010).

2.2   Neurons in the brain, and their representation in neuronal networks

Neurons in the vertebrate brain typically have, extending from the cell body, large dendrites which receive inputs from other neurons through connections called synapses. (The dendrites are the thick tree-like parts of each neuron shown in Fig. 2.1.) The synapses operate by chemical transmission. When a synaptic terminal receives an all-or-nothing action potential from the neuron of which it is a terminal, it releases a chemical transmitter that crosses the synaptic cleft and produces either depolarization (a change of the negative voltage within a cell towards the threshold for firing an action potential, producing excitation) or hyperpolarization (the opposite, producing inhibition) in the postsynaptic neuron, by opening particular ionic channels. (A textbook such as Kandel et al. (2012) gives further information on this process.) Summation of a number of such depolarizations or excitatory inputs within the time constant of the receiving neuron, which is typically 15–25 ms, produces sufficient depolarization that the neuron fires an action potential. There are often 5,000–20,000 synaptic inputs onto each neuron. Examples of cortical neurons are shown in Fig. 2.1, and further examples are shown elsewhere (Shepherd 2004, Rolls and Treves 1998, Rolls 2008c, Shepherd and Grillner 2010). Once firing is initiated in the cell body (or axon initial segment of the cell body), the action potential is conducted in an all-or-nothing way to reach the synaptic terminals of the neuron, whence it may affect other neurons. Any inputs the neuron receives that cause it to become hyperpolarized make it less likely to fire (because the membrane potential is moved away from the critical threshold at which an action potential is initiated), and are described as inhibitory. The neuron can thus be thought of in a simple way as a computational element that sums its inputs within its time constant and, whenever this sum, minus any inhibitory effects, exceeds a threshold, produces an action potential that propagates to all of its outputs. This simple idea is incorporated in many neuronal network models using a formalism of a type described in Section 2.5.

2.3   Information encoded in the brain by neuronal firing rates

2.3.1   Reading the code used by single neurons

A fundamental issue in neuroscience is the question of how information is encoded in the brain (Rolls 2008c, Rolls and Treves 2011). It is fundamental because to discover how the brain processes information we first have to know how it encodes and represents inputs from the environment, and its internal workings. Neuroscience has approached this problem by measuring the level of the spiking activity of single neurons and populations of neurons, as it is

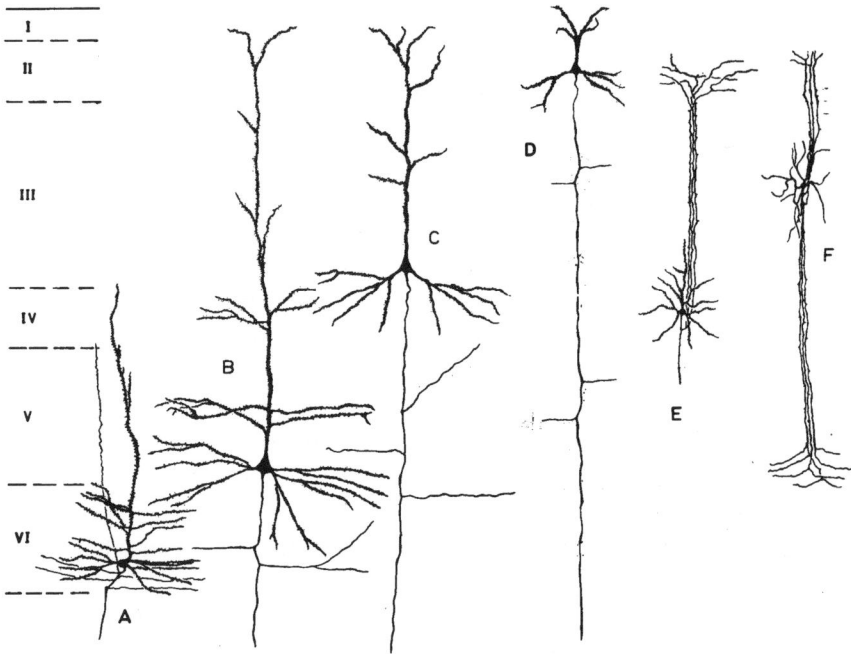

Fig. 2.1 Examples of neurons found in the brain. Cell types in the cerebral neocortex are shown. The different laminae of the cortex are designated I–VI, with I at the surface. Cells A–D are pyramidal cells in the different layers. Cell E is a spiny stellate cell, and F is a double bouquet cell. The surface or top of the cortex is above layer I, and the white matter, which consists of fibre connections, is below layer VI. The whole thickness of the cortex is 2–3 mm. The thick part of each neuron is the dendrite or receiving surface. The thin part of each neuron is the axon, which carries the spiking output of a neuron to other neurons. (After Jones 1981; see Jones and Peters 1984, p. 7.)

by the all-or-none spiking activity that information is accurately carried from one neuron to the thousands of neurons to which it has synaptic connections[1]. E.D. Adrian (1928) working at Cambridge succeeded in measuring the spiking activity of an afferent nerve fibre receiving from a stretch receptor in a muscle of the frog. He found that the nerve fibre generated action potentials or spikes with a firing rate measured in spikes/s that increased when the muscle was stretched. He showed for the first time the characteristic sigmoidal response function of a sensory neuron, i.e. the firing rate of the neuron increased at first slowly, then had a linear portion, and then saturated at a high firing rate as the stimulus applied increased (in this case the weight applied to the muscle).

[1] The electrical spike of a neuron lasts for approximately 1 ms (one thousandth of a second) and is used to transmit information along the axon of a neuron because as an active process it ensures that the spike is all-or-none, either there or not, at the terminals of the axon. If neural transmission was by passive effects of a voltage change being sensed at a distance, it would be subject to unknown degradation influenced by the distance travelled and by the diameter of the axon.

The objective character of the neuronal response function inspired the rate hypothesis, namely that the firing rate of a neuron represents the stimulation, because knowing the firing rate we can decode the strength of the applied stimulation[2].

There are now many experimental demonstrations that the firing rates of neurons indeed do encode information about stimuli. In one example neurons in the primary visual cortex have firing rates that depend on the orientation of edges or bars of light in their receptive field (Hubel and Wiesel 1968)[3]. In another example, neurons in the primary taste cortex have firing rates that depend on the concentration of the taste stimulus in the mouth (Scott, Yaxley, Sienkiewicz and Rolls 1986), as do the activations in this area recorded with functional neuroimaging using functional magnetic resonance imaging (fMRI) of the brain, and the rated subjective intensity of the taste which is proportional to the fMRI signal (Grabenhorst, Rolls and Bilderbeck 2008a, Grabenhorst, Rolls and Parris 2008b).

This type of sensory neuron also conveys information about what stimulus is present, something that goes beyond the orientation of a line in a visual scene, or the intensity of a taste. The information may be for example about what taste is present in the mouth. The neuron shown in Fig. 2.2 recorded in the secondary taste cortex which we discovered in the orbitofrontal cortex (Rolls, Yaxley and Sienkiewicz 1990, Baylis, Rolls and Baylis 1994) increased its firing rate to the sweet taste of glucose, but not to salt, sour or bitter (Rolls et al. 1990). The neuron was quite selective, in that it even had only a small response, a few spikes, to the flavour of fruit juice (BJ) in the mouth. The activity in Fig. 2.2 also illustrates the all-or-none spikes transmitted by single neurons, and the firing rate code being used. What stimulus is present in the mouth is signalled by the firing rate of the neuron, the number of spikes it emits in a given time period.

---

[2] E. D. Adrian was awarded the Nobel Prize in Physiology or Medicine in 1932 (with Sir Charles Sherrington). I remember as a medical student at Cambridge finding one of Lord Adrian's electrometers for measuring neural activity on a rubbish pile outside the Department of Physiology. I asked Fergus Campbell what it was, and when he recognized it, we brought it back into the Physiological Laboratory, where I hope that it is now on display. Physiology at Cambridge continued to develop great expertise in recording and amplifying the small signals from neurons. The father of Richard S. Pumphrey, one of my undergraduate medical student friends at Cambridge, was Professor Richard Julius Pumphrey (http://www.jstor.org/pss/769453), and with his expertise in low-noise neuronal recording, was drafted in the war into the development of radar. I was told that one of the early tests was whether more wobble could be seen on an oscilloscope if it was a woman than a man walking. Even as an undergraduate at Cambridge, there was great expertise in low-noise recording, and I made in the Physiological Laboratory low-noise cathode follower first stage amplifiers for distribution to colleagues internationally. I built on experience I had developed while at school in building an oscilloscope (with a 3.5 inch cathode ray tube and rectifiers as voltage doublers to obtain the 4,000 volts needed for the plate) and amplifiers for recording from the giant axon of the earthworm.

[3] David Hubel and Torsten Wiesel were awarded (with Roger Sperry) the Nobel Prize in Physiology or Medicine in 1981.

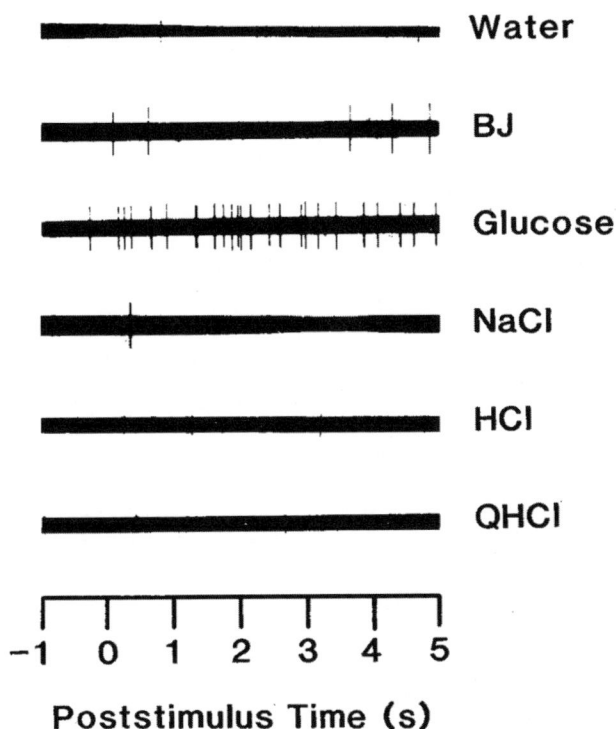

Fig. 2.2 Examples of the responses recorded from one caudolateral orbitofrontal taste cortex neuron to the six taste stimuli, water, 20% blackcurrant juice (BJ), 1 M glucose (sweet), 1 M NaCl (salt), 0.01 M HCl (sour), and 0.001 M quinine HCl (QHCl, bitter). The stimuli were placed in the mouth at time 0. Each vertical line is a voltage spike generated by the action potential of the single neuron being recorded. (From Rolls, Yaxley and Sienkiewicz 1990.)

Although the firing rate of such neurons indicates how strong the stimulus is, which stimulus is present is encoded by which neurons are firing. Each neuron is tuned to respond to a different subset of stimuli. This type of coding, used throughout the cerebral cortex, can also thus be described as a place code, for which neuron is firing defines which stimulus or event is occurring. ('Place' here refers to which neuron is firing, for each neuron is necessarily at a different place in the cortical sheet.) The firing rate in such a 'labelled line' of 'place' representation determines the contribution of whatever is being coded for by that single neuron to the overall representation of a stimulus that is represented and transmitted by a whole population of neurons, as we shall see next.

In another example, the firing rates of the 'face neurons' that we discovered in the inferior temporal visual cortex (IT), amygdala, and orbitofrontal cortex reflect which face has been seen (Sanghera, Rolls and Roper-Hall 1979, Perrett, Rolls and Caan 1982, Rolls, Critchley, Browning and

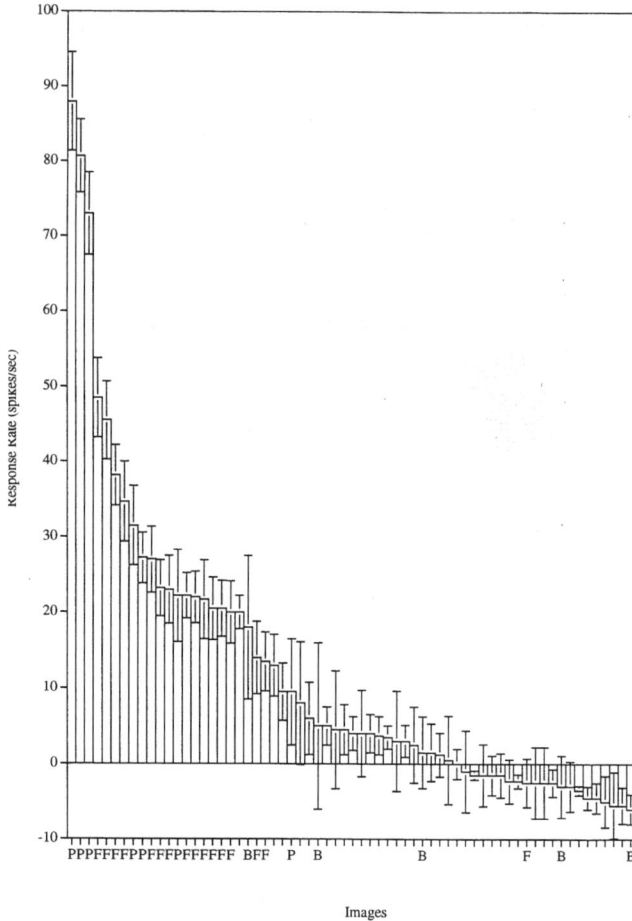

Fig. 2.3 Firing rate distribution of a single neuron in the temporal visual cortex to a set of 23 face (F) and 45 non-face images of natural scenes. The firing rate to each of the 68 stimuli is shown. Faces shown in profile are marked P. B indicates stimuli that included other body parts such as hands. This typical 'face' neuron did not respond to just one of the 68 stimuli. Instead, it responded to a small proportion of stimuli with high rates, to more stimuli with intermediate rates, and to many stimuli with almost no change of firing. This is typical of the distributed representations found in temporal cortical visual areas. (After Rolls and Tovee 1995.)

Inoue 2006a, Rolls 2011d). In particular, each neuron responds with a different firing rate to each different face in a set, as illustrated in Fig. 2.3, so that which face is being seen can be read off or decoded by knowing the firing rates of a set of these neurons, as described below. Face-selective neurons have also been described in humans (Kreiman, Koch and Freid 2000), though in terms of their properties and locations they seem to be involved more in memory-related representations than the perceptual representations described here.

Further, something close to Adrian's 'Sensation' (the title of his 1928 book

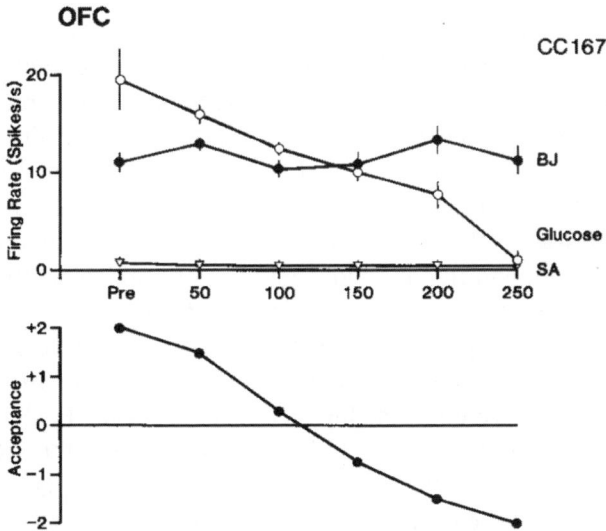

Fig. 2.4 The effect of feeding to satiety with glucose solution on the responses (rate ± sem) of a neuron in the secondary taste cortex to the taste of glucose and of blackcurrant juice (BJ). The spontaneous firing rate is also indicated (SA). Below the neuronal response data, the behavioural measure of the acceptance or rejection of the solution on a scale from +2 (strong acceptance) to −2 (strong rejection) is shown. The solution used to feed to satiety was 20% glucose. The monkey was fed 50 ml of the solution at each stage of the experiment as indicated along the abscissa, until he was satiated as shown by whether he accepted or rejected the solution. Pre is the firing rate of the neuron before the satiety experiment started. (From Rolls, Sienkiewicz and Yaxley 1989.)

was *The Basis of Sensations*) is reflected rather directly in the firing rate code used by neurons. A good example is that the subjective sensation of the thickness of carboxymethylcellulose (a food thickener) in the mouth is proportional to the log of its viscosity (Kadohisa, Rolls and Verhagen 2005), as is the firing rate of neurons (Verhagen, Kadohisa and Rolls 2004) and the signal recorded with functional neuroimaging (using functional magnetic resonance imaging, fMRI) in the human primary taste cortex (De Araujo and Rolls 2004).

In other parts of the brain, the firing rates still encode information, but about events that reflect the internal workings of the brain, rather than events that reflect the firing of sensory receptors. For example, the reward system that we discovered in the orbitofrontal cortex for taste reward (Rolls, Sienkiewicz and Yaxley 1989) as well as many other rewards (Rolls, Burton and Mora 1980, Thorpe, Rolls and Maddison 1983, Rolls 2005a, Rolls and Grabenhorst 2008, Grabenhorst and Rolls 2011) contains neurons that fire fast when a taste is rewarding, and gradually decrease their responses to zero to the food as it is fed to satiety so that the food is no longer rewarding (Rolls, Sienkiewicz and

Yaxley 1989) (Fig. 2.4[4]). Correspondingly, the subjectively rated pleasantness of the food reflects this neuronal firing, for the subjective pleasantness decreases to zero as the food is fed to satiety, as does the signal recorded with fMRI neuroimaging in the human orbitofrontal cortex (Kringelbach, O'Doherty, Rolls and Andrews 2003). Thus the neuronal firing rates, which are reflected in the fMRI signal (Rolls, Grabenhorst and Deco 2010b), encode in this part of the brain the reward value and subjective pleasantness of many stimuli and events in the world, and this is fundamental to our understanding of emotion (Rolls 2005a, Rolls and Grabenhorst 2008) (see Chapter 3).

Another good example of how firing rates can now be read to understand the internal workings of the brain is provided by the spatial view neurons that we discovered in the hippocampus (Georges-François, Rolls and Robertson 1999, Robertson, Rolls and Georges-François 1998, Rolls, Robertson and Georges-François 1997a, Rolls, Treves, Robertson, Georges-François and Panzeri 1998). These neurons increase their firing rates only when one part of a spatial environment is being looked at. These responses occurred when the spatial view was seen from different angles with respect to the body and with many different head directions, so the representation was not egocentric (that is it was not with respect to the body or head axis). The increase in neural firing rate could occur when the spatial view was seen from many places in the environment. These findings indicated that spatial view neurons represent locations 'out there' in space on allocentric or 'world' coordinates. Very interestingly, these neurons had a short-term memory of the allocentric spatial view for their firing, which occurred even in the dark provided that the eyes moved to look at the allocentric location of the spatial view. (This type of memory lasts for one or two minutes, after which one's sense of direction if one is moving around in the dark becomes much less good.) This allocentric representation of space is we believe part of a system for remembering where objects have been seen, for example where one has seen good food, or other good resources (without necessarily ever having visited that place), for some of these neurons respond to a combination of a spatial view and the object that has been shown at that location (Rolls, Xiang and Franco 2005). Some of the neurons also respond during recall of the location in space where an object was last seen (Rolls and Xiang 2006). This type of neuronal activity is prototypical of an episodic memory, the memory for particular past events or episodes, and indeed these neurons are an important component of the theory of episodic memory that we have developed (Treves and Rolls 1994, Rolls 2010c) (see further Section 2.8).

What has been described in this subsection (2.3.1) is that the firing rates of

---

[4]The neuron did not decrease its firing rate to the fruit juice (BJ), which was not being fed to satiety. Thus the reward value decrease was specific to the food eaten to satiety. It was in experiments of this type that we discovered sensory-specific satiety (Rolls 1981, Rolls 2011i).

single neurons provide much evidence about how the world is represented in the brain.

When considering the operation of the brain, and of neuronal networks in the brain, it is found that many useful properties arise if each input to the network (arriving on the set of input axons that make synapses onto a single neuron) is encoded by the firing rate of an ensemble or subset of the axons (distributed encoding), and is not signalled by the activity of a single input or axon, which is called local encoding. Because this firing rate code distributed across a population of neurons each tuned differently is so important to understanding how the brain operates (Rolls 2008c, Rolls and Treves 2011), this neural encoding is considered further in the next section. When taken together with the concepts outlined in Section 2.5, the aim is to show how we at present can go far beyond phenomena found in the brain, towards an understanding of how it works computationally. This is one of the major advances in neuroscience, in the understanding of our own brains.

On a first reading of this chapter non-specialists may wish to advance to Section 2.5.

2.3.2   Understanding the code provided by populations of neurons

I start with some definitions, then summarize some evidence that shows the type of encoding used in some brain regions, and then show how the representation found is advantageous (Rolls 2008c, Rolls and Treves 2011, Deco and Rolls 2011).

2.3.2.1   Definitions

A *local representation* is one in which all the information that a particular stimulus or event occurred is provided by the activity of one of the neurons. In a famous example, a single neuron might be active only if one's grandmother was being seen, and this is sometimes called grandmother cell encoding. (The term was coined by Jerry Lettvin in about 1969 – see Charles Gross (2002).) An implication is that most neurons in the brain regions where objects or events are represented would fire only very rarely (Barlow 1972, Barlow 1995). A problem with this type of encoding is that a new neuron would be needed for every object or event that has to be represented. Another disadvantage is that this type of coding does not generalize easily to similar inputs, so that similarities between perceptions or memories would not be apparent. Another disadvantage is that the system is rather sensitive to brain damage: if a single neuron is lost, the representation may be lost. Another disadvantage of local encoding is that the storage capacity in a memory system in the brain (the number of stimuli that can be stored and recalled) may not be especially high (in the order of the number of synapses onto each neuron) (Rolls 2008c).

A *fully distributed representation* is one in which all the information that a particular stimulus or event occurred is provided by the activity of the full set of neurons. If the neurons are binary (e.g. either active or not), the most distributed encoding is when half the neurons are active (i.e. firing fast) for any one stimulus or event, and half are inactive. Different stimuli are represented by different subsets of the neurons being active.

A *sparse distributed representation* is a distributed representation in which a small proportion of the neurons is active at any one time. In a sparse representation with binary neurons (i.e. neurons with firing rates that are either high or low), less than half of the neurons are active for any one stimulus or event. For binary neurons, we can use as a measure of the sparseness the proportion of neurons in the active state. For neurons with real, continuously variable, values of firing rates, the sparseness a^p of the representation provided by the population can be quantified as described elsewhere (Rolls 2008c). A low value of the sparseness a^p indicates that few neurons are firing for any one stimulus.

2.3.2.2 Encoding provided by neuronal populations

At the time Barlow (1972) wrote, there was little actual evidence on the activity of neurons in the higher parts of the visual and other sensory systems. There is now considerable evidence, some of which is now described (Rolls 2008c, Rolls and Treves 2011).

The representation is distributed as illustrated in Fig. 2.3 by the firing rates produced by each of 68 stimuli in a single neuron in the primate inferior temporal visual cortex. Rather few stimuli produce high firing rates (e.g. above 60 spikes/s), and increasingly large numbers of stimuli produce lower and lower firing rates. The spontaneous firing rate of this neuron, the rate when no stimuli were being shown, was 20 spikes/s (Rolls and Tovee 1995). The histogram bars indicate the change of firing rate from the spontaneous value produced by each stimulus. Stimuli that are faces are marked F, or P if they are in profile. B refers to images of scenes that included either a small face within the scene, sometimes as part of an image that included a whole person, or other body parts, such as hands (H) or legs. The non-face stimuli are unlabelled. The neuron responded best to three of the faces (profile views), had some response to some of the other faces, and had little or no response, and sometimes had a small decrease of firing rate below the spontaneous firing rate, to the non-face stimuli. The representation was thus rather distributed, and this is typical of neurons in the higher order visual cortical areas (Rolls and Tovee 1995, Baddeley, Abbott, Booth, Sengpiel, Freeman, Wakeman and Rolls 1997, Treves, Panzeri, Rolls, Booth and Wakeman 1999, Franco, Rolls, Aggelopoulos and Jerez 2007); of neurons in the taste and flavour cortical areas in the insula and orbitofrontal cortex (Verhagen, Kadohisa and Rolls 2004, Rolls, Verhagen and Kadohisa 2003d, Verhagen, Rolls and Kadohisa 2003, Kadohisa, Rolls and Verhagen

2004, Kadohisa, Rolls and Verhagen 2005); and of neurons tuned to spatial view in the hippocampus (Rolls, Treves, Robertson, Georges-François and Panzeri 1998, Rolls 2008c).

These data provide a clear answer to whether these neurons are grandmother cells: they are not, in the sense that each neuron has a graded set of responses to the different members of a set of stimuli, with the prototypical distribution similar to that of the neuron illustrated in Fig. 2.3. On the other hand, each neuron does respond very much more to some stimuli than to many others, and in this sense is tuned to some stimuli.

With data of this type recorded from single neurons and simultaneously recorded populations of single neurons, and the application of quantitative information theoretic measures described elsewhere (Rolls 2008c, Rolls and Treves 2011), the working hypotheses about neuronal encoding in the visual, hippocampal, olfactory and taste cortical systems are as follows (Rolls 2008c, Rolls and Treves 2011):

1. Much information is available about the stimulus presented in the number of spikes emitted by single neurons in a fixed time period, the firing rate. Importantly, just knowing the firing rate of a single neuron provides quite a lot of information (evidence) about which particular stimulus was shown, as illustrated by the neuronal responses shown in Fig. 2.3.

2. Much of this firing rate information is available in short periods, with a considerable proportion available in as little as 20 ms. This rapid availability of information enables the next stage of processing to read the information quickly, and thus for multistage processing to operate rapidly. This time is the order of time over which a receiving neuron might be able to utilize the information, given its synaptic and membrane time constants. In this time, a sending neuron is most likely to emit 0, 1, or 2 spikes.

3. This rapid availability of information is confirmed by population analyses, which indicate that across a population of neurons, much information is available in short time periods.

4. More information is available using this rate code in a short period (of e.g. 20 ms) than from just the first spike (Rolls, Franco, Aggelopoulos and Jerez 2006b).

5. Little information is available by time variations within the spike train of individual neurons for static visual stimuli (in periods of several hundred milliseconds), apart from a small amount of information from the onset latency of the neuronal response (Tovee, Rolls, Treves and Bellis 1993, Rolls and

Treves 2011, Deco and Rolls 2011). For a time-varying stimulus, clearly the firing rate will vary as a function of time.

6. Across a population of neurons, the firing rate information provided by each neuron tends to be independent; that is, the information increases approximately linearly with the number of neurons. The outcome is that the number of stimuli that can be encoded rises exponentially (i.e. very rapidly) with the number of neurons in the ensemble (Rolls, Treves and Tovee 1997b). An implication of the independence is that the response profiles to a set of stimuli of different neurons are uncorrelated (Franco et al. 2007).

7. The information in the firing rate across a population of neurons can be read moderately efficiently by a decoding procedure as simple as a synapti-cally weighted sum of the inputs (a dot product) (Rolls et al. 1997b, Rolls and Treves 2011), which is what is shown in Equation 2.1 on page 25. This is the simplest type of processing that might be performed by a neuron, as it involves taking a dot (or inner) product of the incoming firing rates with the receiving synaptic weights to obtain the activation (e.g. depolarization) of the neuron. This type of information encoding ensures that the simple emergent proper-ties of associative neuronal networks such as generalization, completion, and graceful degradation (see Section 2.3.2.3) can be realized very naturally and simply. This type of encoding is also what makes the responses of an individual neuron interpretable: it is tuned to respond better to some stimuli than others, with a graded firing rate representation of the type illustrated in Fig. 2.3. This decoding principle is illustrated in Fig. 2.5.

8. There is little additional information to the great deal available in the firing rates from any stimulus-dependent cross-correlations or synchronization bet-ween the spikes of different neurons that may be present (Aggelopoulos, Franco and Rolls 2005, Rolls 2008c, Rolls and Treves 2011, Deco and Rolls 2011). Stimulus-dependent synchronization might in any case only be useful for group-ing different neuronal populations, and would not easily provide a solution to the binding problem in vision. Instead, the binding problem in vision may be solved by the presence of neurons that respond to combinations of features in a given spatial position with respect to each other (Rolls 2008c, Rolls 2009d, Rolls and Treves 2011).

9. There is little information available in the order of the spike arrival times of different neurons for different stimuli that is separate or additional to that provided by a rate code (Rolls, Franco, Aggelopoulos and Jerez 2006b). The presence of spontaneous activity in cortical neurons facilitates rapid neuronal responses, because some neurons are close to threshold at any given time, but

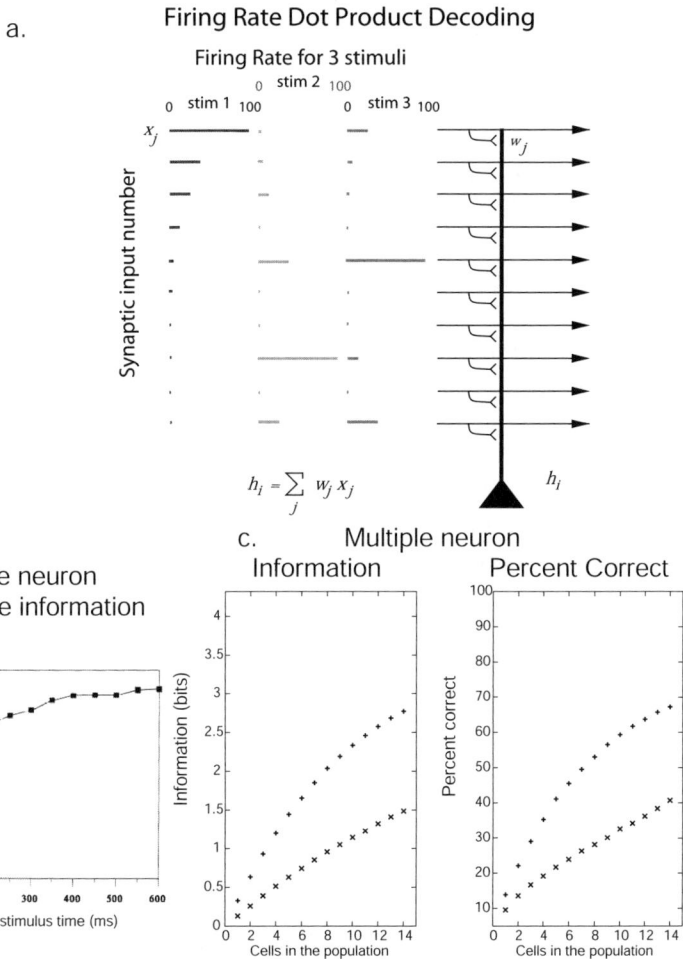

a. **Firing Rate Dot Product Decoding**

Firing Rate for 3 stimuli

$$h_i = \sum_j w_j x_j$$

b. Single neuron cumulative information

c. Multiple neuron

Information — Percent Correct

Fig. 2.5 Information encoding by firing rates. (a). Each stimulus is encoded by an approximately exponential firing rate distribution of a population of neurons. The distribution is ordered to show this for stimulus 1, and other stimuli are represented by similar distributions with each neuron tuned independently of the others. The code can be read by a dot (inner) product decoding performed by any receiving neuron of the firing rates with the synaptic weights, and it is the almost linear increase in information with the number of neurons illustrated in (c) which shows that the tuning profiles to the set of stimuli are almost independent. (b). Much information is available from single neurons, and from populations of neurons, using the number of spikes in short time windows of e.g. 50 ms. This is shown by the cumulative information from the firing rates for different time periods from single neurons in the cortex in the superior temporal sulcus to a set of 10 face and non-face stimuli, and is remarkable in that the neurons do not start to respond to the visual stimuli until 80–90 ms, so that the data point at 100 ms shows the information from 20 ms or less of firing. (c). The information available about which of 20 faces had been seen that is available from the responses measured by the firing rates in a time period of 500 ms (+) or a shorter time period of 50 ms (x) of different numbers of temporal cortex cells. The corresponding percentage correct from different numbers of cells is also shown. Decoding with dot product decoding reveals similar principles. (See colour plates Appendix B.)

this also would make a spike order code difficult to implement.

10. Analysis of the responses of single neurons to measure the sparseness of the representation indicates that the representation is distributed, and not grandmother cell like (or local) (Rolls and Tovee 1995, Franco et al. 2007, Rolls 2008c, Rolls and Treves 2011).

11. The representation is not very sparse in the perceptual systems studied (as shown for example by the values of the single cell sparseness), and this may allow much information to be represented. At the same time, the responses of different neurons to a set of stimuli are decorrelated, in the sense that the correlations between the response profiles of different neurons to a set of stimuli are low. Consistent with this, the neurons convey independent information, at least up to reasonable numbers of neurons. The representation may be more sparse in memory systems such as the hippocampus, and this may help to maximize the number of memories that can be stored in associative networks (Rolls 2008c, Rolls and Treves 2011).

12. Even when temporal order must be encoded and stored in the brain, recent evidence indicates that a firing rate code is being used (see Section 2.7).

13. Because neurons can convey almost independent information, and each neuron is tuned in a different way to a set of stimuli (the most important property of cortical encoding), measures that take the average of the activity of many neurons (or synapses (Logothetis 2008)) cannot reveal how information is encoded in the brain. For example, a typical $3 \times 3 \times 3$ mm voxel (volume) with fMRI would contain 810,000 neurons if the neuronal density is taken as 30,000 neurons/mm^3 (Rolls 2008c). The result is that the activation of a voxel with fMRI reveals little about exactly which stimulus was presented (e.g. whose face), and even the information reflected by a voxel that it is in a category (such as that it is pleasant, is predicted to be chosen, or is a face) is less than that of a typical neuron, and moreover does not increase linearly with the number of voxels (Rolls, Grabenhorst and Franco 2009) a fundamental property of neuronal encoding. The same argument applies also to other measures that group together the effects of a large number of neurons, such as local field potentials (LFP), and magnetoencephalography (MEG). Such measures, including fMRI and positron emission topography (PET), are useful in analyzing what categories of information may be represented in a brain region (e.g. faces, houses, spatial scenes), but do not reveal which particular instance is being represented, or the details of the neuronal code. (Insofar as anything is revealed about the neural code, these measures are consistent with the evidence that firing rates rather than properties such as stimulus-dependent neuronal

synchronization are being used (Rolls, Grabenhorst and Franco 2009, Rolls, Grabenhorst and Deco 2010b, Rolls and Treves 2011, Deco and Rolls 2011).

Because understanding neuronal encoding in a brain region, and how this is related to the detailed neuronal network functional architecture of a region is so crucial to understanding neuronal computation (Rolls 2008c), functional neuroimaging in humans can never replace neuronal recordings, possible mainly in animals, and these are among the reasons why research with animals continues to be important in understanding how the brain actually works. For some brain regions, including the visual system, the prefrontal including orbitofrontal cortex, the hippocampus (given the types of representation found in it), and even the taste system which one might think is evolutionarily old, the human brain is so far developed and so different from that in rodents that in such systems it is important to rely on findings in non-human primates. Further, information is conveyed between the computing elements of the brain, the neurons, by action potentials of single cells travelling along their axons, and it is therefore only by analyzing at this single and multiple simultaneously recorded single neuron level that we will understand the encoding being used in the brain to transfer information between its computing elements, or what information is being encoded and transmitted (Rolls 2008c, Rolls and Treves 2011).

2.3.2.3 Advantages of different types of coding

One advantage of distributed encoding is that the similarity between two representations can be reflected by the correlation between the two patterns of activity that represent the different stimuli. I describe in Section 2.4 and Fig. 2.5 the idea that the input to a neuron is represented by the activity on its set of input synapses with each input weighted by the synaptic strength of the synapse. There are in the order of 10,000 such excitatory synaptic terminals carrying the input spikes, and 10,000 corresponding synapses, on each neuron. In this way, a single cortical neuron receives in the order of 10,000 synaptic inputs, each one from a different cortical neuron. At each synapse, a current is injected into the neuron that is weighted by the synaptic strength. The weighting corresponds mathematically to the operation of multiplication. The total input to the neuron is then the sum of these 10,000 currents. The more similar the input set of firing rates is to the set of synaptic weights, the larger will be the output of the neuron. Thus neurons compute the similarity between the firing rates on each input and the strengths of the corresponding synaptic weights on a neuron, which might represent a previously stored input. In this sense, neurons compute the similarity or correlation between a set of input firings and their synaptic weights, and thus can compare a new input with an input previously stored as modified synaptic strengths of weights (see further Fig. 2.6 and Equation 2.1 in Section 2.4). The correlation will be high if the activity of each axon in the two representations is similar; and will become more and more different as the firing rate activity of more and more of the axons differs in the two representations. Thus the similarity of two inputs can be represented in a graded or continuous way if (this type of) distributed encoding is used. This enables generalization to similar stimuli, or to incomplete versions of a stimulus (if it is, for example, partly seen or partly remembered), to occur. With a local representation, either one stimulus or another is represented, each by its own neuron firing, and similarities between different stimuli are not encoded.

Another advantage of distributed encoding is that the number of different stimuli that can be represented by a set of C components (e.g. the firing rates applied to the C synaptic inputs to a neuron) can be very large. It can be much larger than C, as shown in Section 2.5.3 (Treves and Rolls 1991, Rolls 2008c) (see Section 2.3). Put the other way round, even if a neuron has only a limited number of inputs (e.g. a few thousand), it can nevertheless receive a great deal of information about which stimulus was present. This ability of a neuron with a limited number of inputs to receive information about which of potentially very many input events is present is probably one factor that makes computation by the brain possible. With local encoding, the number of stimuli that can be encoded increases only linearly with the number C of axons or components (because a different component is needed to represent each new stimulus).

In the real brain, there is now good evidence that in a number of brain systems, including the high-order visual and olfactory cortices, and the hippocampus, distributed encoding with the properties described above, of representing similarity, and of exponentially increasing encoding capacity as the number of neurons in the representation increases, is found. For example, as we have seen, in the primate inferior temporal visual cortex, the number of faces or objects that can be represented increases approximately exponentially with the number of neurons in the population (Rolls and Tovee 1995, Abbott, Rolls and Tovee 1996, Rolls, Treves and Tovee 1997b, Rolls, Treves, Robertson, Georges-François and Panzeri 1998, Rolls, Franco, Aggelopoulos and Reece 2003b, Rolls, Aggelopoulos, Franco and Treves 2004, Franco, Rolls, Aggelopoulos and Treves 2004, Aggelopoulos, Franco and Rolls 2005, Rolls, Franco, Aggelopoulos and Jerez 2006b, Rolls 2008c). A similar result has been found for the encoding of position in space by the primate hippocampus (Rolls, Treves, Robertson, Georges-François and Panzeri 1998), and for the encoding of olfactory stimuli in the orbitofrontal cortex (Rolls, Critchley, Verhagen and Kadohisa 2010a).

2.4   What neurons do: neuronal computation

The computing elements of the brain are the neurons. To understand how the brain works, we must consider how neurons work.

A first important working concept is that each neuron in the cerebral cortex has a large number of excitatory synaptic connections, in the order of 10,000. Each synapse receives spikes of activity from one other neuron at a rate typically between 0 and 100 spikes/s. Each time a spike arrives, it acts through synaptic receptors to open ion channels in the postsynaptic neuron. Let us consider the set of excitatory synaptic inputs where these inputs tend to depolarize the neurons, bringing it from a negative voltage (typically −70 mV) towards the firing threshold (typically −55 mV) at which the neuron emits an action potential or spike of activity. Each synapse produces a small effect, which depends on the strength of the synaptic connection, which can be modified by learning. The neuron then sums all these products of an input firing rate and synaptic strength in its membrane potential. Mathematically, the depolarization of the neuron can be described as the sum of the products of the input firing multiplied by the relevant synaptic weight. Mathematically this is a dot product or inner product operation, where the change of membrane potential is a product between the set (or vector) of input firings and the set (or vector) of synaptic weights. The more similar the set of input firings is to the set of strong synapses, the larger will be the effect produced in the neuron. If the input does not match the set of strengthened synapses, the excitation of the neuron will be low.

A neuron can thus be seen as performing an analogue computation between the set of inputs and its synaptic weights, and producing a firing rate output that depends on this product. A neuron can thus be seen as performing a similarity operation. Its output depends on the similarity between its set of inputs and its synaptic weights. Mathematically, this operation is close to a correlation, and thus a neuron can be thought of as computing a correlation between its input set of firings, and its synaptic weights (Rolls 2008c). Inhibitory interneurons receive from the excitatory neurons, and send back inhibition to them, so that it is the neurons that receive the most similar input set of firings to their set of synaptic weights that end up firing fastest.

For those who would like to see this a little more formally (and this paragraph might be skipped), let us consider a neuron i as shown in Fig. 2.6, which receives inputs from axons that we label j through synapses of strength w_{ij}. The first subscript (i) refers to the receiving neuron, and the second subscript (j) to the particular input. j counts from 1 to C, where C is the number of synapses or connections received. The firing rate of the ith neuron is denoted as y_i, and that of the jth input to the neuron as x_j. To express the idea that the neuron makes a simple linear summation of the inputs it receives, we can write the activation of neuron i, denoted h_i, as

$$h_i = \sum_j x_j w_{ij} \qquad (2.1)$$

where $\sum_j$ indicates that the sum is over the C input axons (or connections) indexed by j to each neuron. The multiplicative form here indicates that activ-

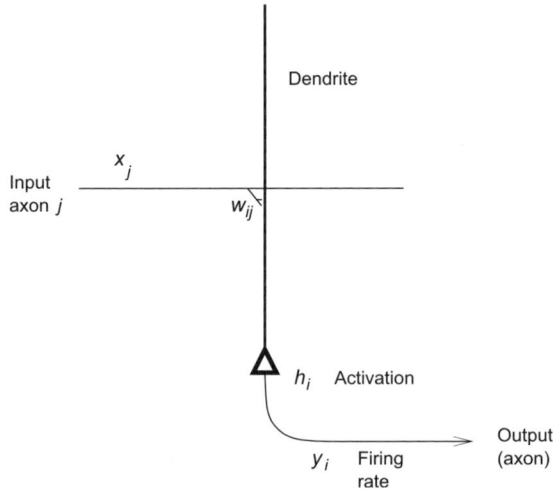

Fig. 2.6 Notation used to describe an individual neuron in a network model. By convention, we generally represent the dendrite as thick, and vertically oriented (as this is the normal way that neuroscientists view cortical pyramidal cells under the microscope); and the axon as thin. The cell body or soma is indicated between them. The firing rate we also call the activity of the neuron. The firing rate of the jth input axon is x_j, and it synapses at a synapse with weight w_{ij} onto neuron i to produce activation h_i. The activation, if sufficient to exceed the threshold for firing of the neuron, elicits an action potential or spike. The firing rate of these action potentials is y_i, and is the output of the neuron. There are many such inputs to a neuron, and their effects are summed within a short time period in the order of 20 ms.

ation should be produced by an axon only if it is firing, and depending on the strength of the synapse w_{ij} from input axon j onto the dendrite of the receiving neuron i. Equation 2.1 indicates that the strength of the activation reflects how fast the axon j is firing (that is x_j), and how strong the synapse w_{ij} is. The sum of all such activations expresses the idea that summation (of synaptic currents in real neurons) occurs along the length of the dendrite, to produce activation at the cell body. (The operation shown in Equation 2.1 is inner product or dot product multiplication of two vectors, a firing rate and a synaptic weight vector, to produce a scalar output, the activation of the neuron. The operation is identical to computing the correlation between two vectors each with a mean of zero (see further Rolls (2008c)).)

The activation of the neuron, or its depolarization, h_i, is then converted into a firing rate y_i, by a spike being produced whenever the firing threshold is received. Because a threshold is a non-linear operation (in that an output occurs only if the threshold is exceeded), a neuron after the linear summation then performs a non-linear operation, and this non-linearity is important in many ways, where the activation h_i is converted into a spike firing rate y_i. The non-linearity is important for example in only allowing relatively large inputs to produce an output, and this is helpful in removing the effects of interfering inputs to the neuron from for example other memory patterns stored in a network so that recall can be perfect, and in many computations for which non-linearity is essential (Rolls 2008c).

This is a fundamental way of thinking about brain computation. It applies directly to understanding memory retrieval in the brain. If a set of synaptic weights was increased to store a memory, then the neuron performs memory retrieval by firing fast when the recall cue exactly matches the synaptic weights on that neuron that stored the memory. It applies to perception, in that if a set of synaptic weights was increased to represent a feature in the input (such as a vertical bar, or a face), then that neuron will fire fastest to a stimulus that matches the stimulus feature (the vertical bar, or the face) stored by the modified synaptic weights.

2.5 Learning implemented in different neuronal network architectures

We now consider three architectural principles for neuronal networks found in the brain, and examine what functions each type of architecture performs. This provides a foundation for understanding many important functions performed by the brain in particular brain regions, which are described later in this neuroscience chapter. First I describe the rules that set the strengths of the synaptic connections between the excitatory cortical pyramidal cells in these architectures.

2.5.1 Learning implemented by associative synaptic modification

The main learning principle established for learning in the brain is an increase in synaptic strength if the presynaptic neuron is firing fast (x_j is high), and at the same time the postsynaptic neuron is strongly activated (y_i is high). The change in the synaptic strength δw_{ij} is associative in that both the presynaptic firing and the postsynaptic firing must be high at the same time (or with the presynaptic firing starting just before the postsynaptic firing). This associative effect was guessed at by Donald Hebb (Hebb 1949), and is now established in many studies of long-term synaptic potentiation (LTP) which also implicate this mechanism in many types of learning (Martin, Grimwood and Morris 2000, Rolls 2008c).

Because *both* presynaptic *and* postsynaptic activity are required at the same time for synaptic modification, we can express this mathematically by multiplication, so that we can formalize the change of synaptic strength by what has become known as the Hebb rule (although Hebb himself did not formalize it):

$$\delta w_{ij} = \alpha y_i x_j \qquad (2.2)$$

where δw_{ij} is the change of the synaptic weight w_{ij} which results from the simultaneous (or conjunctive) presence of presynaptic firing x_j and postsynaptic firing y_i (or strong depolarization), and α is a learning rate constant that specifies how much the synapse alters on any one pairing. The presynaptic and postsynaptic activity must be present approximately simultaneously (to within perhaps 100–500 ms in the real brain). In practice, NMDA (N-methyl-D-aspartate) receptors appear to be required for this type of synaptic modification, and they have an interesting non-linearity such that only a strongly activated postsynaptic neuron will support synaptic modification, and this may be a way for the brain not to store small, possibly noise-related or interference-related, effects, but instead to store information when there is strong co-activity between neurons (Rolls 2008c).

An important property is that the effects are synapse-specific, that is the only synapses that increase in strength are those with conjunctive presynaptic and postsynaptic activity. Part of the importance of this is that the capacity of the network, for example the number of memories it can store, or the number of associations that can be formed, is in the order of the number of synapses onto any one neuron in the network. In the cortex this number is large, in the order of 10,000 or more synapses onto each neuron, and thus any one network in the brain can store and recall correctly in the order of 10,000 different items (Rolls 2008c). Such a network might involve just 100,000 neurons (given that every neuron in a small region is not connected to every other neuron), and might occupy an area just 1–2 mm in diameter of the cerebral cortex. The size of such a module or computational cortical column in the cortex is determined by the fact that the excitatory connections that run back from one neuron to connect to other cortical neurons (the recurrent collateral axons, which are branches of the axon that continues on to another cortical area) have this connectivity limited to a small diameter of the cortex, typically with a diameter of 1–2 mm in a column that occupies the depth of the cortex, 2–3 mm.

This storage capacity of a single small network in the brain occupying just one computational cortical column of the cerebral cortex is quite remarkable, given that the number of words in a person's vocabulary is typically less than 10,000 (and this fact may be no coincidence). To be clear: all the nouns in a person's vocabulary might require just one cortical network 1–2 mm in diameter to store them.

I believe that spatially separate representations in the cortex of this modular type, possibly in different cortical areas, have this ability to recall their contents separately from other computational cortical columns, but that the location in the brain defines the role. For example, in language, one network in one part of the cortex might specify 10,000 nouns used as subjects in a sentence, and another network in another cortical module 10,000 nouns used as objects. In this way, computational cortical columns each containing the same set of nouns would specify, depending on their place in the cortex, whether the columns represent the subjects or objects in sentences. Nearby cortical columns to each of these subject or object computational columns, connected by stronger forward than backward connections to ensure a smooth forward dynamical trajectory in time during the production of language (Rolls 2008c, Rolls and Deco 2010), would specify adjectives related to the subjects or to the objects that need to be represented. This proposal is for a **cortical place code for the syntax required by language**. The proposal is that the place code is similar to that used for other cortical functions, where neuronal firing in one part of the cortex defines it as being related to visual inputs, and in another part of the cortex as being related to auditory inputs. However, we know little about how language and especially syntax (grammar) is computed in the cortex. This is perhaps the last great computational mystery of brain function to be unravelled, given that we do have principles that appear to be capable of accounting for many brain functions including long-term memory, short-term memory, perception, attention, emotion, and decision-making (Rolls 2005a, Rolls 2008c, Rolls and Deco 2010).

2.5.2 Pattern association memory

One example of associative learning is association of a stimulus pattern (set of neurons firing) such as a sound or visual stimulus which when paired with a reward or punisher such as food or shock comes to predict that food or shock. The prediction may lead to a response such as salivation or fear, which is an example of classical or Pavlovian conditioning (Rolls 2005a, Rolls 2008c). In this case the sound of the bell becomes a conditioned stimulus that elicits salivation as a conditioned response when paired with food, which is the unconditioned stimulus that elicits the unconditioned response of salivation.

This type of pattern association learning is also that which is fundamental to emotions, in which a stimulus paired for example with a loved one, or a recalled memory of that stimulus, can come to elicit emotional states, as described in Chapter 3. Another example is that the sight of a food that has become associated by learning with its flavour can elicit some of the same states that are elicited by the flavour of the food, including wanting the food (an effect called the salted nut phenomenon or incentive motivation, in which a salted nut given away in a market may make one want to buy some (Rolls 2005a)). More generally, this type of pattern association learning occurs whenever two stimuli, events, or memories are present at approximately the same time, when synaptic modification between the neurons that represent each may allow them to be associated together, and one to recall the other.

The neuronal network architecture for pattern association learning is shown in Fig. 2.7a and b. The output neurons are driven by the unconditioned stimulus

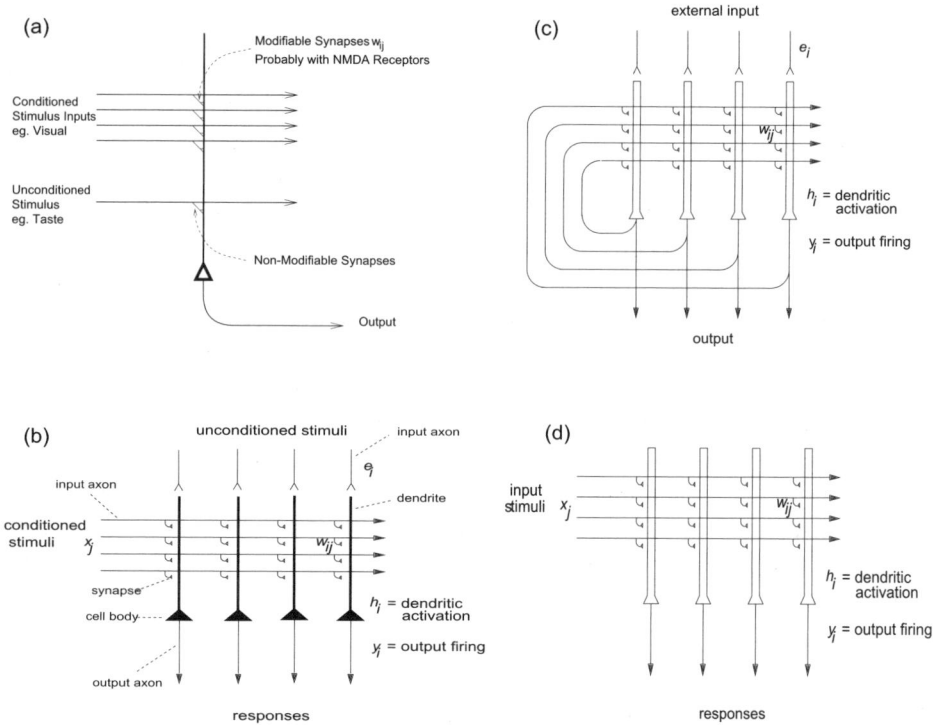

Fig. 2.7 Three neuronal network architectures. (a) Pattern association introduced with a single output neuron. (b) Pattern association network. (c) Autoassociation network. e_i is the external input to each output neuron i, and y_j is the presynaptic firing rate to synapse w_{ij}. The triangles represent the pyramidal cell bodies, and the thick lines or open rectangles above each cell the dendrites of the cell. The thin lines represent the axons. (d) Competitive network.

(or by the stimulus to which the association should be made). The conditioned stimulus reaches the output neurons by associatively modifiable synapses w_{ij}. If the conditioned stimulus is paired during learning with activation of the output neurons produced by the unconditioned stimulus, then later, after learning, due to the associative synaptic modification, the conditioned stimulus alone will produce the same output as the unconditioned stimulus. This class of network shows generalization provided that distributed representations are used. A more detailed account is provided by Rolls (2008c).

2.5.3 Autoassociation or attractor memory

Another type of associative learning occurs when the memory of a particular episode is formed, such as where one was at breakfast yesterday, what was eaten, and who was present, in which the different parts of the memory (typically involving a place, a time, and objects or people present) become associated together because different synapses carrying the different types of information all increase in strength on the same postsynaptic neuron or neurons, using the architecture of an autoassociative or attractor memory (Rolls 2008c, Rolls 2010c) (see Fig. 2.7c). The same architecture provides a short-term memory (Rolls 2008c).

Formal models of neural networks are needed in order to provide a basis for understanding the memory, attention, and decision-making functions performed by real neuronal networks in the brain. The aims of this section are to describe formal models of attractor networks in the brain and their dynamics as implemented by neurons, to introduce the concept of stability in these networks, and to introduce the concept of noise produced by statistical fluctuations of the spiking of neurons, and how this noise influences the stability of these networks. More detailed descriptions of some of the quantitative aspects of storage in autoassociation networks are provided in the Appendices of Rolls and Treves (1998) *Neural Networks and Brain Function*. Another book that provides a clear and quantitative introduction to some of these networks is Hertz, Krogh and Palmer (1991) *Introduction to the Theory of Neural Computation*, and other useful sources include Dayan and Abbott (2001), Amit (1989) (for attractor networks), Koch (1999) (for a biophysical approach), Wilson (1999) (on spiking networks), Gerstner and Kistler (2002) (on spiking networks), Rolls and Deco (2002), Rolls (2008c) *Memory, Attention, and Decision-Making*, and Rolls and Deco (2010) *The Noisy Brain: Stochastic Dynamics as a Principle of Brain Function*.

Autoassociative memories, or attractor neural networks, store memories, each one of which is represented by a pattern of neural activity. The memories are stored in the recurrent synaptic connections between the neurons of the network, for example in the recurrent collateral connections between cortical pyramidal cells. Indeed, the presence of recurrent collateral excitatory connections between pyramidal cells is perhaps the most important characteristic of the cerebral neocortex, for they allow the formation of short-term memory, and hence attention which requires a short-term memory to hold the object of attention in mind, planning (which requires several steps of a plan to be held in mind), and language (which probably requires several parts of each sentence to be held online during processing).

This is a key architecture in the design of the neocortex, which is most of the cortex that overlies the rest of the brain, and in which the connections between the excitatory neurons spread over just a few mm. The same design is used in the hippocampal cortex, in the CA3 region, but here the connections range widely between the pyramidal cells.

2.5.3.1 Architecture and operation

The prototypical architecture of an autoassociation memory is shown in Fig. 2.7c. The external input e_i is applied to each neuron i by unmodifiable synapses.

This produces firing y_i of each neuron (or a vector **y** of the firing of the output neurons where each element in the vector is the firing rate of a single neuron). Each output neuron i is connected by a recurrent collateral connection to the other neurons in the network, via connection weights w_{ij}. The synaptic connections can be increased associatively, that is their strength will increase if the postsynaptic neuron and the presynaptic neuron are simultaneously active. This architecture effectively enables the output firing to be associated during learning with itself, hence the term autoassociation memory (Kohonen 1977, Kohonen, Oja and Lehtio 1981). Later on, during recall, presentation of part of the external input will force some of the output neurons to fire, but through the recurrent collateral axons and the modified synapses, other neurons can be brought into activity. This process can be repeated a number of times, and recall of a complete pattern may be perfect. Effectively, a pattern can be recalled or recognized because of associations formed between its parts.

The network is a positive feedback system, and its activity is controlled by inhibitory interneurons. The connectivity is that the pyramidal cells have collateral axons that excite the inhibitory interneurons, which in turn connect back to the population of pyramidal cells to inhibit them. However, the positive feedback architecture is inherently unstable, and the price paid for the valuable property of short-term and long-term memory, and all the computational processes including decision-making that are related to this, is the potential for instability, evident in states such as epilepsy.

During recall, a part of one of the originally learned stimuli can be presented as an external input. The resulting firing is allowed to iterate repeatedly round the recurrent collateral system, gradually on each iteration recalling more and more of the originally learned pattern. *Completion* thus occurs. If a pattern is presented during recall that is similar but not identical to any of the previously learned patterns, then the network settles into a stable recall state in which the firing corresponds to that of the previously learned pattern. The network is thus attracted into a state that represents one of the previously learned memory patterns, and this is why they are called *attractor networks*. An attractor network can store many different memory patterns, and during recall when the whole of one of the stored memories is recalled, an *attractor state* arises in which the neurons that are active are those that are in one of the stored patterns, and the other neurons in the network are inhibited by the inhibitory neurons. During recall, there is thus a form of competition between the possible attractor states, with one winning.

Once a high firing rate attractor state is present, the positive feedback round the recurrent collaterals keeps the neurons firing, and this is how short-term memory is implemented.

The main factors that determine the maximum number of memories that can be stored in an autoassociative network are the number of connections on each neuron devoted to the recurrent collaterals, and the sparseness of the representation which can be thought of as the proportion of neurons that are active for any one memory pattern. For example, for 12,000 recurrent collateral connections onto each neuron, a number found in the rat hippocampus and possibly higher in humans, and the sparseness = 0.02, the number of memories that can be stored is calculated to be approximately 36,000 (Treves and Rolls 1991, Rolls and Treves 1998, Rolls 2008c).

2.5.3.2 Attractor networks, energy landscapes, and stochastic dynamics

The attractor network system formally resembles spin glass systems of magnets analysed quantitatively in the area of physics known as statistical mechanics. This has led to the analysis of (recurrent) autoassociative networks as dynamical systems made up of many interacting elements, in which the interactions are such as to produce a large variety of basins of attraction of the dynamics. Each basin of attraction corresponds to one of the originally learned patterns, and once the network is within a basin it keeps iterating until a recall state is reached that is the learned pattern itself or a pattern closely similar to it. (Interference effects may prevent an exact identity between the recall state and a learned pattern.) The states reached within each basin of attraction are called attractor states, and the analogy between autoassociator neural networks and physical systems with multiple attractors was drawn by John Hopfield (1982) in a very influential paper. He was able to show that the recall state can be thought of as the local minimum in an energy landscape, where the energy would be defined as

$$E = -\frac{1}{2} \sum_{i,j} w_{ij}(y_i - \langle y \rangle)(y_j - \langle y \rangle). \qquad (2.3)$$

This equation can be understood in the following way. If two neurons are both firing above their mean rate (denoted by $\langle y \rangle$), and are connected by a weight with a positive value, then the firing of these two neurons is consistent with each other, and they mutually support each other, so that they contribute to the system's tendency to remain stable. If across the whole network such mutual support is generally provided, then no further change will take place, and the system will indeed remain stable. If, on the other hand, either of our pair of neurons was not firing, or if the connecting weight had a negative value, the neurons would not support each other, and indeed the tendency would be for the neurons to try to alter ('flip' in the case of binary units) the state of the other. This would be repeated across the whole network until a situation in which most mutual support, and least 'frustration', was reached. What makes it possible to define an energy function and for these points to hold is that the synaptic connection matrix is symmetric (Hopfield 1982, Hertz, Krogh and Palmer 1991, Amit 1989).

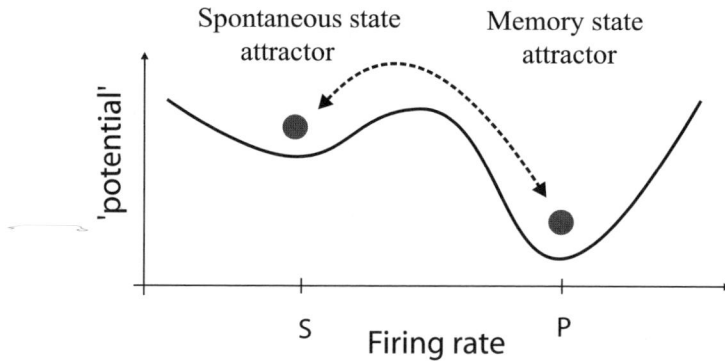

Fig. 2.8 Energy landscape. The noise influences when the system will jump out of the spontaneous firing stable (low energy) state S, and whether it jumps into the high firing rate state labelled P (with persistent or continuing firing in a state which is even more stable with even lower energy), which might correspond to a short-term memory, or to a recalled long-term memory, or to a decision.

Autoassociation attractor systems (described in Section 2.5.3) have two types of stable fixed points: a spontaneous state with a low firing rate, and one or more attractor states with high firing rates in which the positive feedback implemented by the recurrent collateral connections maintains a high firing rate. We sometimes refer to this latter state as the persistent state, because the high firing normally persists to maintain a set of neurons active, which might implement a short-term memory.

The stable points of the system can be visualized in an energy landscape (see Fig. 2.8). The area in the energy landscape within which the system will move to a stable attractor state is called its basin of attraction. The attractor dynamics can be pictured by energy landscapes, which indicate the basins of attraction by valleys, and the attractor states or fixed points by the bottom of the valleys (see Fig. 2.8).

The stability of an attractor is characterized by the average time in which the system stays in the basin of attraction under the influence of noise. The noise provokes transitions to other attractor states. One source of noise results from the interplay between the Poissonian character of the spikes (the fact that they occur at nearly random times for a given mean firing rate) and the finite-size effect due to the limited number of neurons in the network (Rolls and Deco 2010).

Two factors determine the stability. First, if the depths of the attractors are shallow (as in the left compared to the right valley in Fig. 2.8), then less force is needed to move a ball from one valley to the next. Second, high noise will make it more likely that the system will jump over an energy boundary from one state to another. We envisage that the brain as a dynamical system has characteristics of such an attractor system including statistical fluctuations (see Rolls and Deco (2010), where the effects of noise are defined quantitatively). The noise could arise not only from the probabilistic spiking of the neurons which has significant effects in finite size integrate-and-fire networks (Deco and Rolls 2006), but also from any other source of noise in the brain or the environment (Faisal, Selen and Wolpert 2008), including the effects of distracting stimuli. The reader is referred to one of the sources above for a more quantitative description of these processes.

Attractor networks in the cortex are fundamental to the modern understanding we have developed (Rolls 2008c) of short-term memory (Section 2.6), of attention (Rolls and Deco 2002, Rolls 2008c) (Section 2.9), of long-term memory and the recall of long-term memory (Rolls 2010c) (Section 2.8), of decision-making (Rolls and Deco 2010) (Section 2.12), of how instability or noise in these process is important in creativity (Rolls and Deco 2010) (Section 2.12.9) and in neuropsychiatric disorders such as schizophrenia (Rolls, Loh, Deco and Winterer 2008d) (Section 10.1), and of how overstability in some cortical systems is implicated in obsessive-compulsive disorder (Rolls, Loh and Deco 2008c, Rolls 2011f) (Section 10.2).

2.5.4 Competitive networks

In the third architecture, the main input to the output neurons is received through associatively modifiable synapses w_{ij} (see Fig. 2.7d). Because of the initial values of the synaptic strengths, or because every axon does not contact every output neuron, different input patterns tend to activate different output neurons. When one pattern is being presented, the most strongly activated neurons tend (via lateral inhibition) to inhibit the other neurons. For this reason the network is called competitive. During the presentation of that pattern, associative modification of the active axons onto the active postsynaptic neuron takes place. Later, that or similar patterns will have a greater chance of activating that neuron or set of neurons. Other neurons learn to respond to other input patterns. In this way, a network is built that can categorize patterns, placing similar patterns into the same category.

This type of computation is fundamental as a processor of sensory information, for the connections in the architecture self-organize to produce feature analyzers in which each neuron responds to a different combination of perceptual features, such as two lines that intersect at a given angle, or a particular combination of eyes, mouth and hair to form a particular face (Rolls 2008c, Rolls 2009d, Rolls 2011d). This is a key building block of perception.

This architecture is frequently found at each stage of the hierarchy of cortical processing areas for each sensory system (vision, taste, smell etc). In such a hierarchy, there is frequently convergence of inputs to a given part of one stage from a small region of the preceding stage. This allows feature combinations to be formed at early stages between quite local feature, such as two lines that form a corner of an eye, to combinations of many separate features such as eyes, mouth, nose, forehead and hair to form a particular combination that represents the face of a particular person (see Chapter 4 of *Memory, Attention, and Decision-Making* (Rolls 2008c), and Fig. 2.31).

All three architectures require inhibitory interneurons, which receive inputs from the principal neurons in the network (usually the pyramidal cells shown in Fig. 2.7) and implement feedback inhibition by connections to the pyramidal cells. The inhibition is usually implemented by GABA (gamma-amino-butyric acid) neurons, and maintains a small proportion of the pyramidal cells active (Rolls 2008c).

These fundamental building blocks can be combined together in single networks. For example, an attractor network in the cortex may act as a competitive network for learning new representations when it is being driven by forward inputs synapsing onto another part of the dendrite than that devoted to the recurrent collateral connections (Rolls 2008c). Such a network can then operate as a building block of perceptual representations, but can also implement a short-term memory for what is represented by the neurons, which can be extremely useful (Fig. 2.31) (Rolls 2008c).

2.5.5 Stochastic dynamics: integrate-and-fire neuronal networks

So far, we have been thinking of neurons as having average firing rates, measured for example by y_i. The firing rates to which we refer are of action potentials or spikes elicited by a neuron. These spikes all have the same height, and can be carried faithfully without signal loss to the ends of the axons, because as they travel the spikes self-regenerate. This provides for lossless signal transmission. The rate of arrival of the spikes carries the information (Section 2.3). However, the exact time at which each spike of a neuron is emitted is rather variable and random, even when it is firing at a given average firing rate measured over say 1 s. The randomness is sometimes described as noise (because the hiss sound that one hears when a radio is not tuned into a station is effectively a random process), and the spiking is described as stochastic, meaning that it jumps into action at rather random times.

It turns out that this stochasticity is important in brain function, in for example decision-making (Rolls and Deco 2010). For example, this stochasticity of neurons accounts for why the particular choices we make when the odds are equal are often essentially random, for example in a choice of heads or tails for the toss of a coin.

To understand how noise becomes important in brain function, we need to understand how neurons operate in more detail, how their spiking times can be almost random for a given mean rate, how this contributes to probabilistic

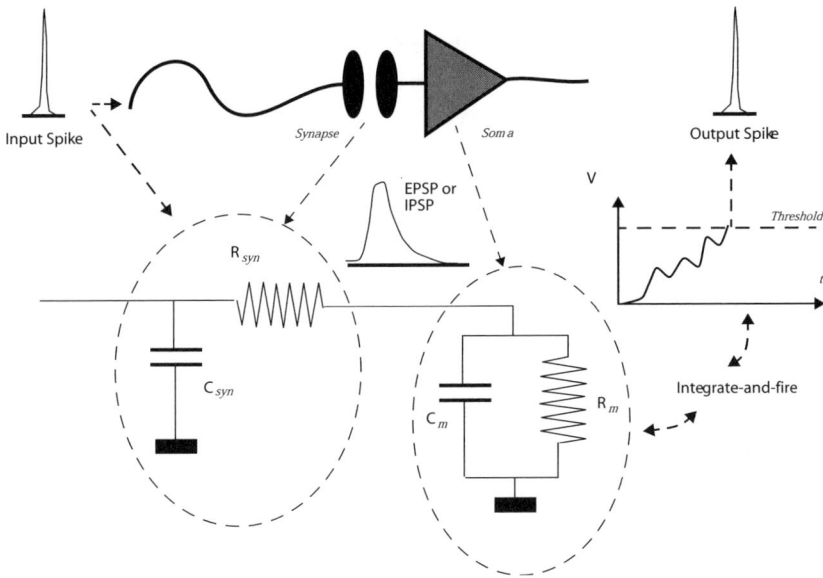

Fig. 2.9 Integrate-and-fire neuron. The basic circuit of an integrate-and-fire model consists of the neuron's membrane capacitance C_m in parallel with the membrane's resistance R_m (the reciprocal of the membrane conductance g_m) driven by a synaptic current with a conductance and time constant determined by the synaptic resistance R_{syn} (the reciprocal of the synaptic conductance g_j) and capacitance C_{syn} shown in the figure. These effects produce excitatory or inhibitory postsynaptic potentials, EPSPs or IPSPs. These potentials are integrated by the cell, and if a threshold V_{thr} is reached a δ-pulse (spike) is fired and transmitted to other neurons, and the membrane potential is reset. (After Deco and Rolls 2003.)

effects in networks of neurons, and how we can implement this in our models of brain activity.

A relatively simple way to understand and model neurons is the integrate-and-fire approach. When an action potential arrives at a synapse, it opens ion channels for a short time, typically 10 ms, and this allows current to flow across the cell membrane. This process has a synaptic time constant (characteristic time for the effect to decay) determined by the synaptic resistance R_{syn} and capacitance C_{syn} shown in Fig. 2.9. The dendrites of the neuron sum or integrate all these currents to alter the voltage across the cell membrane which decays with a neuronal time constant typically in the order of 20 ms determined by the neuron's membrane capacitance C_m in parallel with the membrane's resistance R_m. As spikes arrive at the neuron through excitatory synapses, the neuron's membrane potential then becomes depolarized and moves from approximately -70 mV, the resting potential, towards -50 mV, as illustrated in Fig. 2.9. When the firing threshold is reached, which is typically -50 mV, an action potential or spike is generated, and is transmitted along the axon of the neuron. After the action potential, which is typically 1 ms long, the neuron's membrane potential is reset to approximately −70 mV, there is a short refractory period of typically 1–2 ms, and then the whole process starts again.

The time of emission of the spikes is almost random (almost a Poisson process) because the membrane potential typically moves to be quite close to the firing threshold and then stays close to it for some time, so that a small amount of extra input, caused by random effects described below and the random firing times of neurons that connect to it, can cause the firing threshold to be suddenly exceeded at a rather unpredictable time.

2.5.6 Reasons why the brain is inherently noisy and stochastic

We must therefore address the conceptually important issue of why neuronal spiking, and the operation of what is effectively memory retrieval, is probabilistic. Even when a fully connected recurrent attractor network has as many as 4,000 neurons, the operation of the network is still probabilistic (Deco and Rolls 2006). Under these conditions, the probabilistic spiking of the excitatory (pyramidal) cells in the recurrent collateral firing, rather than variability in the external inputs to the network, is what makes the major contribution to the noise in the network (Deco and Rolls 2006). Thus, once the firing in the recurrent collaterals is spike-implemented by integrate-and-fire neurons, the probabilistic behaviour seems inevitable, even up to quite large attractor network sizes.

We may then ask why the spiking activity of any neuron is probabilistic, and what the advantages are that this may confer. The answer suggested (Rolls 2008c) is that the spiking activity is approximately Poisson-like (as if generated by a random process with a given mean rate, both in the brain and in the integrate-and-fire simulations we describe), because the neurons are held close to (just slightly below) their firing threshold, so that any incoming input can rapidly cause sufficient further depolarization to produce a spike. It is this ability to respond rapidly to an input, rather than having to charge up the cell membrane from the resting potential to the threshold, a slow process determined by the time constant of the neuron and influenced by that of the synapses, that enables neuronal networks in the brain, including attractor networks, to operate and retrieve information so rapidly (Treves 1993, Rolls and Treves 1998, Battaglia and Treves 1998, Rolls 2008c, Panzeri, Rolls, Battaglia and Lavis 2001). The spike trains are essentially Poisson-like because the cell potential hovers noisily close to the threshold for firing (Dayan and Abbott 2001), the noise being generated in part by the Poisson-like firing of the other neurons in the network (Jackson 2004), and partly by randomness (noise) in the ion channels opened by synapses and in the neuron (Faisal et al. 2008). The noise and spontaneous firing help to ensure that when a stimulus arrives, there are always some neurons very close to threshold that respond rapidly, and then communicate their firing to other neurons through the modified synaptic weights, so that an attractor process can take place very rapidly (Rolls 2008c).

The implication of these concepts is that the operation of networks in the brain is inherently noisy because of the Poisson-like timing of the spikes of the neurons, which itself is related to the mechanisms that enable neurons to respond rapidly to their inputs (Rolls 2008c). Factors that cause the neurons to be noisy include synaptic noise or variability (Faisal et al. 2008).

We will see in Section 2.12 how this effect at a neuronal level is translated into probabilistic effects including probabilistic decision-making at the neuronal network level.

We now consider briefly the operation of a number of different brain systems, to illustrate how we understand now quite a lot about how the brain works. This conceptual and computational understanding for how a number of brain systems may actually operate, including systems involved in short-term memory, attention, long-term memory, decision-making, and perception provides a foundation for some of the topics considered in the other chapters of this book. More detailed descriptions of the operation of these systems are provided elsewhere (Rolls 2008c, Rolls and Deco 2010).

2.6 Short-term memory

We have seen that an attractor or autoassociation network can provide a short-term memory. To illustrate this, and to clarify the concept of stability, we show examples of trials of integrate-and-fire attractor network simulations in which the statistical fluctuations have different impacts on the temporal dynamics (Rolls, Loh and Deco 2008c). Figure 2.10 shows the possibilities, as follows.

In the spontaneous state simulations, no cue was applied, and we are interested in whether the network remains stably in the spontaneous firing state, or whether it is unstable and on some trials due to statistical fluctuations enters one of the attractors, thus falsely retrieving a memory. Figure 2.10 (top) shows an example of a trial on which the network correctly stayed in the low spontaneous firing rate regime, and (bottom) another trial (labelled spontaneous unstable) in which statistical spiking-related fluctuations in the network caused it to enter a high activity state, moving into one of the attractors even without a stimulus.

In the persistent state simulations (in which the short-term memory was implemented by the continuing neuronal firing), a strong excitatory input was given to the S1 neuronal population between 0 and 500 ms. Two such trials are shown in Fig. 2.10. In Fig. 2.10 (top), the S1 neurons (correctly) keep firing at approximately 30 Hz after the retrieval cue is removed at 500 ms. However, due to statistical fluctuations in the network related to the spiking activity, on the trial labelled persistent unstable the high firing rate in the attractor for S1 was not stable, and the firing decreased back towards the spontaneous level, in

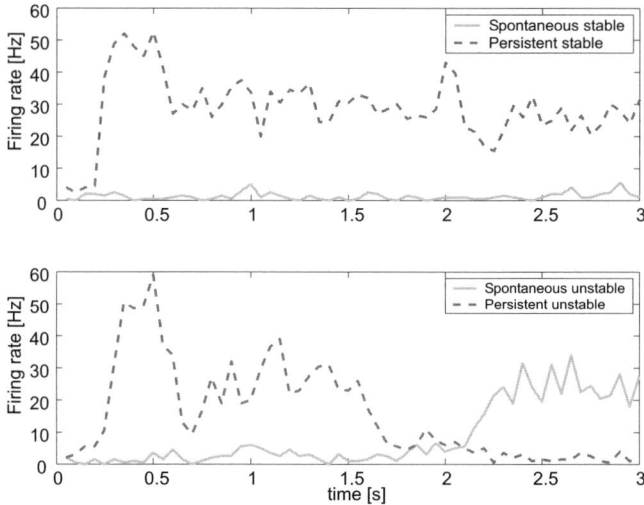

Fig. 2.10 Example trials of the integrate-and-fire attractor network simulations of short-term memory. The average firing rate of all the neurons in the S1 pool is shown. Top. Normal operation. On a trial in which a recall stimulus was applied to S1 at 0–500 ms, firing continued normally until the end of the trial in the 'persistent' simulation condition. On a trial on which no recall stimulus was applied to S1, spontaneous firing continued until the end of the trial in the 'spontaneous' simulation condition. Bottom: Unstable operation. On this persistent condition trial, the firing decreased during the trial as the network fell out of the attractor because of the statistical fluctuations caused by the spiking dynamics. On the spontaneous condition trial, the firing increased during the trial because of the statistical fluctuations. In these simulations the network parameter was w_+=2.1. (After Rolls, Loh, and Deco 2008b.)

the example shown starting after 1.5 s (Fig. 2.10 bottom). This trial illustrates a failure to maintain a stable short-term memory state.

When an average was taken over many trials, for the persistent run simulations, in which the cue triggered the attractor into the high firing rate attractor state, the network was still in the high firing rate attractor state in the baseline condition on 88% of the runs. The noise had thus caused the network to fail to maintain the short-term memory on 12% of the runs.

The spontaneous state was unstable on approximately 10% of the trials, that is, on 10% of the trials the spiking noise in the network caused the network run in the condition without any initial retrieval cue to end up in a high firing rate attractor state. This is of course an error that is related to the spiking noise in the network.

I emphasize that the transitions to the incorrect activity states illustrated in Fig. 2.10 are caused by statistical fluctuations (noise) in the spiking activity of the integrate-and-fire neurons. Indeed, we used a mean-field approach with an equivalent network with the same parameters but without noise to establish parameter values where the spontaneous state and the high firing rate short-term memory ('persistent') state would be stable without the spiking noise, and then in integrate-and-fire simulations with the same parameter values examined the effects of the spiking noise (Loh, Rolls and Deco 2007a, Rolls, Loh and Deco 2008c). These attractor states are independent of any simulation protocol of the spiking simulations and represent the behaviour of the network by mean firing rates to which the system would converge in the absence of statistical fluctuations caused by the spiking of the neurons and by external changes. Therefore the mean-field technique is suitable for tasks in which temporal dynamics and fluctuations are negligible. It also allows a first assessment of the attractor landscape and the depths of the basin of attraction which then need to be investigated in detail with stochastic spiking simulations. Part of the utility of the mean-field approach is that it allows the parameter region for the synaptic strengths to be investigated to determine which synaptic strengths will on average produce stable activity in the network, for example of persistent activity in a delay period after the removal of a stimulus. For the spontaneous state, the initial conditions for numerical simulations of the mean-field method were set to 3 Hz for all excitatory pools and 9 Hz for the inhibitory pool. These values correspond to the approximate values of the spontaneous attractors when the network is not driven by stimulus-specific inputs. For the persistent state, the network parameters resulted in a selective pool having a high firing rate value of approximately 30 Hz when in its attractor state (Loh, Rolls and Deco 2007a, Rolls, Loh and Deco 2008c).

The dorsolateral prefrontal cortex (which includes areas 9, 10 and 46 shown in Fig. 3.4) is one brain area in which one or more short-term memory systems are implemented (Fuster 1973, Fuster 1989, Fuster 1995, Goldman-Rakic 1996, Fuster 2000). For example, the neuron shown in Fig. 2.11 fired in the delay period in which a monkey was remembering to move its eyes in the downward direction, but not for other directions (Funahashi, Bruce and Goldman-Rakic 1989, Goldman-Rakic 1996). Given that cortical connections are typically short range, over a few mm, several different items can be held in short-term memory simultaneously, with each item in an independent network. This is important in planning, in which the several possible steps to a plan must be held in mind simultaneously, and possibly re-ordered. (The manipulation of items in short-term memory is referred to as working memory (Baddeley 2002).) Disorders produced by damage to the prefrontal cortex may include disorders of planning in which the order of items is performed wrongly or the items may not even be remembered, and these are key symptoms of the dysexecutive syndrome.

Not only planning, but also reasoning, requires a number of items to be kept in mind simultaneously, and again attractor networks in cortical areas provide a fundamental computational contribution to reasoning. Reasoning involves syntactically (grammatically) organized thoughts, a fundamental aspect of language, and again, a set of attractor networks each capable of holding one item

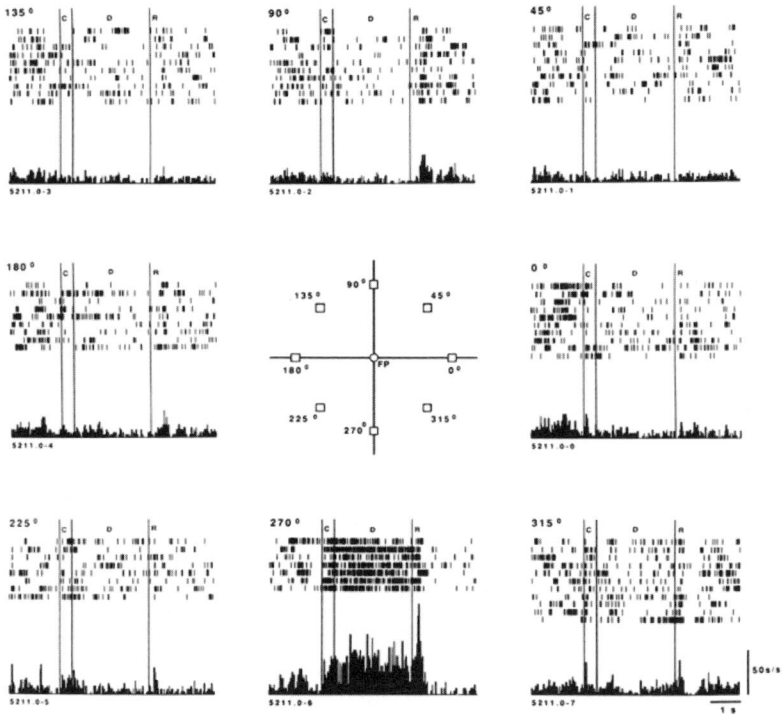

Fig. 2.11 The activity of a single neuron in the dorsolateral prefrontal cortical area involved in remembered saccades. Each row is a single trial, with each spike shown by a vertical line. A cue is shown in the cue (C) period, there is then a delay (D) period without the cue in which the cue position must be remembered, then there is a response (R) period. The monkey fixates the central fixation point (FP) during the cue and delay periods, and saccades to the position where the cue was shown, in one of the eight positions indicated, in the response period. The neuron increased its activity primarily for saccades to position $270°$. The increase of activity was in the cue, delay, and response period while the response was made. The time calibration is 1 s. (Reproduced with permission from Funahashi, Bruce and Goldman-Rakic 1989.)

active in short-term memory is likely to be important in sentence construction and parsing.

2.7 Memory for the order of items in short-term memory

How are multiple items held in short-term memory? How is a sequence of items recalled in the correct order? These questions are closely related in at least forms of short-term memory, and this provides an indicator of possible mechanisms. For example in human auditory-verbal short-term memory, when we remember a set of numbers, we naturally retrieve the last few items in the order in which they were delivered, and this is termed the recency effect (Baddeley 1986). Another fascinating property of short-term memory is that the number of items we can store and recall is in the order of the 'magic number' 7 plus or minus 2 (Miller 1956).

Investigations on the hippocampus are leading to an approach that potentially may answer these questions. Although the hippocampus is involved in spatial memory (see Section 2.8) (Rolls and Kesner 2006, Rolls 2008c, Rolls 2010c), there is also evidence that the hippocampus is involved in associations between objects and time, for example in tasks in which the temporal order of items must be remembered (Rolls and Kesner 2006). Recent neurophysiological evidence shows that neurons in the rat hippocampus have firing rates that reflect which temporal part of the task is current. In particular, a sequence of different neurons is activated at successive times during a time delay period in each trial of a task (see Fig. 2.12) (MacDonald and Eichenbaum 2009, MacDonald, Lepage, Eden and Eichenbaum 2011). The tasks used included an object-odor paired associate non-spatial task with a 10 s delay period between the visual stimulus and the odor. The new evidence also shows that a large proportion of hippocampal neurons fire in relation to individual events in a sequence being remembered (e.g. a visual object or odor), and some to combinations of the event and the time in the delay period (MacDonald et al. 2011).

These interesting neurophysiological findings indicate that rate encoding is being used to encode time, that is, the firing rates of different neurons are high at different times within a trial, delay period, etc., as illustrated in Fig. 2.12. This provides the foundation for a new computational theory of temporal order memory within the hippocampus (and also the prefrontal cortex) which I outline next, and which utilizes the slow transitions from one attractor to another which are a feature that arises at least in some networks in the brain due to the noise-influenced transitions from one state to another (Rolls and Deco 2010).

One hypothesis is that in a recurrently connected system such as the hippocampal CA3 system or local recurrent circuits in the neocortex, there are several attractor states, but that each attractor is connected by slightly stronger forward than reverse synaptic weights to the next. In previous work, we have shown that with an integrate-and-fire implementation with spiking noise this allows slow transitions from one attractor state to the next (Deco and Rolls 2003, Deco, Ledberg, Almeida and Fuster 2005). It will be of interest to investigate whether this system, because of the noise, is limited to transitions between up to perhaps 7 ± 2 different sequential firing rate states with different neuronal subpopulations for each state, and thus provides an account for the

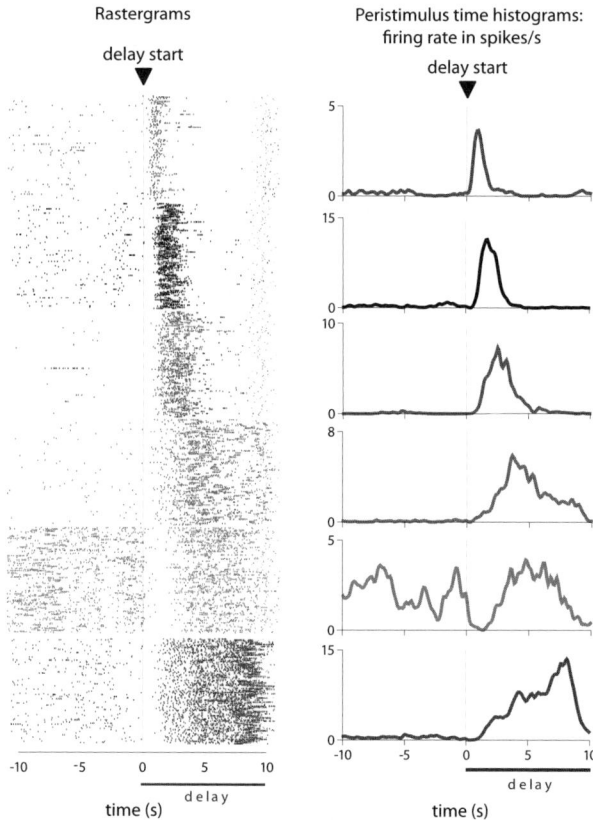

Fig. 2.12 Time encoding neurons. The activity of 6 simultaneously recorded hippocampal neurons each of which fires with a different time course during the 10 s delay in a visual object-delay-odor paired associate task. Each peristimulus time histogram (right, showing the average firing rate in spikes/s as a function of time) and set of rastergrams (left) is for a different neuron. In the rastergrams, each row is a separate trial, and each dot shows the spike time of a single neuron. The onset of the delay is shown. The onset of the delay is shown. The visual stimulus was shown before the delay period; and the odor stimulus was delivered after the delay period. (After MacDonald and Eichenbaum 2009 with permission.) (See colour plates Appendix B.)

limit of the magical number 7 ± 2 on short-term memory and related types of processing (Miller 1956), and for the recency part of short-term memory in which the items are naturally recalled in the order in which they were presented.

A variation on this implementation is to have short-term attractor memories with different time constants (for example of adaptation), but all started at the same time. This could result in some attractors starting early in the sequence and finishing early, and with other attractors starting up a little later, but lasting for much longer in time. The neurons shown in Fig. 2.12 are not inconsistent with this possibility. This type of time-encoding representation could also be used to associate with items, to implement an item-order memory.

This is thus a fundamentally new approach to how noisy networks in the brain could implement the encoding of time using one of several possible mechanisms in which temporal order is an integral component. The approach applies not only to the hippocampus, but also potentially to the prefrontal cortex. It is established that neurons in the prefrontal cortex and connected areas have activity that increases or decreases in a temporally graded and stimulus-specific way during the delay period in short-term memory, sequential decision-making with a delay, and paired associate tasks (Fuster, Bodner and Kroger 2000, Fuster 2008, Brody, Hernandez, Zainos and Romo 2003, Romo, Hernandez and Zainos 2004, Deco, Ledberg, Almeida and Fuster 2005, Deco, Rolls and Romo 2010), and it is likely too that some prefrontal cortex neurons use time encoding of the type illustrated in Fig. 2.12.

Part of the adaptive value of having a short-term memory in for example the prefrontal cortex that implements the order of the items is that short-term memory is an essential component of planning with multiple possible steps each corresponding to an item in a short-term memory, for plans require that the components be performed in the correct order. Similarly, if a series of events that has occurred results in an effect, the ability to remember the order in which those events occur can be very important in understanding the cause of the effect. This has great adaptive value in very diverse areas of human cognition, from interpreting social situations with several interacting individuals, to understanding mechanisms with a number of components.

We can note that with the above mechanism for order memory, it will be much easier to recall the items in the correct order than in any other order. Indeed, short-term memory becomes working memory when the items in the short-term memory need to be manipulated, for example recalled in reverse order, or with the item that was 2-,3-,n-back in a list. Also, in the type of order short-term memory described, recall can be started anywhere in the list by a recall cue. There is also the interesting weaker prediction that the recall will be best if it occurs at the speed with which the items arrived, as during recall the speed of the trajectory through the order state space of short-term memories will be likely to be that inherently encoded in the noisy dynamics of the linked attractors with their characteristic time constants. We can also note that a similar mechanism, involving for example stronger forward than reverse weights between long-term memory representations in attractors, could build a long-term memory with the order of the items implemented.

2.8 Long-term episodic memory and the hippocampus

Investigations of episodic memory reveal how a beautiful anatomical arrangement enables an autoassociation memory implemented by the associatively modifiable recurrent collateral connections between CA3 pyramidal neurons in the hippocampus to store episodic memories, which can then later be recalled to the neocortex. This illustrates how theories are developing on how some parts of the brain operate to implement important functions such as memory. Many of the properties of human memory, such as the ability to remember a whole memory from just a fragment, and the ability to associate together items that co-occur in time, are illustrated by the properties of this memory system. Many people as they get older have an impairment in remembering recent events, and this can also be understood by examining the factors that influence the operation of this memory system, which utilizes the hippocampus. Indeed, it is one of the aims of this brain research that by understanding the changes in the operation of the system during aging, we may be able to develop treatments that will help to maintain the performance of memory systems in the brain during aging (Barnes 2003, Burke and Barnes 2006, Rolls and Deco 2010).

2.8.1 Episodic memory is impaired by damage to the hippocampus

Episodic memory, the memory of a particular episode, requires the ability to remember particular events, and to distinguish them from other events. An event consists of a set of items that occur together, such as seeing a particular object or person's face in a particular place. An everyday example might be remembering where one was for dinner, who was present, what was eaten, what was discussed, and the time at which it occurred. The spatial context is almost always an important part of an episodic memory, and it may be partly for this reason that episodic memory is linked to the functions of the hippocampal system, which is involved in spatial processing and memory.

The most quoted example of the memory deficit produced by damage to the hippocampal system is that of the patient HM, who after bilateral surgical removal for epilepsy of part of the medial temporal lobe of the brain in which the hippocampus is located developed an inability to learn about new events (Scoville and Milner 1957, Corkin 2002). He could remember events that happened prior to the surgery, so that memory recall was little affected, but he could remember very little about events that happened after the temporal lobe surgery but more than a few minutes earlier. After the surgery, he could still find the house in which he lived before the surgery, but could not remember the way to a house to which he moved after the surgery. Semantic information (concepts and meanings) learned before the surgery was retained. The location of the hippocampus in the brain is shown in Fig. 2.13.

We can note that another type of long-term memory, recognition memory (tested for example by the ability to recognize a sample visual stimulus after a delay in which other stimuli are shown), is not impaired by hippocampal damage, produced for example by neurotoxic lesions, but depends on the

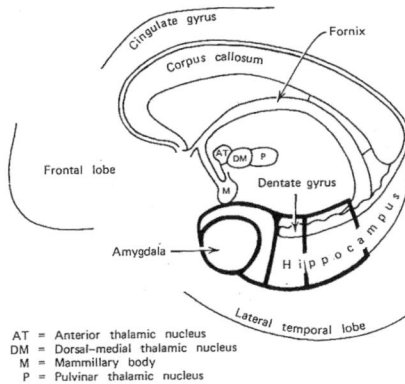

Fig. 2.13 Schematic diagram showing the hippocampus (near the base of the brain in the medial temporal lobe) which if damaged bilaterally can produce amnesia in humans. The front of the brain is on the left, the top extends above this diagram, and the back of the brain extends to the right.

perirhinal cortex, a cortical area that overlies the hippocampus (Buckley and Gaffan 2006), and in which we discovered that neuronal activity is related to the long-term familiarity memory for visual stimuli (Hölscher, Rolls and Xiang 2003, Rolls 2008c).

2.8.2 Hippocampal anatomy and episodic memory

The hippocampus receives inputs from many of the higher cortical areas of the cerebral cortex (those far from the primary cortical areas that receive sensory inputs), and these provide the hippocampus with the information that is to be remembered, such as what object is being seen, and where it is in space. After the information has been stored in the hippocampus, it must be recalled within the hippocampus by typically just a fragment of the original memory. Then the whole memory must be recalled back to the neocortex, and this is provided for by the return connections from the hippocampus back to the areas of the cerebral cortex from which it receives inputs. The hippocampus allows all these different representations (the face, the object, the location in space, etc.), each represented in a different part of the cerebral cortex, to be associated together in a single network, the CA3 network in the hippocampus, if they occur at the same time as part of an event to be remembered. This process is described to illustrate our understanding of how some parts of the brain work, and the properties of our human memory systems, in the rest of this section (2.8). A fuller description and citations to the literature are provided by Rolls (2008c) and Rolls (2010c).

Figure 2.14 shows that the primate hippocampus receives major inputs

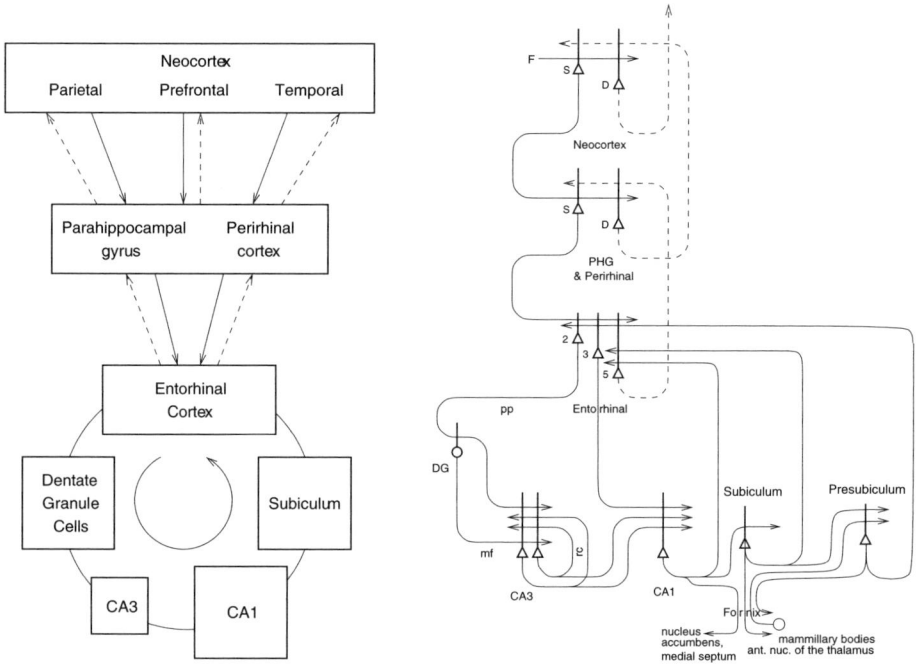

Fig. 2.14 Forward connections (solid lines) from areas of cerebral association neocortex via the parahippocampal gyrus and perirhinal cortex, and entorhinal cortex, to the hippocampus; and backprojections (dashed lines) via the hippocampal CA1 pyramidal cells, subiculum, and parahippocampal gyrus to the neocortex. There is great convergence in the forward connections down to the single network implemented in the CA3 pyramidal cells; and great divergence again in the backprojections. Left: block diagram. Right: more detailed representation of some of the principal excitatory neurons in the pathways. Abbreviations: D, deep pyramidal cells; DG, dentate granule cells; F, forward inputs to areas of the association cortex from preceding cortical areas in the hierarchy. mf: mossy fibres; PHG, parahippocampal gyrus and perirhinal cortex; pp, perforant path; rc, recurrent collaterals of the CA3 hippocampal pyramidal cells; S, superficial pyramidal cells; 2, pyramidal cells in layer 2 of the entorhinal cortex; 3, pyramidal cells in layer 3 of the entorhinal cortex; 5, 6, pyramidal cells in the deep layers of the entorhinal cortex. The thick lines above the cell bodies represent the dendrites.

via the entorhinal cortex (area 28). These come from the highly developed parahippocampal gyrus (areas TF and TH) as well as the perirhinal cortex, and thereby from the ends of many processing streams of the cerebral association cortex, including the visual and auditory temporal lobe association cortical areas, the prefrontal cortex, and the parietal cortex. The hippocampus is thus by its connections potentially able to associate together object representations (from the temporal lobe visual and auditory areas) and spatial representations. In addition, the entorhinal cortex receives inputs from the amygdala, and the orbitofrontal cortex, which could provide reward-related information to the hippocampus.

The primary output from the hippocampus to neocortex originates in CA1

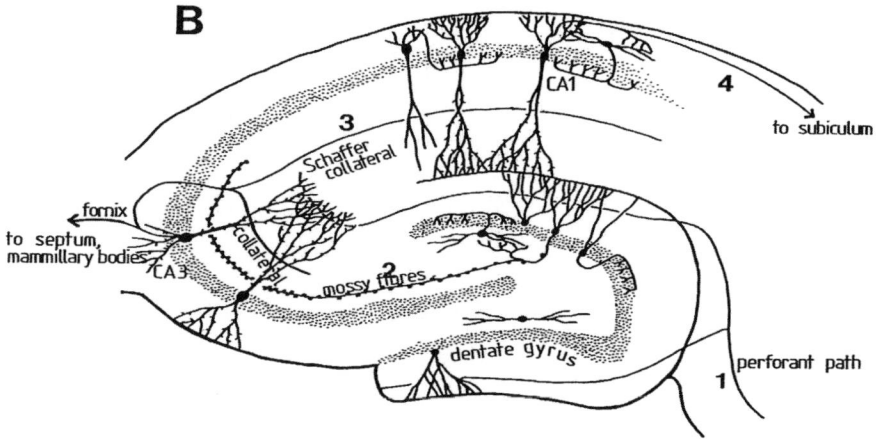

Fig. 2.15 Representation of connections within the hippocampus. Inputs reach the hippocampus through the perforant path (1) which makes synapses with the dendrites of the dentate granule cells and also with the apical dendrites of the CA3 pyramidal cells. The dentate granule cells project via the mossy fibres (2) to the CA3 pyramidal cells. The well-developed recurrent collateral system of the CA3 cells is indicated. The CA3 pyramidal cells project via the Schaffer collaterals (3) to the CA1 pyramidal cells, which in turn have connections (4) to the subiculum.

and projects to subiculum, entorhinal cortex, and parahippocampal structures (areas TF-TH) as well as prefrontal cortex.

The circuitry within the hippocampus is shown in Fig. 2.14 and in more detail in Fig. 2.15. Projections from the entorhinal cortex reach the granule cells (of which there are 1,000,000 in the rat) in the dentate gyrus (DG), via the perforant path (pp) (Witter 1993). In the dentate gyrus there is a set inhibitory interneurons that provide feedback inhibition between the granule cells. The dentate granule cells project to the CA3 cells (of which there are 300,000 in the rat) via the mossy fibres (mf). Each CA3 cell receives approximately 50 mossy fibre inputs, as shown in Fig. 2.16. By contrast, there are many more (probably much weaker) direct perforant path inputs also from the entorhinal cortex onto each CA3 cell (in the rat in the order of 4,000). The largest number of synapses (about 12,000 in the rat) on the dendrites of CA3 pyramidal cells is, however, provided by the (recurrent) axon collaterals of CA3 cells themselves (rc) (see Fig. 2.16). It is remarkable that the recurrent collaterals are distributed to other CA3 cells throughout the hippocampus (Amaral and Witter 1989, Amaral and Witter 1995, Amaral, Ishizuka and Claiborne 1990, Ishizuka, Weber and Amaral 1990), so that effectively the CA3 system provides a single network (with a connectivity of approximately 2% between the different CA3 neurons given that the connections are bilateral). The neurons that comprise CA3, in turn, project to CA1 neurons via the Schaffer collaterals.

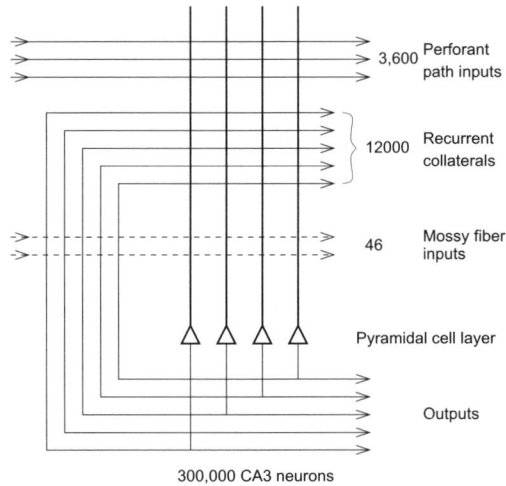

Fig. 2.16 The numbers of connections onto each CA3 cell from three different sources in the rat. (After Treves and Rolls 1992, and Rolls and Treves 1998.)

2.8.3   A theory of the operation of hippocampal circuitry as a memory system

The next step is to consider how using its internal connectivity and synaptic modifiability the hippocampus could store and retrieve many memories, and how retrieval within the hippocampus could lead to retrieval of the activity in the neocortex that was present during the original learning of the episode. To develop understanding of how this is achieved, we have developed a computational theory of the operation of the hippocampus (Rolls 1989a, Rolls 1989b, Rolls 1989d, Rolls 1989e, Rolls 1987, Treves and Rolls 1994, Rolls 1990a, Rolls 1990b, Treves and Rolls 1991, Treves and Rolls 1992, Rolls 1996, Rolls and Treves 1998, Rolls, Stringer and Trappenberg 2002, Stringer, Rolls and Trappenberg 2005, Rolls and Stringer 2005, Rolls and Kesner 2006, Rolls, Stringer and Elliot 2006c, Rolls 2008c, Rolls 2010c).

2.8.3.1   The hippocampal CA3 network as an autoassociation memory

I have proposed that the CA3 stage acts as an autoassociation memory which enables episodic memories to be formed and stored in the CA3 network, and that subsequently the extensive recurrent collateral connectivity allows for the retrieval of a whole representation to be initiated by the activation of some small part of the same representation (the cue) (Rolls 1987, Rolls 1989a, Rolls 1989b, Rolls 1989e, Rolls 1990a, Rolls 1990b). The crucial synaptic modification for this is in the recurrent collateral synapses. The hypothesis is that because the CA3 operates effectively as a single network, given that the recurrent collateral connections of the CA3 neurons to other CA3 neurons extend widely throughout the CA3 region, it can allow arbitrary associations between inputs originating from very different parts of the cerebral cortex to be formed. Empirical support for this proposal has been obtained (Rolls and Kesner 2006), including evidence supporting the theory from Tonegawa[5] and colleagues that genetic knock-out of just the CA3 NMDA receptors implicated in the CA3 recurrent collateral autoassociative network impairs spatial learning and recall of the type for which the hippocampus is needed (Nakazawa, Quirk, Chitwood, Watanabe, Yeckel, Sun, Kato, Carr, Johnston, Wilson and Tonegawa 2002, Nakazawa, Sun, Quirk, Rondi-Reig, Wilson and Tonegawa 2003).

An important point here in unravelling the brain is that the local circuitry of the networks provides important evidence about how the networks operate (Rolls 2008c, Rolls 2011c). The CA3 recurrent collateral system that forms a single network is the only single network I know in the brain in which any neuron has a reasonable probability (in this case 0.02) of connecting to any other neuron in the whole network. This is the important property, of being a single network, so that any neurons firing for example to a spatial location can become associated by co-activity-dependent (associative) synaptic modification with any other input to the CA3 system, for example the representation of an object. Then, later, the presentation of just the spatial location (e.g. the city where a meeting was held) can enable recall of who was there (the participants); or equally, vice versa. This is what is provided for by the single autoassociation or attractor network in CA3 (Figs. 2.15 and 2.16).

The number of episodic memories that can be stored in this CA3 system is reasonably large. Given that the number of recurrent collateral excitatory modifiable synapses onto any one CA3 neuron in the rat is approximately 12,000 (see Fig. 2.16), and the sparseness of the representation, we predict that the number of different memories that can be stored is in the order of 12,000. The number of memories is likely to be higher in humans with larger numbers of recurrent collateral synapses onto each CA3 neuron.

However, the memory capacity of any attractor network (such as the hippocampal CA3 system) is limited. Exceeding the capacity can lead to a loss of much of the information retrievable from the network. It is therefore necessary to have some form of forgetting in this store, and this is a deep computational neuroscience reason why forgetting is a characteristic of human memory, for it serves an important purpose (see Section 2.8.4).

2.8.3.2 Mossy fibre inputs and perforant path inputs to the CA3 cells

Another remarkable fact about the functional architecture of the hippocampus is shown in Figs. 2.15 and 2.16. Although there are a large number of recurrent collateral synapses onto each CA3 cell, there are only approximately 48 inputs to each CA3 neuron from the mossy fibres, and these are large and strong

---

[5] Susumu Tonegawa is a molecular biologist who received the Nobel Prize for Physiology or Medicine in 1987 for his discovery of the genetic mechanism that produces antibody diversity.

synapses. We have theorized that (because each CA3 cell receives from only 48 selected randomly from the approximately 1,000,000 dentate gyrus neurons that send the mossy fibre inputs), every time a new memory is formed in CA3, a randomly selected set of a few CA3 neurons is activated. This ensures that each new memory stored in CA3 is as different as possible by randomization from any other memory stored by recurrent collateral synaptic modification between CA3 neurons that have been forced into activity by the mossy fibre inputs. This is important for an episodic memory system in which each memory must be kept as separate as possible from the other memories. We refer to this as a pattern separation effect. There is empirical support for this theory, including the finding that memory storage is impaired by damage to the dentate granule cell / mossy fibre system, especially when pattern separation is required (Rolls and Kesner 2006).

In fact, following another clue from the functional architecture, the dentate granule cells have the architecture of a competitive network, and I have theorized (with empirical support now available) that they do act as a competitive network to form the sparse representations that help the pattern separation effect just referred to (Rolls 1989b, Rolls et al. 2006c, Rolls 2008c).

The functional architecture indicates that there is a second set of inputs to the CA3 neurons, the perforant path fibres from the entorhinal cortex, and these we have shown are sufficiently numerous (1,200 onto each CA3 neuron in the rat), given their associative modifiability, to initiate recall from the hippocampus, which the mossy fibres are insufficient in number to do (Treves and Rolls 1992). This is also now supported by experimental evidence (Rolls and Kesner 2006, Rolls 2008c).

2.8.3.3 CA1 cells

The CA1 cells start the return path back to the cortex for recall. The associative modifiability of the CA3 to CA1 connections helps information to be maintained. Their architecture makes them appear as a competitive network, and indeed it is proposed that this allows recoding of the parts of a memory which necessarily must be represented as separate parts in CA3 for arbitrary associations to be formed into a single new representation of the whole memory which is then more efficient as a retrieval cue (Rolls and Kesner 2006, Rolls 2008c).

2.8.3.4 Backprojections to the neocortex for memory recall

It is suggested above that the modifiable connections from the CA3 neurons to the CA1 neurons allow the whole episode in CA3 to be produced in CA1. The CA1 neurons would then activate, via their termination in the deep layers of the entorhinal cortex, at least the pyramidal cells in the deep layers of the entorhinal cortex (see Fig. 2.14). These entorhinal cortex layer 5 neurons would then, by virtue of their backprojections to the parts of cerebral cortex that originally provided the inputs to the hippocampus, terminate in the superficial layers of those neocortical areas, where synapses would be made onto the distal parts of the dendrites of the cortical pyramidal cells (Rolls 1989b, Rolls 2008c). The areas of cerebral neocortex in which this recall would be produced could include multimodal cortical areas, and also areas of unimodal association cortex (e.g. inferior temporal visual cortex). The backprojections, by recalling previous episodic events, could provide information useful to the neocortex in the building of new representations in the multimodal and unimodal association cortical areas, which by building new long-term representations can be considered as a form of memory consolidation (Rolls 1989b, Rolls 2008c).

The hypothesis of the architecture with which this would be achieved is shown in Fig. 2.14. The feedforward connections from association areas of the cerebral neocortex (solid lines in Fig. 2.14) show major convergence as information is passed to CA3, with the CA3 autoassociation network having the smallest number of neurons at any stage of the processing. The backprojections allow for divergence back to neocortical areas. The way in which we suggest that the backprojection synapses are set up to have the appropriate strengths for recall is by associative synaptic modification of the backprojection synapses at one or more stages during the original learning (Treves and Rolls 1994, Rolls 1995a).

A fascinating issue is how many synaptic input connections would be needed onto each neocortical neuron at each of the backprojection stages for the recall to operate. We have been able to show that this must be large, in the order of the number of connections onto each neuron in the CA3 recurrent collateral system, weighted by the ratio of the sparseness in the neocortical backprojecting neurons vs CA3 neurons (Treves and Rolls 1994). Thus the number of backprojection synapses onto each neuron should be as high as possible (of order 12,000 if possible). This provides a theory of why the number of backprojections between cortical areas in the neocortex is as high as the number of forward projections between cortical areas, and it is the only theory I know of this.

2.8.3.5 Neuronal activity in the hippocampus related to memory

An important part of the approach to understanding memory formation in the hippocampus is to understand what the neurons encode for in the hippocampus, that could be associated together by the memory processes described above. There are place cells in rodents that fire when a rodent is in a particular place in an environment (O'Keefe 1976, Wills, Lever, Cacucci, Burgess and O'Keefe 2005), but these are not useful for a human who can just by looking at different locations in an environment (without ever being there which is the requirement for a place cell to fire), remember who or what was seen there[6]. It was therefore a useful discovery that there are spatial view neurons in the primate hippocampus (Georges-François, Rolls and Robertson 1999, Robertson, Rolls and Georges-François 1998, Rolls, Robertson and Georges-François 1997a, Rolls, Treves, Robertson, Georges-François and Panzeri 1998) (see Section 2.3.1). These neurons increase their firing rates when one part of a spatial environment is being looked at. The allocentric (world based) representation of space is, we have proposed, an important part of the system for remembering where objects have been seen, for example where one has seen good food, or other good resources (without necessarily having visited that place), for some of these neurons respond to a combination of a spatial view and the object that has been shown at that location (Rolls, Xiang and Franco 2005). Some of the neurons also respond during recall of the location in space where an object was last seen (Rolls and Xiang 2006). This type of neuronal activity is prototypical of an episodic memory, the memory for particular past events or episodes, and thus these neurons are an important component of the theory of episodic memory that we have developed (Treves and Rolls 1994, Rolls 2010c), for they represent information of the type that must be associated together in a human episodic memory (Rolls 2008c, Rolls 2010c).

2.8.4 Forgetting

Forgetting is an important feature of associative neural networks and the brain, and is important in their successful operation. There are a number of different mechanisms for forgetting, and a number of different reasons why forgetting is important in particular classes of network (Rolls 2008c).

Consider attractor, that is autoassociation, networks, which are used for short-term memory, episodic memory, etc. These networks have a critical storage capacity, which is in the order of the number of synapses on each neuron (Hopfield 1982, Treves and Rolls 1991), and if this is exceeded, most of the memories in the network become unretrievable. It is therefore crucial to have a mechanism for forgetting in these networks.

One mechanism is decay of synaptic strength. The simple forgetting mechanism is just a decay of the synaptic value back to its baseline, which may be exponential in time or in the number of learning changes incurred (Nadal, Toulouse, Changeux and Dehaene 1986, Rolls 2008c).

A second mechanism is that synapses may also become weaker as a result of active processes more closely related to learning. We have seen that if the presynaptic firing to a synapse is high at the same time as the postsynaptic

---

[6]The famous example is the Greek orator Simonides of Ceos who was at a banquet in the 5th century B.C. to give a speech. He stepped outside for a moment to receive a message whereupon the banquet hall collapsed, leaving no survivors inside. Simonides was able to identify the bodies of the killed guests based on the location where he had last seen them sitting or standing before he left the building.

neuron is activated, then that synapse increases in strength by the associative learning process described as long-term synaptic potentiation (LTP) (Section 2.5.1). However, the synapse may get weaker if for example the presynaptic terminal is not active, yet the postsynaptic neuron is active, a process known as long-term synaptic depression (LTD). (This is heterosynaptic LTD, in that the synapse that becomes weaker is *other than* a synapse through which the neuron is being activated.) This process is very useful in a number of ways in associative and competitive neuronal networks, including keeping the total strength of all the synapses on a neuron limited (see Rolls (2008c)), but the effect in the present context is that this can result in some loss of memories previously stored in these associative networks, if a particular synapse had been strengthened as part of a previous memory.

Forgetting is thus to some extent a useful property in its own right to ensure that the memory capacity of an attractor network is not exceeded, and is also a consequence of other processes important in neuronal learning systems, including preventing all the synaptic weights on a neuron from growing too large (Rolls 2008c).

We should not complain too much about forgetting, but learn to live with it, by using the items that we wish to remember, and not re-accessing old useless memories, such as one's room number in a hotel that one visited recently.

Another type of forgetting is important in short-term memory systems. Remember that the memory is maintained in a short-term memory attractor network by the continuing firing of a subset of the neurons, reactivating themselves by positive feedback by the strengthened synapses that couple that subset of neurons (Section 2.6). Also remember that such an attractor network can maintain the activity of only one subset of neurons active at any one time. It is then important that if another item is to be placed in the short-term memory, then the ongoing firing must be interrupted by a strong new input which forces the network into a new attractor state. This effect could be facilitated by synaptic or neuronal adaptation (that is, decreasing efficacy) happening over several seconds, which would make the first item held in the short-term memory gradually be held less stably, so that over time a new input might more easily enter the short-term memory. Given that such short-term memories are also used in the brain to maintain our attention on something (which is held in the short-term memory), we also become somewhat more distractible over time due to these adaptation effects. In fact, adaptation in the brain may serve a useful function, by helping us to forget things previously held in our short-term memories, so that we can use them for new things in our changing environment. Some drugs such as nicotine and caffeine may help to reduce this adaptation, and may therefore help concentration.

2.8.5 Memory reconsolidation

An unexpected phenomenon discovered in recent years is reconsolidation. This refers to a process in which after a memory has been stored, it may be weakened or lost if recall of the memory is performed during the presence of a protein synthesis inhibitor (Debiec, LeDoux and Nader 2002, Debiec, Doyere, Nader and LeDoux 2006, Nader and Einarsson 2010). The implication that has been drawn is that whenever a memory is recalled, some reconsolidation process requiring protein synthesis may be needed.

One possible function of reconsolidation is that it may allow some restructuring of a memory: when an item is recalled, it might be linked to other memories to help form a semantic (meaning-related) memory, and then re-stored. If for example we know that birds fly, and then we are told that secretary birds do not fly, during the processing in which the semantic meaning of birds as flying is being recalled, that memory may be restructured to include the exception, and then stored again in the restructured form.

A second possible computational function is that reconsolidation might be useful as a mechanism to ensure that whenever a memory is retrieved, additional LTP (long-term potentiation of synaptic strength) is not added to the existing LTP. This could be achieved if during the recall process the memory strength is reset to a low value from which it is then strengthened.

A third possible computational function of reconsolidation is that it could enable the selective retention of 'useful' memories (or in fact memories being used, that is recalled), in a situation where strengthened synapses otherwise decay in strength over time to implement a type of forgetting (Rolls 2008c).

Therapies have been proposed for removing traumatic memories by requiring memory recall in the presence of a treatment that prevents the memory reconsolidation (Nader and Einarsson 2010, Schiller, Monfils, Raio, Johnson, LeDoux and Phelps 2010).

2.8.6 Memory and aging

During normal aging, short-term memory, attention, and episodic memory (the memory for recent events) tend to become less good (Grady 2008), while semantic memory (the memory for concepts, ideas, and knowledge) tends to be less impaired. What is happening? Can modern neuroscience do anything to elucidate, and potentially to treat, these changes? A number of different factors are relevant.

2.8.6.1 Synaptic strength and aging

Memories stored early in life may be stored better, and later recalled better, than those stored later in life. There are a number of possible reasons for this. One is that the neurotransmitters that generally facilitate the synaptic modification involved in memory formation (Section 2.5.1), such as acetylcholine and noradrenaline (Rolls 2008c), may become depleted with aging.

Another mechanism, not necessarily independent, is that new synaptic modification, as assessed by long-term potentiation (LTP), appears to be less long-lasting with aging (Burke and Barnes 2006). Another mechanism may be that storing memories in a flat energy landscape (i.e. without much prior synaptic modification) may help these memories to stand out from those added later. While this would not be a natural property of the type of autoassociation palimpsest memory described above, it could be a property of the way in which an episodic memory stored in the hippocampus may be retrieved into the neocortex where it can be incorporated into a semantic memory (see Section 2.8), the relevant example of which in this case would be an autobiographical memory.

In semantic memories, it could be that the first stored links tend to provide the framework around which other information is structured. Also, the structure of a semantic memory, with many different associations linked to each concept, may be more resistant to degradation than an episodic memory, where events must be kept separate and can rely on support or associations with other related memories.

Another factor in the apparent strength of early memories may be that some may be stored with an affective component, and this may not only make the memory strong by activation of the cholinergic and related systems (Rolls 2008c), but may also mean that part at least of the memory is stored in different brain structures such as the amygdala which may have relatively more persistent and less flexible or reversible memories than other memory systems.

Another factor in the importance and stability of synaptic modification early in life arises in perceptual systems, in which it is important to allow neurons to become tuned to the statistics of for example the visual environment, but once feature analyzers have been formed, stability of the feature analyzers in early cortical processing layers may be important so that later stages in the hierarchy can perform reliable object recognition which achieves stability only if the input filters to the system do not keep changing (Rolls 2008c, Rolls 2012b). This could be the importance of a critical period for learning early on in perceptual development.

2.8.6.2 Memory dynamics and aging

We have seen that short-term memory is maintained by recirculating activity in a population of strongly interconnected cortical neurons (Section 2.6). With high firing of the neurons, and strong synapses, the positive feedback keeps the network stable, and resistant to interruption by small inputs, or by internal noise in the brain due to the spiking of the neurons. We can describe the network as being in a low-energy, that is stable, state (Section 2.5.3.2). Under these conditions, short-term memory will be maintained well, and will be good.

However, if either the firing rates become reduced, or the synapses are less strong, then the stability of the network will be reduced, and the short-term memory will be more likely to be destabilized by a distracting stimulus, or by the spiking-related noise caused by the neurons. This is illustrated in Fig. 2.10 on page 40. The quantitative description of this is in Equation 2.3 on page 33. (These processes are described further in Sections 10.1 and 10.3.)

We have proposed that a number of changes in the brain that occur during normal aging act to reduce the stability of these short-term memory networks, and thus contribute to the short-term memory (and attentional) difficulties that occur in normal aging (Rolls and Deco 2010, Rolls, Deco and Loh 2012), and some of these processes are considered in more detail in Section 10.3.

One change is that NMDA receptor functionality (one of the excitatory receptors in the recurrent circuitry) tends to decrease with aging (Kelly, Nadon, Morrison, Thibault, Barnes and Blalock 2006). This would act to reduce the depth of the basins of attraction, both by reducing the firing rate of the neurons in the active attractor, and effectively by decreasing the strength of the potentiated synaptic connections that support each attractor as the currents passing through these potentiated synapses would decrease.

Another change in aging is that dopaminergic function in the prefrontal cortex may decline with aging (Sikström 2007). This change can reduce NMDA receptor activated ion channel conductances (Seamans and Yang 2004) (see further Section 10.1), with the effects described above. Thus part of the role of dopamine in the prefrontal cortex in short-term memory can be accounted for by a decreased depth in the basins of attraction of prefrontal attractor networks (Loh, Rolls and Deco 2007a, Loh, Rolls and Deco 2007b, Rolls, Loh, Deco and Winterer 2008d). The decrease in dopamine could thus contribute to the reduced short-term memory and attention in aging.

Another factor is acetylcholine. Acetylcholine in the neocortex has its origin largely in the few thousand cholinergic neurons in the basal forebrain. The correlation of clinical dementia ratings with the reductions in a number of cortical cholinergic markers such as choline acetyltransferase, muscarinic and nicotinic acetylcholine receptor binding, as well as levels of acetylcholine, suggested an association of cholinergic hypofunction with cognitive deficits, which led to the formulation of the cholinergic hypothesis of memory dysfunction in senescence and in Alzheimer's disease (Bartus 2000, Schliebs and Arendt 2006). The cholinergic neurons are activated by emotion-related and novel stimuli, and may by releasing acetylcholine in the cortex facilitate synaptic modification (LTP) during learning (Rolls 2008c). Further, if acetylcholine is reduced, cortical neurons tend to adapt, that is to reduce their firing rates over periods of a few seconds, and the reduced firing rates by decreasing the stability of short-term memory networks is also likely to contribute to decreased short-term memory and attention in aging (Rolls 2008c, Rolls et al. 2012).

In view of these changes, boosting glutamatergic transmission is being explored as a means of enhancing cognition and minimizing its decline in aging. Several classes of AMPA glutamate receptor potentiators have been explored in rodent models and are now entering clinical trials (Lynch and Gall 2006, O'Neill and Dix 2007). These treatments might increase the depth of the basins of attraction. Agents that activate the glycine or serine modulatory sites on the NMDA receptor (Coyle 2006) would also be predicted to be useful. The stochastic neurodynamics approach (Rolls and Deco 2010) suggests that there could be a whole range of treatments that may increase the firing of cortical pyramidal cells that could be useful to ameliorate the cognitive changes in aging such as reduced short-term memory, attention, and episodic memory, and the computational neuroscience approach allows possible new combinations of treatments to be considered and explored in computer simulations (Rolls and Deco 2010, Rolls, Deco and Loh 2012). Further, this approach may lead to predictions for effective treatments that need not necessarily restore the particular change in the brain that caused the symptoms, but may find alternative routes to restore the stability of the dynamics (see further Section 10.3).

To conclude this section on memory, I wish to make the point that we now have the computational tools and approaches to understand how some parts of the brain actually operate. Thus we are beyond the era of phenomenology, metaphor, and an un-understood 'miracle' of memory. Instead, we can understand just how miraculous and elegant memory is in its workings. This is thus not only a reductionist approach, for we now understand the emergent properties of the system. The new approach shows how we can understand the workings of our memories, for example why it may take time to recall an item as attractor processes settle in the brain under the influence of noise caused by neuronal firing; why forgetting is so useful in fact to brain function; and, an important goal that drives much of this research, how to improve memory, and to minimize its deterioration in aging or disease.

2.9 Attention

Here I show that what might appear as a somewhat nebulous concept, attention, can now be understood at a fully mechanistic level, which by no means diminishes its interest. Indeed, it opens a way to understanding the cognitive symptoms of some psychiatric disorders, and potentially new treatments for them (Chapter 10).

One type of attentional process operates when salient features in a visual scene attract attention (Itti and Koch 2001). This visual processing is described as feedforward and bottom-up, in that it operates forward in the visual pathways from the visual input (see Fig. 2.17 in which the solid lines show the forward pathways). Salient features such as a moving visual stimulus produce strong inputs which dominate selection in the system.

A second type of selective attentional process, with which we are concerned in this section, involves actively maintaining in short-term memory a location or object as the target of attention, and using this by top-down processes to influence earlier cortical processing. Some of the top-down or backprojection pathways in the visual system are shown in Fig. 2.17 by the dashed lines.

Observations from a number of cognitive neuroscience experiments lead to an account of attention termed the 'biased competition hypothesis', which aims to explain the computational algorithms governing top-down visual attention and their implementation in the brain's neural circuits and neural systems (Duncan and Humphreys 1989, Desimone and Duncan 1995, Rolls and Deco 2002, Rolls and Deco 2010). According to this hypothesis, attentional selection operates by top-down biasing of an underlying competitive interaction between multiple stimuli in the visual field. The massively feedforward and feedback connections that exist in the anatomy and physiology of the cortex implement the attentional mutual biasing interactions that result in a neurodynamical account of visual perception consistent with the seminal constructivist theories of Helmholtz (1867) and Gregory (1970) which argue that visual perception involves constructing a perceptual representation that is guided by but goes beyond what is actually on the retina.

In this section, I show how biased competition can operate as a dynamical process that enables multiple interacting brain systems with bottom-up and top-down effects to be analyzed, and show how noise can affect the stability of these attentional processes. More extensive accounts of the empirical data on attention that can be accounted for by the biased competition approach are provided by Rolls and Deco (2002) in *Computational Neuroscience of Vision* and Rolls (2008c) in *Memory, Attention, and Decision-Making*. Then I show how this approach has been extended to a 'biased activation hypothesis of attention', and show how this has implications for how our brains process information by emotional or more perceptual routes, which in turn has many implications for understanding sensory testing, and how we take decisions.

The **biased competition hypothesis** of attention proposes that multiple stimuli in the visual field activate populations of neurons that engage in competitive interactions. Attending to a stimulus at a particular location or with a particular feature biases this competition in favour of neurons that respond to the feature or location of the attended stimulus. This attentional effect is produced by generating signals within areas outside the visual cortex that are then fed back to extrastriate areas such as V4 and V2 (see Fig. 2.17), where they bias the competition such that when multiple stimuli appear in the visual

ventral, what   dorsal, where and motion

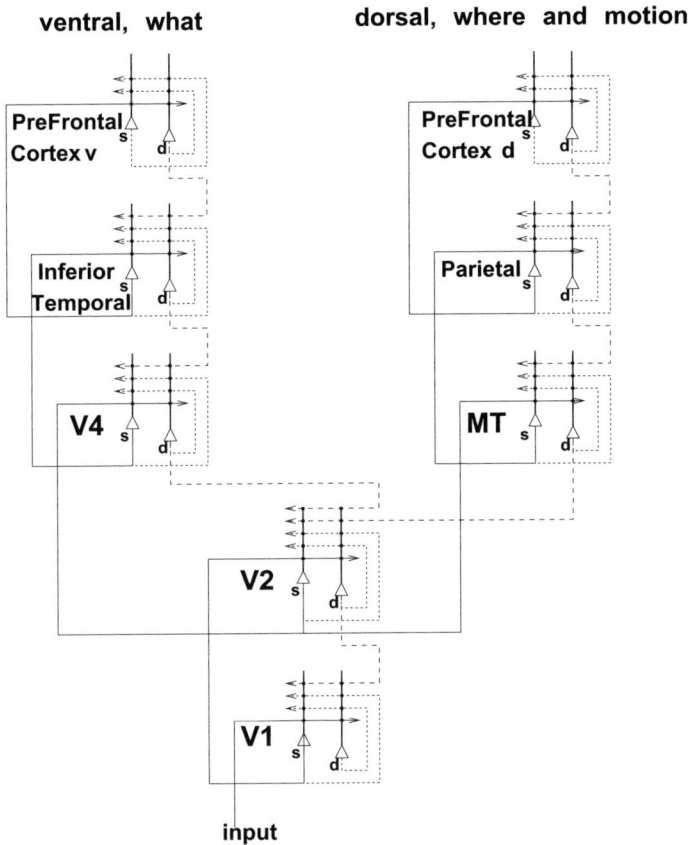

Fig. 2.17 The overall architecture of a model of object and spatial processing and attention, including the prefrontal cortical areas that provide the short-term memory required to hold the object or spatial target of attention active. Forward connections are indicated by solid lines; backprojections, which could implement top-down processing, by dashed lines; and recurrent connections within an area by dotted lines. The triangles represent pyramidal cell bodies, with the thick vertical line above them the dendritic trees. The cortical layers in which the cells are concentrated are indicated by s (superficial, layers 2 and 3) and d (deep, layers 5 and 6). V1 is the primary visual cortex, V2 the second visual cortical area, etc. V4 is in the ventral visual stream which processes 'what' object is present. MT is an area in the dorsal visual stream, which processes 'where' objects are and their motion. The prefrontal cortical areas most strongly reciprocally connected to the inferior temporal cortex 'what' processing stream are labelled v to indicate that they are in the more ventral part of the lateral prefrontal cortex, area 46, close to the inferior convexity in macaques. The prefrontal cortical areas most strongly reciprocally connected to the parietal visual cortical 'where' processing stream are labelled d to indicate that they are in the more dorsal part of the lateral prefrontal cortex, area 46, in and close to the banks of the principal sulcus in macaques (see Rolls and Deco 2002, and Rolls 2008c).

field, the cells representing the attended stimulus 'win', thereby suppressing cells representing distracting stimuli (Duncan and Humphreys 1989, Desimone and Duncan 1995, Duncan 1996).

Biased Competition

Short Term Memory
Top-Down Bias Source

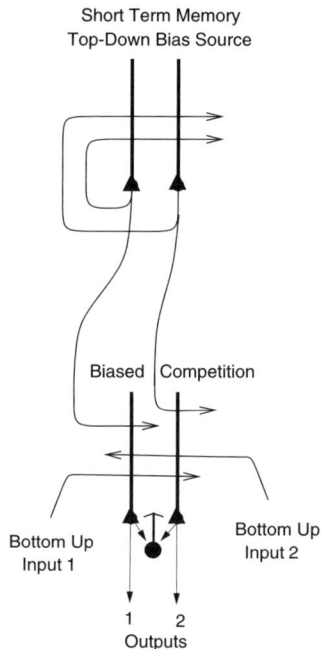

Fig. 2.18 Biased competition theory of attention. There is usually a single attractor network that can enter different attractor states to provide the source of the top-down bias (as shown). If it is a single network, there can be competition within the short-term memory attractor states, implemented through the local GABA inhibitory neurons. The top-down continuing firing of one of the attractor states then biases in a top-down process some of the neurons in a cortical area to respond more to one than the other of the bottom-up inputs, with competition implemented through the GABA inhibitory neurons (symbolized by a filled circle) which make feedback inhibitory connections onto the pyramidal cells (symbolized by a triangle) in the cortical area. The thick vertical lines above the pyramidal cells are the dendrites. The axons are shown with thin lines and the excitatory connections by arrow heads. (After Grabenhorst and Rolls, 2010.)

We have seen in Fig. 2.7d on page 30 how given forward (bottom-up) inputs to a network, competition between the principal (excitatory) neurons can be implemented by inhibitory neurons which receive from the principal neurons in the network and send back inhibitory connections to the principal neurons in the network. This competition can be biased by a top-down input to favour some of the populations of neurons, and this describes the biased competition hypothesis.

The network and mechanisms that implement the biased competition hypothesis of attention are shown in more detail in Fig. 2.18. Populations or pools of neurons are shown as a single neuron in this figure. The way that we think of top-down biased competition as operating in for example visual selective attention (Desimone and Duncan 1995) is that within an area, e.g. a cortical region, some neurons receive a weak top-down input that increases their response to the bottom-up stimuli, potentially supralinearly if the bottom-up stimuli are weak (Deco and Rolls 2005a, Rolls 2008c, Rolls and Deco 2002). The enhanced firing of the biased neurons then, via the local inhibitory neurons, inhibits the other neurons in the local area from responding to the bottom-up stimuli. This is a local mechanism, in that the inhibition in the neocortex is primarily local, being implemented by cortical inhibitory neurons that typically have inputs and outputs over no more than a few mm (Rolls 2008c).

Each population or pool of neurons is tuned to a different stimulus property, e.g. spatial location, or feature, or object. The more pools of the module that are active, the more active the common inhibitory pool will be and, consequently, the more feedback inhibition will affect the pools in the module, such that only the most excited group of pools will survive the competition. The external top-down bias can shift the competition in favour of a specific group of pools. Some non-linearity in the system, implemented for example by the fact that neurons have a threshold non-linearity below which they do not fire, can result in a small top-down bias having quite a large attentional effect, especially for relatively weak bottom-up inputs, as shown in a realistic spiking neuron simulation by Deco and Rolls (2005a).

The architecture of Fig. 2.18 is that of a competitive network (described in more detail by Rolls and Deco (2002) and Rolls (2008c)) but with top-down backprojections. However, the network could equally well include associatively modifiable synapses in recurrent collateral connections between the excitatory (principal) neurons in the network, making the network into an autoassociation or attractor network capable of short-term memory, as described in Section 2.5.3. The competition between the neurons in the attractor network is again implemented by the inhibitory feedback neurons.

The '**biased activation hypothesis of attention**' (Grabenhorst and Rolls 2010) is illustrated in Fig. 2.19. Here, in contrast to the locally implemented biased competition, we have facilitation of processing in a whole cortical area or even cortical processing stream. For example, when the instructions were 'pay attention to pleasantness' a whole cortical processing stream including the orbitofrontal cortex and pregenual cingulate cortex (which might be the left stream in Fig. 2.19) had its activations to a standard test taste (umami, monosodium glutamate) increased relative to the activations in another linked cortical set of areas, the anterior insular primary taste cortex, and the mid-insular cortex (which might be the right-hand stream in Fig. 2.19). On the other hand, when the instructions were 'pay attention to intensity' a whole cortical processing stream including the anterior insular primary taste cortex, and the mid-insular cortex had its activations to a standard test odor of umami increased relative to the activations in another linked cortical set of areas, the orbitofrontal cortex and pregenual cingulate cortex (Grabenhorst and Rolls 2010). Moreover, we showed that the short-term memory systems that hold the object of attention

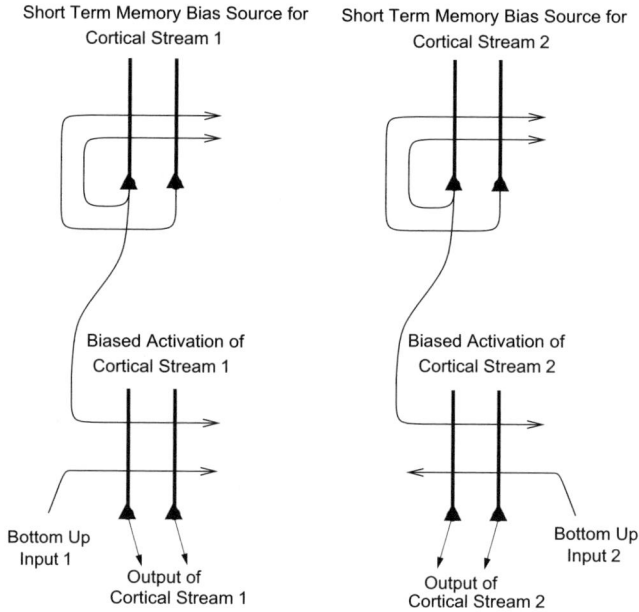

Fig. 2.19 Biased activation theory of attention. The short-term memory systems that provide the source of the top-down activations may be separate (as shown), or could be a single network with different attractor states for the different selective attention conditions. The top-down short-term memory systems hold what is being paid attention to active by continuing firing in an attractor state, and bias separately either cortical processing system 1, or cortical processing system 2. This weak top-down bias interacts with the bottom-up input to the cortical stream and produces an increase of activity that can be supralinear. Thus the selective activation of separate cortical processing streams can occur. In the example, stream 1 might process the affective value of a stimulus, and stream 2 might process the intensity and physical properties of the stimulus. The outputs of these separate processing streams then must enter a competition system, which could be for example a cortical attractor decision-making network that makes choices between the two streams, with the choice biased by the activations in the separate streams (see text). (After Grabenhorst and Rolls, 2010.)

in mind during a trial (attention to pleasantness or to intensity) were separate (as illustrated in Fig. 2.19), with the lateral prefrontal cortex region that provide the bias for attention to pleasantness more anterior than the region that provides the source of the top-down bias for attention to intensity. So the attentional mechanism might more accurately be described in this case as biased activation, without local competition being part of the effect. This biased activation theory and model of attention, illustrated in Fig. 2.19, is a rather different way to implement attention in the brain than biased competition, and each mechanism may apply in different cases, or both mechanisms in some cases.

The implications of the biased activation theory are of considerable interest, for they show that the way in which whole brain systems process taste (and probably most other) stimuli depends on exactly what instructions are given. Thus the task instructions, or more generally environmental context, can influence how brain areas process tastes with respect to affective value, and taking this into account is likely to have important implications for understanding the effects produced by sensory stimuli not only in the laboratory, but also in daily life. For example, when developing a new perfume or food flavour, it may be important to engage processing in brain areas concerned with processing the pleasantness of the stimuli by using appropriate attentional facilitation directed towards the pleasantness of stimuli. Similarly, when individuals are aiming to control their responses to for example a delicious food, and how much they eat, it may be important to control the aspect of the stimulus to which they are paying attention (Rolls 2011i). Further, cognitive descriptions of a food, such as that it is 'rich and delicious', may influence processing in whole cortical streams for odour (De Araujo, Rolls, Velazco, Margot and Cayeux 2005), taste and flavour (Grabenhorst, Rolls and Bilderbeck 2008a), and touch (McCabe, Rolls, Bilderbeck and McGlone 2008, Rolls 2010a).

Biased activation is predicted to apply in systems where different attributes of the same stimulus are processed in separate cortical areas and streams. A good example is the ventral vs dorsal visual streams, in which a model described by Deco and Rolls predicts that activation in a whole series of dorsal stream areas including parietal cortex and MT will be higher when attention is directed to the spatial location of the stimulus, and that activation in a whole series of ventral stream areas including the inferior temporal visual cortex and V4 will be higher when attention is being paid to the identity of the stimulus (Deco and Rolls 2004, Rolls 2008c) (Fig. 2.17).

2.10 Mechanisms of visual perception

The aim of this section is to show how we are starting to understand something as apparently bafflingly complex as perception. How can we recognise an object or face as being the same object or face from different angles, when it may look very different in shape? Even when it is seen in different sizes, positions on the retina, and in-plane rotations, there is still a massive computational problem. How do we go from a retinal image of an object to something that we can recognise invariantly with respect to all these transforms, when the image on the retina for these different transforms may be so different, and activate different sets of receptor cells, the cones? This is such an enormous problem that it has not been solved yet in computer vision. But we are starting to understand how the brain performs this wonderful function, and it is very interesting to see the approaches it uses (Rolls 2008c, Rolls 2012b).

An important early advance was made when David Hubel and Torsten Wiesel discovered that single neurons in the primary visual cortex, V1 (visual area 1, at the back of the brain), respond to edges or bars shown at a particular angle or orientation on the retina (Hubel and Wiesel 1962, Hubel and Wiesel 1968). The story of the scientific discovery is interesting. They were using slides in a projector to produce small circles of light as visual stimuli, and were getting little response from single neurons. They were then startled when a neuron did respond, when they were changing the slide. It turned out that it was the straight edge visual stimulus as the slide was dropped into the projector that was what was needed to make the cell fire. This does illustrate the role of serendipity in science – but also of careful observation, and being willing to follow up the unexpected.

However, this did not solve the major issues of visual perception. As Martha Farah, the distinguished cognitive neuroscientist at the University of Pennsylvania, put it "a miracle occurs" (Farah 2000) in order for the responses of the face-selective neurons that we discovered in the inferior temporal visual cortex (IT) and amygdala (Sanghera, Rolls and Roper-Hall 1979, Perrett, Rolls and Caan 1982, Rolls 1984, Rolls 2011f) to be accounted for. These neurons, and similar object-selective neurons in IT, have selective responses to a face, or an object, that are invariant with respect to position on the retina, size, rotation, and even in some cases viewing angle (Rolls and Baylis 1986, Hasselmo, Rolls, Baylis and Nalwa 1989, Tovee, Rolls and Azzopardi 1994, Booth and Rolls 1998, Rolls 2008c, Rolls 2011f, Rolls 2012b). (As described in Section 2.3, the neurons are quite selective, but do use a sparse distributed code with its advantages for generalization to similar stimuli, etc.)

Information in the *'ventral or what'* visual cortical processing stream projects after the primary visual cortex, area V1, to the secondary visual cortex (V2), and then via area V4 to the posterior and then to the anterior inferior temporal visual cortex (see Figs. 2.20, 2.21, 2.22, 2.17 and 2.25).

Information processing along this stream is primarily unimodal, as shown by the fact that inputs from other modalities (such as taste or smell) do not anatomically have significant inputs to these regions, and by the fact that neurons in these areas respond primarily to visual stimuli, and not to taste or olfactory stimuli, etc. (Rolls 2000a, Baylis, Rolls and Leonard 1987, Ungerleider 1995, Rolls and Deco 2002). The representation built along this pathway is mainly about what object is being viewed, independently of exactly where it is on the retina, of its size, and even of the angle with which it is viewed (see Rolls (2008c) and Rolls and Deco (2002)), and for this reason it is frequently referred to as the 'what' visual pathway. The representation is also independent of whether the object is associated with reward or punishment; that is the representation is about objects per se (Rolls, Judge and Sanghera 1977). The computation that must be performed along this stream is thus primarily to build a representation of objects that shows invariance. After this processing, the visual representation is interfaced to other sensory systems in areas in which simple associations must

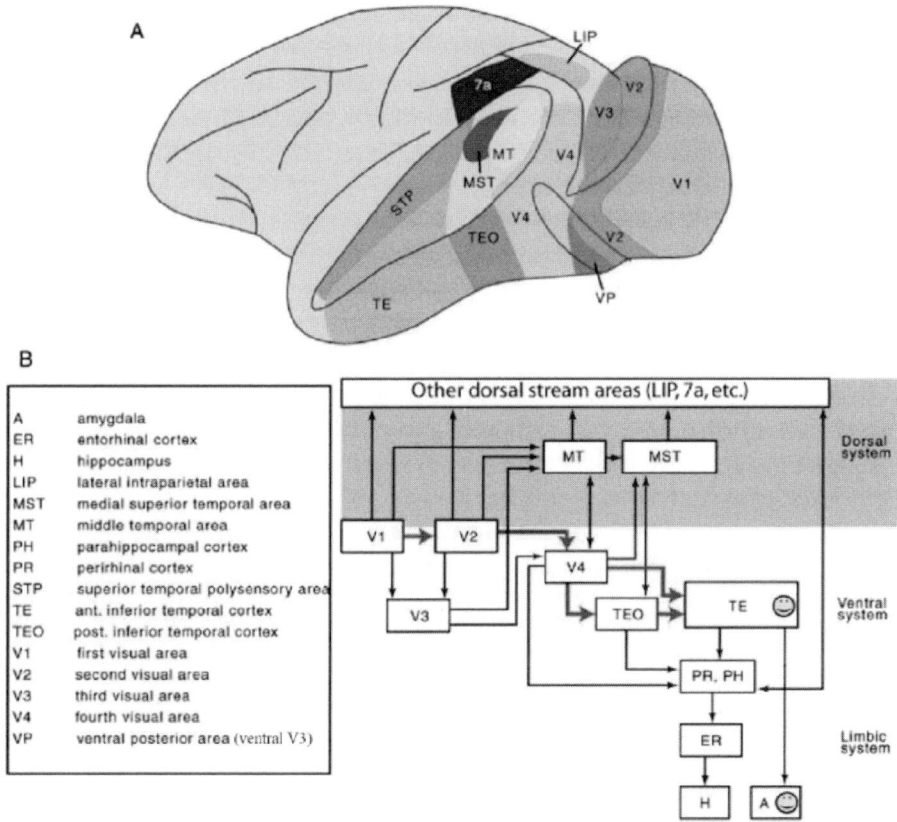

Fig. 2.20 (A). A lateral view of the monkey brain illustrating the multiplicity of functional areas within both processing streams. (B). Some of the pertinent connections of the inferior temporal cortex with other cortical areas and medial temporal lobe structures. Red lines indicate the main afferent pathway to area TE, which includes areas V1, V2, V4, and TEO. For simplicity, only projections from lower order to higher order areas are shown, but each of these feedforward projections is reciprocated by a feedback projection. Faces indicate areas in which neurons selectively responsive to faces have been found. (Adapted from Gross et al. 1993.) (See colour plates Appendix B.)

be learned between stimuli in different modalities (see Sections 2.8 and 2.11, and Chapter 3). The representation must thus be in a form in which the simple generalization properties of associative networks can be useful. Given that the association is about what object is present (and not where it is on the retina), the representation computed in sensory systems must be in a form that allows the simple correlations computed by associative networks to reflect similarities between objects, and not between their positions on the retina. Rolls' theory of invariant visual object recognition which shows how such invariant sensory representations could be built in the brain is the subject of Chapter 4 of Rolls (2008c) and Rolls (2012b), with its beginnings in Rolls (1992).

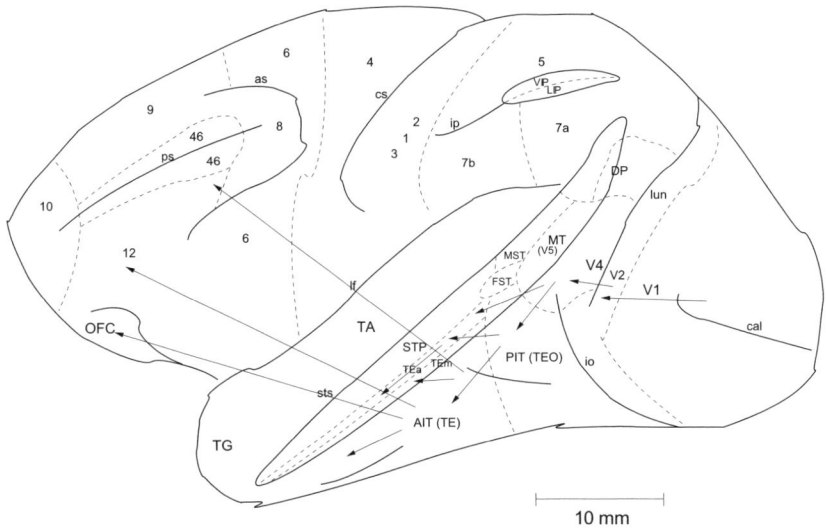

Fig. 2.21 Lateral view of the macaque brain showing the connections in the 'ventral or what visual pathway' from V1 to V2, V4, the inferior temporal visual cortex, etc., with some connections reaching the amygdala and orbitofrontal cortex. as, arcuate sulcus; cal, calcarine sulcus; cs, central sulcus; lf, lateral (or Sylvian) fissure; lun, lunate sulcus; ps, principal sulcus; io, inferior occipital sulcus; ip, intraparietal sulcus (which has been opened to reveal some of the areas it contains); sts, superior temporal sulcus (which has been opened to reveal some of the areas it contains). AIT, anterior inferior temporal cortex; FST, visual motion processing area; LIP, lateral intraparietal area; MST, visual motion processing area; MT, visual motion processing area (also called V5); OFC, orbitofrontal cortex; PIT, posterior inferior temporal cortex; STP, superior temporal plane; TA, architectonic area including auditory association cortex; TE, architectonic area including high order visual association cortex, and some of its subareas TEa and TEm; TG, architectonic area in the temporal pole; V1–V4, visual areas 1–4; VIP, ventral intraparietal area; TEO, architectonic area including posterior visual association cortex. The numbers refer to architectonic areas, and have the following approximate functional equivalence: 1, 2, 3, somatosensory cortex (posterior to the central sulcus); 4, motor cortex; 5, superior parietal lobule; 7a, inferior parietal lobule, visual part; 7b, inferior parietal lobule, somatosensory part; 6, lateral premotor cortex; 8, frontal eye field; 12, inferior convexity prefrontal cortex; 46, dorsolateral prefrontal cortex.

The ventral visual stream converges with other mainly unimodal information processing streams for taste, olfaction, touch, and hearing in a number of areas, particularly the amygdala and orbitofrontal cortex (see Figs. 2.25, 2.21, and 2.22). These areas appear to be necessary for learning to associate sensory stimuli with other reinforcing (rewarding or punishing) stimuli. For example, the amygdala is involved in learning associations between the sight of food and its taste. (The taste is a primary or innate reinforcer.) The orbitofrontal cortex is especially involved in rapidly relearning these associations, when environmental contingencies change (see Rolls (2005a) and Rolls (2000b)). They thus are brain regions in which the computation at least includes simple pattern association (e.g. between the sight of an object and its taste). In the orbitofrontal cortex, this association learning is also used to produce a repres-

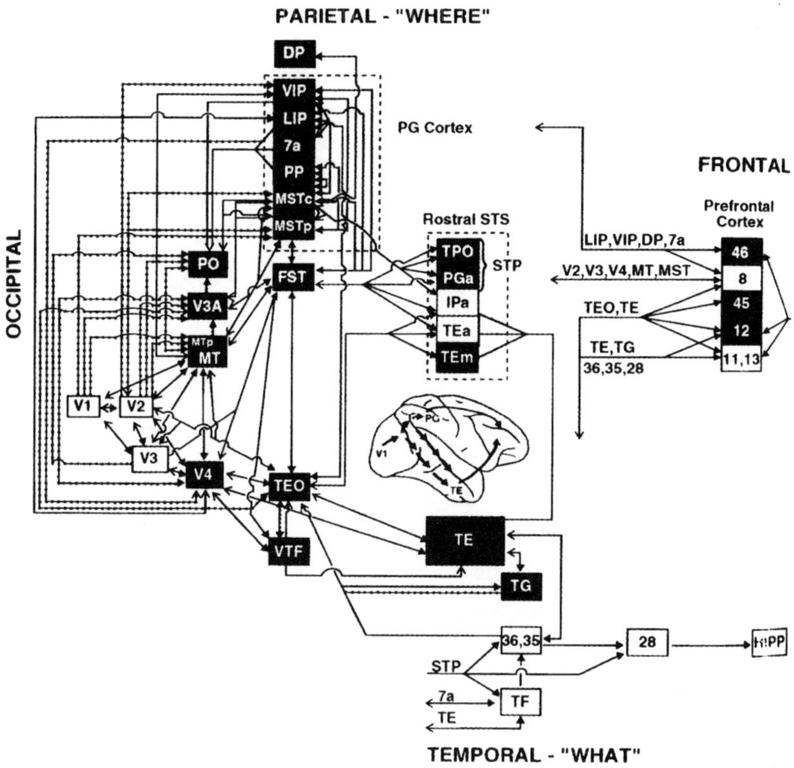

Fig. 2.22 Visual processing pathways in monkeys. Solid lines indicate connections arising from both central and peripheral visual field representations; dotted lines indicate connections restricted to peripheral visual field representations. Shaded boxes in the 'ventral (lower) or what' stream indicate visual areas related primarily to object vision; shaded boxes in the 'dorsal or where' stream indicate areas related primarily to spatial vision; and white boxes indicate areas not clearly allied with only one stream. Abbreviations: DP, dorsal prelunate area; FST, fundus of the superior temporal area; HIPP, hippocampus; LIP, lateral intraparietal area; MSTc, medial superior temporal area, central visual field representation; MSTp, medial superior temporal area, peripheral visual field representation; MT, middle temporal area; MTp, middle temporal area, peripheral visual field representation; PO, parieto-occipital area; PP, posterior parietal sulcal zone; STP, superior temporal polysensory area; V1, primary visual cortex; V2, visual area 2; V3, visual area 3; V3A, visual area 3, part A; V4, visual area 4; and VIP, ventral intraparietal area. Inferior parietal area 7a; prefrontal areas 8, 11 to 13, 45 and 46 are from Brodmann (1925). Inferior temporal areas TE and TEO, parahippocampal area TF, temporal pole area TG, and inferior parietal area PG are from Von Bonin and Bailey (1947). Rostral superior temporal sulcal (STS) areas are from Seltzer and Pandya (1978) and VTF is the visually responsive portion of area TF. (Reprinted with permission from Ungerleider 1995.)

entation of flavour, in that neurons are found in the orbitofrontal cortex that are activated by both olfactory and taste stimuli (Rolls and Baylis 1994), and in that the neuronal responses in this region reflect in some cases olfactory to taste

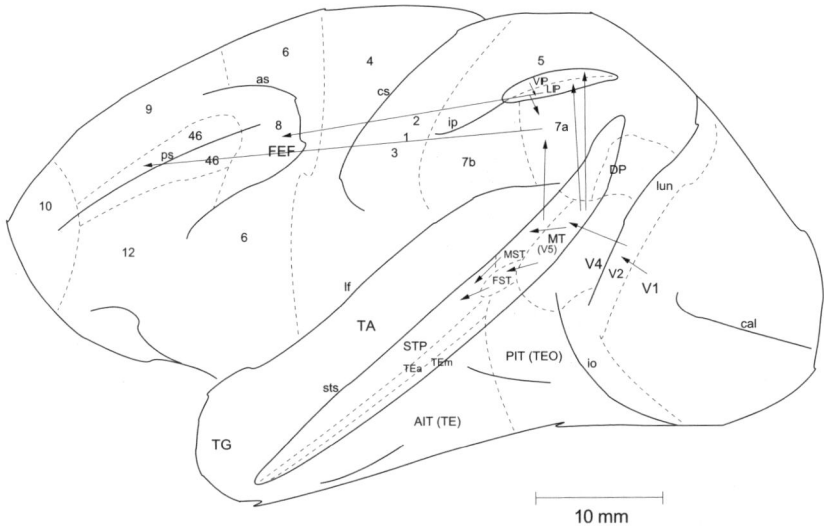

Fig. 2.23 Lateral view of the macaque brain showing the connections in the 'dorsal or where visual pathway' from V1 to V2, MST, LIP, VIP, and parietal cortex area 7a, with some connections then reaching the dorsolateral prefrontal cortex. Abbreviations as in Fig. 2.21. FEF - frontal eye field.

association learning (Rolls, Critchley, Mason and Wakeman 1996b, Critchley and Rolls 1996b). In these regions too, the representation is concerned not only with what sensory stimulus is present, but for some neurons, with its hedonic or reward-related properties, which are often computed by association with stimuli in other modalities. For example, many of the visual neurons in the orbitofrontal cortex respond to the sight of food only when hunger is present. This probably occurs because the visual inputs here have been assoc-iated with a taste input, which itself only produces neuronal responses in this region if hunger is present, that is when the taste is rewarding (see Section 3.13) (Rolls 2005a, Rolls and Grabenhorst 2008, Rolls 2011i). The outputs from these associative memory systems, the amygdala and orbitofrontal cor-tex, project onwards to structures such as the hypothalamus, through which they control autonomic and endocrine responses such as salivation and insulin release to the sight of food; and to the striatum, including the ventral striatum, through which behaviour to learned reinforcing stimuli is produced (Fig. 2.25).

The *'dorsal or where' visual processing stream* shown in Figs. 2.23, 2.22 and 2.25, is that from V1 to MT, MST and thus to the parietal cortex (Ungerleider 1995, Ungerleider and Haxby 1994, Rolls and Deco 2002, Rolls 2008c)). This 'where' pathway for primate vision is involved in representing where stimuli are relative to the animal (i.e. in egocentric space), and the motion of these stimuli. Neurons here respond, for example, to stimuli in visual space around the animal, including the distance from the observer, and also respond to optic flow or to moving stimuli. Outputs of this system control eye movements to

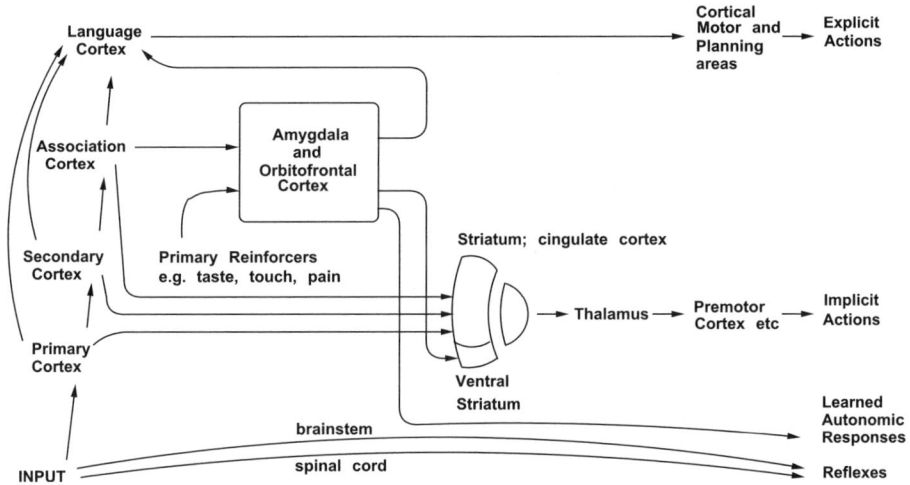

Fig. 2.24 Dual routes to the initiation of actions in response to rewarding and punishing stimuli. The inputs from different sensory systems to brain structures such as the orbitofrontal cortex and amygdala allow these brain structures to evaluate the reward- or punishment-related value of incoming stimuli, or of remembered stimuli. The different sensory inputs enable evaluations within the orbitofrontal cortex and amygdala based mainly on the primary (unlearned) reinforcement value for taste, touch, and olfactory stimuli, and on the secondary (learned) reinforcement value for visual and auditory stimuli. In the case of vision, the 'association cortex' that outputs representations of objects to the amygdala and orbitofrontal cortex is the inferior temporal visual cortex. One route for the outputs from these evaluative brain structures is via projections directly to structures such as the basal ganglia (including the striatum and ventral striatum) and cingulate cortex to enable implicit, direct behavioural responses based on the reward- or punishment-related evaluation of the stimuli to be made. The second route is via the language systems of the brain, which allow explicit (verbalizable) decisions involving multistep syntactic planning to be implemented.

visual stimuli (both slow pursuit and saccadic eye movements). These outputs proceed partly via the frontal eye fields, which then project to the striatum, and then via the substantia nigra reach the superior colliculus (Goldberg 2000). Other outputs of these regions are to the dorsolateral prefrontal cortex, area 46, which is important as a short-term memory for where fixation should occur next, as shown by the effects of lesions to the prefrontal cortex on saccades to remembered targets, and by neuronal activity in this region (Goldman-Rakic 1996). The dorsolateral prefrontal cortex short-term memory systems in area 46 with spatial information received from the parietal cortex play an important role in attention, by holding on-line the target being attended to, as described in Section 2.9.

2.11 Multiple learning systems in the brain

There are many learning systems in the brain for different functions, with activity in some available to conscious report, and others not. The learning systems do not necessarily produce consistent outputs, and this leads to complexity on behaviour. Moreover, the different learning systems are subject, as is the rest of the brain, to variation between individuals as part of the process of Darwinian evolution, and this leads to individual differences in different types of learning. To understand human behaviour, and how this shapes our culture and vice versa, I next outline some of the different types of learning in the brain, using Fig. 2.24 to show some of the systems.

First, **perceptual learning** takes place in the hierarchy of cortical areas for each sensory modality (vision, hearing, taste, olfaction and touch) using competitive learning of the type illustrated in Fig. 2.7d. An outline has been described for the visual system in Section 2.10. The competitive learning enables neurons to learn to respond to complex combinations of features, such as faces which include a combination of eyes, nose, mouth, hair etc. in the correct relative spatial positions, and allowing the particular details of the features, and the distances between them, to provide information about which individual's face is shown (Perrett, Rolls and Caan 1982, Rolls 2008c, Rolls 2011d). For vision, the representations of individual objects and faces must be invariant with respect to the position of the object on the retina, its size, and even its view (e.g. a face seen from the front or in profile), and we have shown that such invariant representations of objects and faces are present in the inferior temporal visual cortex (Rolls and Baylis 1986, Hasselmo, Rolls, Baylis and Nalwa 1989, Gross, Rodman, Gochin and Colombo 1993, Tovee, Rolls and Azzopardi 1994, Booth and Rolls 1998, Rolls, Aggelopoulos and Zheng 2003a). We have produced a theory of how the visual system learns these invariant representations, using competitive learning in the cortical hierarchy with spatial convergence from stage to stage. Part of the key to the success of the system is that the competitive learning rule involves a short-term memory trace for the previous few seconds, which enables neurons to learn to categorize stimuli as similar over periods of a few seconds (Földiák 1991, Rolls 1992, Wallis and Rolls 1997, Rolls and Stringer 2006, Rolls 2008c, Rolls 2012b). This allows the system to learn about objects from the statistics of the natural world, for we typically look at one object for one to a few seconds during which time the object may be seen in different transforms, then move our eyes to look at another object (see Chapter 4 of Rolls (2008c)).

A key property of this perceptual learning that we have discovered is that the brain learns a representation of objects that is independent of their reward or affective (emotional, aesthetic) value, and only in later cortical areas relates this to the reward value of the object (Rolls 2005a, Rolls and Grabenhorst 2008). In vision, because invariant object representations are learned first, anything about the reward value of that object that is learned by a subsequent stage such as the orbitofrontal cortex generalizes automatically to other transforms of the object. This is a crucial and fundamental property of brain design. It also enables us to learn about objects, for example where they are located, independently of whether we want them, i.e. whether they are rewarding, at the time. A similar principle has been shown to operate for taste, in that what the taste is (sweet, salt, bitter, sour or umami) is represented in the primary taste cortex (in the anterior insula) independently of its reward value (as modulated for example by hunger), and the pleasantness of taste (its affective or reward value) is represented in the orbitofrontal cortex (see Fig. 2.25) (Chapter 3). This enables us again to learn about for example the location of the source of the taste, and to report about the identity and intensity of the taste, independently of how rewarding it is at the time.

Second, **reward-related learning** can take place at several levels of the system shown in Fig. 2.25.

One level allows the autonomic system to learn responses such as salivation to the sight of food (Figs. 2.24 and 2.25). Here the sight of food becomes the conditioned stimulus, the taste of the food is the unconditioned stimulus, and the salivation is the conditioned response in what is known as classical or Pavlovian conditioning. Similarly, an increased heart rate might become a conditioned response to a visual stimulus such as the sight of a face associated with a pleasant or unpleasant unconditioned (i.e. unlearned) stimulus such as pleasant touch or an unpleasant stimulus such as a spike[7]. This system operates implicitly, by which we mean unconsciously[8]. Brain regions such as the amygdala and orbitofrontal cortex are involved in this type of learning (Blair, Schafe, Bauer, Rodrigues and LeDoux 2001, Pare, Quirk and LeDoux 2004, Davis 2006, Rolls 2005a, Rolls 2008c).

A second level involving the striatum and other parts of the basal ganglia learns about automatic responses or habits (Figs. 2.24 and 2.25). This is sometimes called stimulus-response learning, and becomes established after much training. It operates typically implicitly (unconsciously), and is used for many types of skills. During acquisition of the skills the actions may be directed by the goal-directed or more consciously by the reasoning system. However, the reasoning system that can be guided by verbal instruction is too slow to perform the skills as well as the habit system once the skills have been learned by the habit system. This is why for example thinking consciously about one's moves, in for example golf, tennis, piano playing, or windsurfing, may disrupt performance after it has been learned by the implicit habit system. Learning in the habit system is slow, and much practice makes perfect in this system.

When humans or animals perform an action to get a reward (termed an instrumental action), and this is performed by the stimulus-response or habit system, the behaviour may be initiated, even if the goal is no longer desirable

---

[7] *Stimulus* is the Latin word for a sharpened post set into the ground at an angle to deter an enemy (Lewis and Short 1879).

[8] *Autonomic* is Greek for self-ruling (Liddell and Scott 1891).

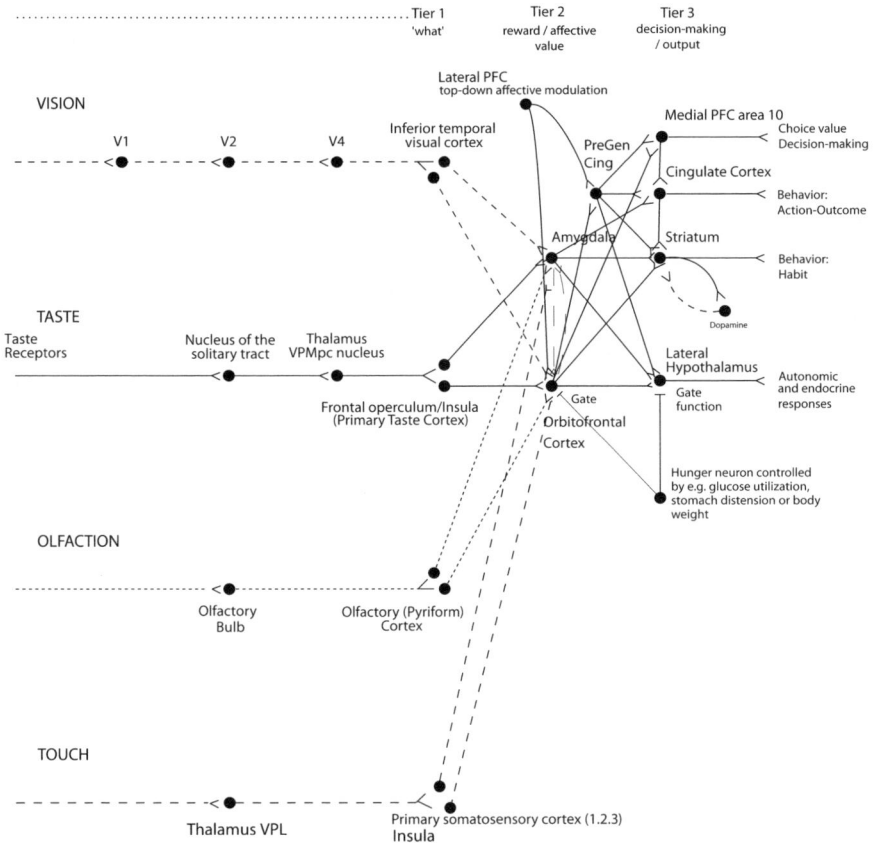

Fig. 2.25 Schematic diagram showing some of the gustatory, olfactory, visual and somatosensory pathways to the orbitofrontal cortex, and some of the outputs of the orbitofrontal cortex, in primates. The secondary taste cortex, and the secondary olfactory cortex, are within the orbitofrontal cortex. V1 - primary visual cortex. V4 - visual cortical area V4. PreGen Cing - pregenual cingulate cortex. "Gate" refers to the finding that inputs such as the taste, smell, and sight of food in some brain regions only produce effects when hunger is present (Rolls, 2005). Tier 1 is the column of brain regions including and below the inferior temporal visual cortex represents brain regions in which what stimulus is present is made explicit in the neuronal representation. In Tier 2 the reward or affective value is made explicit in the representation, in regions including the orbitofrontal cortex, pregenual cingulate cortex, and amygdala. Tier 3 shows areas where choice decisions are made between the values of stimuli. The top pathway, also shown in Fig. 2.21, shows the connections in the 'ventral or what visual pathway' from V1 to V2, V4, the inferior temporal visual cortex, etc., with some connections reaching the amygdala and orbitofrontal cortex. The taste pathways project after the primary taste cortex to the orbitofrontal cortex and amygdala. The olfactory pathways project from the primary olfactory, pyriform, cortex to the orbitofrontal cortex and amygdala. The bottom pathway shows the connections from the primary somatosensory cortex, areas 1, 2 and 3, to the mid-insula, orbitofrontal cortex, and amygdala. Somatosensory areas 1, 2 and 3 also project via area 5 in the parietal cortex, to area 7b.

because for example it has been devalued or made unpleasant in another context. This has been taken to imply that we may 'want' something when we do not 'like' it (Berridge and Robinson 1998, Berridge and Robinson 2003), but my view (Rolls 2008c) is that it just reflects the operation of the habit system which once set up and until it is retrained implements implicit or automatic actions or responses that are independent of the goal value because of the overlearning in the habit system.

The third level involves regions such as the cingulate cortex which perform goal-based learning of instrumental actions (Fig. 2.25). The process involves learning of associations between behavioral actions and reinforcers (reward and or punishment outcomes, hence 'action-outcome' learning) (Rushworth, Walton, Kennerley and Bannerman 2004, Rushworth, Noonan, Boorman, Walton and Behrens 2011). This may often involve trial-and-error learning or what is required to obtain a reward. Once learned, if training continues, the habit system may gradually take over control. If however the habit system suddenly fails, perhaps because the contingencies have changed, the goal based system will switch back in and start to relearn whatever is required.

The stimulus-reward learning in brain regions such as the orbitofrontal cortex and amygdala is probably implemented by pattern association networks of the type illustrated in Fig. 2.7a and b. However, if an expected reward is not obtained, a population of neurons in the orbitofrontal cortex responds (Thorpe, Rolls and Maddison 1983, Kringelbach and Rolls 2003, Rolls 2009b), to help the system correct itself. These neurons are called negative reward prediction error neurons, for they respond when the error or difference between the prediction and the reward outcome obtained is negative. With practice, it is possible to learn to switch very rapidly the stimulus chosen when the contingency reverses ('serial reversal learning set'). This capability may involve a short-term memory network which holds the current rule in mind, and can be rapidly reset so that the neurons representing the other contingency then become active as a result of the activity of the negative reward prediction error neurons. The rule short-term memory neurons then bias other neurons termed conditional reward neurons to reset the choices of the stimuli (Deco and Rolls 2005c, Rolls 2009b).

A fourth level allows multistep reasoning to possibly defer an immediate reward, and to obtain a more valuable reward (as judged by this rational system), as indicated in Fig. 2.24, and as described in Chapter 5.

We have already seen descriptions of other learning systems that use attractor networks (Fig. 2.7c and Section 2.5.3), for example those involved in short-term memory (Section 2.6) and episodic memory (Section 2.8). We now show that the same architecture, prototypical of the cerebral neocortex, can be used for decision-making.

2.12 Decision-making

It turns out that with great elegance and simplicity of design, the brain uses the attractor network architecture described in Section 2.5.3 not only for functions such as short-term memory (Section 2.6) and long-term memory (Section 2.8 and Rolls (2008c)), but also for decision-making. Indeed, we can understand memory recall as a special case of decision-making, in which sometimes multiple and conflicting recall cues lead to a single recalled memory. We can also understand creative thought as multiple step associative memory recall, driven by the same internal neuronal spiking-related noise that makes decision-making probabilistic (Section 2.12.9) (Rolls and Deco 2010).

2.12.1 Decision-making by an attractor network

To understand the mechanisms by which the brain takes decisions (Wang 2002, Deco and Rolls 2006, Wang 2008, Rolls and Deco 2010), let us consider the attractor network architecture again, but this time as shown in Fig. 2.26a with two competing inputs λ_1 and λ_2, each encouraging the network to move from a state of spontaneous activity into the attractor corresponding to λ_1 or to λ_2. These are separate attractor states that have been set up by associative synaptic modification, one attractor for the neurons that are co-active when λ_1 is applied, and a second attractor for the neurons that are co-active when λ_2 is applied. When λ_1 and λ_2 are both applied simultaneously, each attractor competes through the inhibitory interneurons (not shown), until one attractor state wins the competition, and the network falls into one of the high firing rate attractors that represents the decision. The noise in the network caused by the random spiking of the neurons means that on some trials, for given inputs, the neurons in the decision 1 attractor are more likely to win, and on other trials the neurons in the decision 2 attractor are more likely to win. This makes the decision-making probabilistic, for, as shown in Fig. 2.26b, the noise influences when the system will jump out of the spontaneous firing stable (low energy) state S, and whether it jumps into the high firing state for decision 1 or decision 2 (D).

2.12.2 Confidence in the brain

We can report after we have taken a decision how confident we are in the decision, even before the outcome of the decision becomes available. Our subjective confidence is higher as the easiness of the decision increases, where the easiness is measured by $\Delta\lambda$, the difference between the two input stimuli between which a choice is made (Vickers 1979, Vickers and Packer 1982). We can now understand how confidence is an emergent property of the attractor decision-making network, and this illustrates how now that we are staring to understand the mechanisms of cortical function, further properties can be

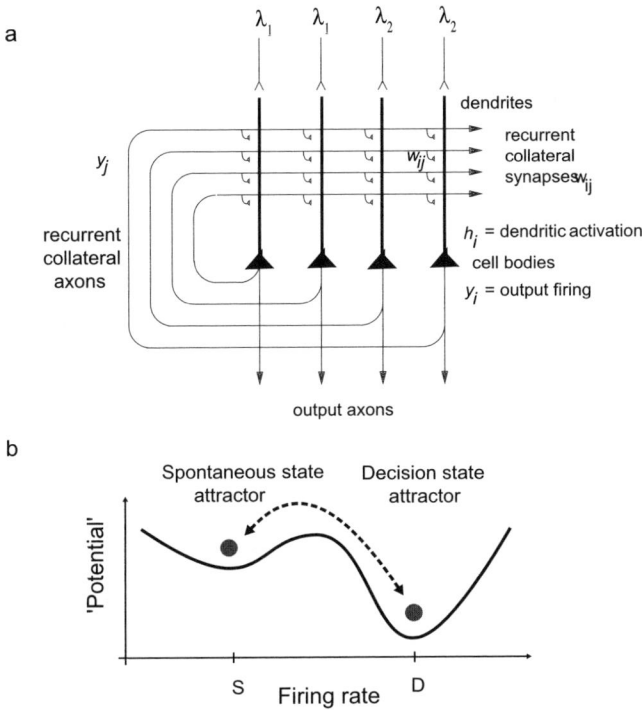

Fig. 2.26 (a) Attractor or autoassociation network architecture for decision-making. The evidence for decision 1 is applied via the λ_1 inputs, and for decision 2 via the λ_2 inputs. The synaptic weights w_{ij} have been associatively modified during training in the presence of λ_1 and at a different time of λ_2. When λ_1 and λ_2 are applied, each attractor competes through the inhibitory interneurons (not shown), until one wins the competition, and the network falls into one of the high firing rate attractors that represents the decision. The noise in the network caused by the random spiking of the neurons means that on some trials, for given inputs, the neurons in the decision 1 attractor are more likely to win, and on other trials the neurons in the decision 2 attractor are more likely to win. This makes the decision-making probabilistic, for, as shown in (b), the noise influences when the system will jump out of the spontaneous firing stable (low energy) state S, and whether it jumps into the high firing state for decision 1 or decision 2 (D).

accounted for without extra postulates or ad hoc machinery. In this case, let us assume that our attractor network (Fig. 2.26a) has made a decision, which is influenced by the internal spiking-related noise. Let us assume that decision population 1 wins the competition. If that is the correct decision, then the larger λ_1 is relative to λ_2, the greater will be the firing of the population of neurons that represent choice 1, because the continuing presence of λ_1 adds to the input from the recurrent collateral synaptic connections that is maintaining the firing rate of the neurons for decision 1 high. This is illustrated in Fig. 2.27 (Rolls, Grabenhorst and Deco 2010c, Rolls et al. 2010b).

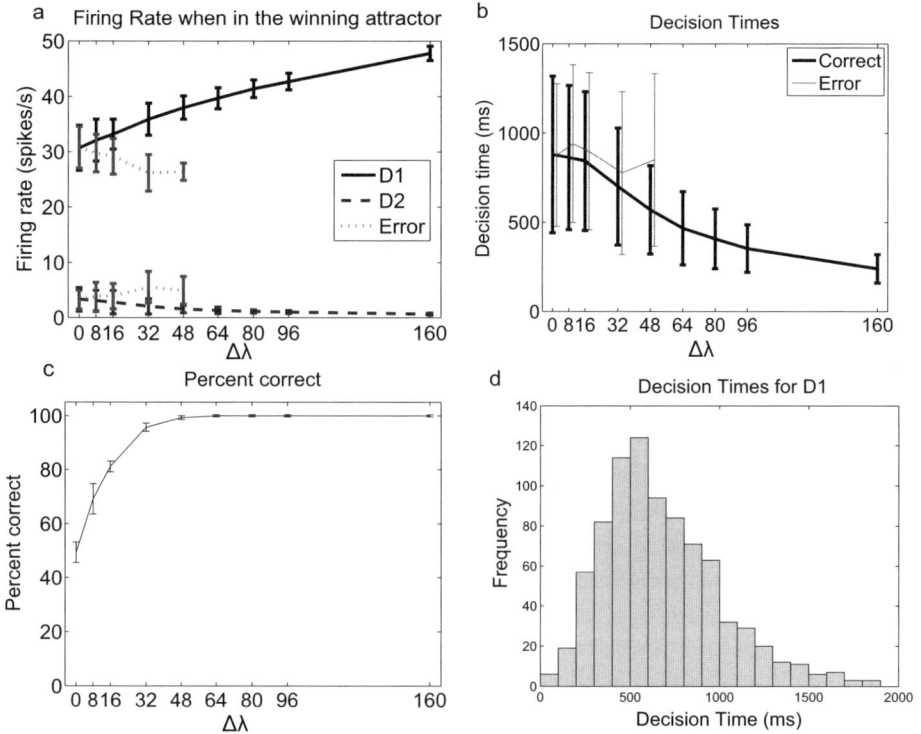

Fig. 2.27 (a) Firing rates (mean ± sd) on correct trials when in the D1 attractor as a function of ΔI. $\Delta\lambda$=0 corresponds to difficult, and $\Delta\lambda$=160 spikes/s corresponds to easy. The firing rates on correct trials for the winning population D1 are shown by solid lines, and for the losing population D2 by dashed lines. All the results are for 1000 simulation trials for each parameter value, and all the results shown are statistically highly significant. The results on error trials are shown by the dotted lines, and in this case the D2 attractor wins, and the D1 attractor loses the competition. (b) Decision or reaction times (mean ± sd) for the D1 population to win on correct trials as a function of the difference in inputs $\Delta\lambda$ to D1 and D2. (c) Per cent correct performance, i.e. the percentage of trials on which the D1 population won, as a function of the difference in inputs $\Delta\lambda$ to D1 and D2. (d) The distribution of decision times for the model for $\Delta\lambda$=32 illustrating the long tail of slow responses. Decision times are shown for 837 correct trials, the level of performance was 95.7% correct, and the mean decision time was 701 ms. (After Rolls, Grabenhorst and Deco, 2010c.) (See colour plates Appendix B.)

Conversely, on error trials the firing rate for the winning attractor decreases as $\Delta\lambda$ increases (Fig. 2.27, Rolls, Grabenhorst and Deco (2010c)). The reason for this is that the continuing input is now working against the noise-influenced decision that has been taken by the attractor decision-making network. This accounts for why our confidence (before the outcome of the action chosen is known) is lower on trials when we make an error. The predictions of this theory of decision-making about brain activations on error trials in parts of the brain implicated in value-based decision-making have been confirmed (Rolls et al. 2010c).

Thus we have a theory of how confidence, which can be subjective and conscious, is an emergent property of the type of decision-making network that we believe is implemented in the brain.

2.12.3 Decision-making with multiple alternatives

This framework can also be extended very naturally to account for the probabilistic decision taken when there are multiple, that is more than two, choices. One such extension models choices between continuous variables in a continuous or line attractor network (Furman and Wang 2008, Liu and Wang 2008) to account for the responses of lateral intraparietal cortex neurons in a 4-choice random dot motion decision task (Churchland, Kiani and Shadlen 2008). In another approach, a network with multiple discrete attractors (Albantakis and Deco 2009) can account well for the same data.

2.12.4 The matching law

Another potential application of this model of decision-making is to probabilistic decision tasks. In such tasks, the proportion of choices reflects, and indeed may be proportional to, the expected value of the different choices. This pattern of choices is known as the matching law (Sugrue, Corrado and Newsome 2005). An example of a probabilistic decision task in which the choices of the human participants in the probabilistic decision task clearly reflected the expected value of the choices is described by Rolls, McCabe and Redoute (2008e).

A network of the type described here in which the biasing inputs λ_1 and λ_2 to the model are the expected values of the different choices alters the proportion of the decisions it makes as a function of the relative expected values in a way similar to that shown in Fig. 2.27, and provides a model of this type of probabilistic reward-based decision-making (Marti, Deco, Del Giudice and Mattia 2006). It was shown for example that the proportion of trials on which one stimulus was shown over the other was approximately proportional to the difference of the two values between which choices were being made. In setting the connection weights to the two attractors that represent the choices, the returns (the average reward per choice), rather than the incomes (the average reward per trial) of the two targets, are relevant (Soltani and Wang 2006).

This type of model also accounts for the observation that matching is not perfect, and the relative probability of choosing the more rewarding option is often slightly smaller than the relative reward rate ('undermatching'). If there were no neural variability, decision behaviour would tend to get stuck with the more rewarding alternative; stochastic spiking activity renders the network more exploratory and produces undermatching as a consequence (Soltani and Wang 2006).

2.12.5 Symmetry-breaking

It is of interest that the noise that contributes to the stochastic dynamics of the brain through the spiking fluctuations may be behaviourally adaptive, and that the noise should not be considered only as a problem in terms of how the brain works. This is the issue raised for example by the donkey in the medieval Duns Scotus paradox, in which a donkey situated between two equidistant food rewards might never make a decision and might starve.

The problem raised is that with a deterministic system, there is nothing to break the symmetry, and the system can become deadlocked. In this situation, the addition of noise can produce probabilistic choice, which is advantageous. We have shown here that stochastic neurodynamics caused for example by the relatively random spiking times of neurons in a finite sized cortical attractor network can lead to probabilistic decision-making, so that in this case the stochastic noise is a positive advantage.

2.12.6 The evolutionary utility of probabilistic choice

Probabilistic decision-making can be evolutionarily advantageous in another sense, in which sometimes taking a decision that is not optimal based on previous history may provide information that is useful, and which may contribute to learning. Consider for example a probabilistic decision task in which choice 1 provides rewards on 80% of the occasions, and choice 2 on 20% of the occasions. A deterministic system with knowledge of the previous reinforcement history would always make choice 1. But this is not how animals including humans behave. Instead (especially when the overall probabilities are low and the situation involves random probabilistic baiting, and there is a penalty for changing the choice), the proportion of choices made approximately matches the outcomes that are available, as described by the matching law (Sugrue, Corrado and Newsome 2005, Corrado, Sugrue, Seung and Newsome 2005, Rolls, McCabe and Redoute 2008e) (Section 2.12.4). By making the less favoured choice sometimes, the organism can keep obtaining evidence on whether the environment is changing (for example on whether the probability of a reward for choice 2 has increased), and by doing this approximately according to the matching law minimizes the cost of the disadvantageous choices in obtaining information about the environment.

This probabilistic exploration of the environment is very important in trial-and-error learning, and indeed has been incorporated into a simple reinforcement algorithm in which noise is added to the system, and if this improves outcomes above the expected value, then changes are made to the synaptic weights in the correct direction (in the associative reward-penalty algorithm) (Sutton and Barto 1981, Barto 1985, Rolls 2008c).

In perceptual learning, probabilistic exploratory behaviour may be part of the mechanism by which perceptual representations can be shaped to have

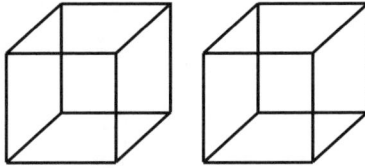

Fig. 2.28 Two Necker cubes. (It may be helpful to increase the viewing distance.)

appropriate selectivities for the behavioural categorization being performed (Sigala and Logothetis 2002, Szabo, Deco, Fusi, Del Giudice, Mattia and Stetter 2006).

Another example is in food foraging, which probabilistically may reflect the outcomes (Krebs and Davies 1991, Kacelnik and Brito e Abreu 1998), and is a way optimally in terms of costs and benefits to keep sampling and exploring the space of possible choices.

Another sense in which probabilistic decision-making may be evolutionarily advantageous is with respect to detecting signals that are close to threshold, as described by Rolls and Deco (2010) in their section on stochastic resonance. In this case, adding a little noise to the signal brings the state above threshold more frequently, allowing more of the signal to be revealed.

Intrinsic indeterminacy may be essential for unpredictable behaviour (Glimcher 2005). For example, in interactive games like matching pennies or rock–paper–scissors, any trend that deviates from random choice by an agent could be exploited to his or her opponent's advantage. This again is an emergent property of this noise-influenced decision-making system implemented in the brain (Rolls and Deco 2010).

2.12.7 Perceptual decision-making and rivalry

Another application is to changes in perception. Perceptions can change 'spontaneously' from one to another interpretation of the world, even when the visual input is constant. A good example is the Necker cube, in which visual perception flips occasionally to make a different edge of the cube appear nearer

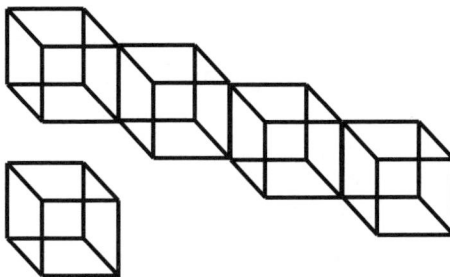

Fig. 2.29 Linked Necker cubes, and a companion.

Fig. 2.30 The Rubin vase is on the right. Noise in the brain influences the transitions from seeing this as a vase, or as two faces in profile looking at each other. (See colour plates Appendix B.)

to the observer (Fig. 2.28). We hypothesize that the switching between these multistable states is due in part to the statistical fluctuations in the network due to the Poisson-like spike firing that is a form of noise in the system. It will be possible to test this hypothesis in integrate-and-fire simulations. (This may or may not be supplemented by adaptation effects (of the synapses or neurons) in integrate-and-fire networks.)

You may observe interesting effects in Fig. 2.28, in which when one cube flips which face appears closer, the other cube performs a similar flip, so that the two cubes remain for most of the time in the same configuration. This effect can be accounted for by short-range cortico-cortical excitatory connections between corresponding depth feature cue combination neurons that normally help to produce a consistent interpretation of 3D objects. The underlying mechanism is that of attractor dynamics linking in this case corresponding features in different objects. When the noise in one of the attractors makes the attractor flip, this in turn applies a bias to the other attractor, making it very likely that the attractor for the second cube will flip soon under the influence of its internal spiking-related noise. The whole configuration provides an interesting perceptual demonstration of the important role of noise in influencing attractor dynamics, and of the cross-linking between related attractors which helps the whole system to move towards energy minima, under the influence of noise. Another interesting example is shown in Fig. 2.29.

The same approach should provide a model of pattern and binocular rivalry and ambiguous figures, where one image is seen at a time even though two images are presented simultaneously (Fig. 2.30), and indeed an attractor-based noise-driven model of perceptual alternations has been described (Moreno-Bote, Rinzel and Rubin 2007). When these are images of objects or faces, the system that is especially important in the selection is the inferior temporal visual cortex (Blake and Logothetis 2002, Maier, Logothetis and Leopold 2005), for it is here that representations of whole objects are present (Rolls 2008c, Rolls and Stringer 2006, Rolls 2009d), and the global interpretation of one object can compete with the global interpretation of another object. These simulation models are highly feasible, in that the effects in integrate-and-fire simulations to influence switching between stable states not only of noise, but also of synaptic adaptation and neuronal adaptation which may contribute, have already been investigated (Deco and Rolls 2005c, Deco and Rolls 2005b, Moreno-Bote et al. 2007).

2.12.8   Memory recall as a stochastic process

The theory is effectively a model of the stochastic dynamics of the recall of a memory in response to a recall cue. The memory might be a long-term memory, but the theory applies to the retrieval of any stored representation in the brain. The way in which the attractor is reached depends on the strength of the recall cue, and inherent noise in the attractor network performing the recall because of the spiking activity in a finite size system. The recall will take longer if the recall cue is weak. Spontaneous stochastic effects may suddenly lead to the memory being recalled, and this may be related to the sudden recovery of a memory which one tried to remember some time previously. These processes are considered further by Rolls (2008c).

The theory applies to a situation where the representation may be being 'recalled' by a single input, which is perceptual detection as described by Rolls and Deco (2010).

The theory also applies to a situation where the representation may be being 'recalled' by two or more competing inputs λ, which is computationally a type of decision-making.

The theory also applies to short-term memory, in which the continuation of the recalled state as a persistent attractor is subject to stochastic noise effects, which may knock the system out of the short-term memory attractor, as described in Section 2.6.

The theory also applies to attention, in which the continuation of the recalled state as a persistent attractor is subject to stochastic noise effects, which may knock the system out of the short-term memory attractor that is normally stable because of the non-linear positive feedback implemented in the attractor network by the recurrent collateral connections, as described in Section 2.9 and Chapter 10.

2.12.9   Creative thought

Another way in which probabilistic decision-making may be evolutionarily advantageous is in creative thought, which is influenced in part by associations between one memory, representation, or thought, and another. If the system were deterministic, i.e. for the present purposes without noise, then the trajectory through a set of thoughts would be deterministic and would tend to follow the same furrow each time. However, if the recall of one memory or thought from another were influenced by the statistical noise due to the random spiking of neurons, then the trajectory through the state space would be different on different occasions, and we might be led in different directions on different occasions, facilitating creative thought (Rolls 2008c).

Of course, if the basins of attraction of each thought were too shallow, then the statistical noise might lead one to have very unstable thoughts that were too loosely and even bizarrely associated to each other, and to have a short-term memory and attentional system that is unstable and distractible, and indeed this is an account that we have proposed for some of the symptoms of schizophrenia (Rolls 2005a, Rolls 2008c, Loh, Rolls and Deco 2007a, Loh, Rolls and Deco 2007b, Rolls, Loh, Deco and Winterer 2008d) (see Section 10.1).

The stochastic noise caused by the probabilistic neuronal spiking plays an important role in these hypotheses, because it is the noise that destabilizes the attractors when the depth of the basins of attraction is reduced. If the basins of attraction were too deep, then the noise might be insufficient to destabilize attractors, and this leads to an approach to understanding obsessive-compulsive disorders (Rolls, Loh and Deco 2008c) (see Section 10.2).

2.12.10   Unpredictable behaviour

An area where the spiking-related noise in the decision-making process may be evolutionarily advantageous is in the generation of unpredictable behaviour, which can be advantageous in a number of situations, for example when a prey is trying to escape from a predator, and perhaps in some social and economic situations in which organisms may not wish to reveal their intentions (Maynard Smith 1982, Maynard Smith 1984, Dawkins 1995). We note that such probabilistic decisions may have long-term consequences. For example, a probabilistic decision in a 'war of attrition', such as staring down a competitor in dominance hierarchy formation, may fix the relative status of the two individual animals involved, who then tend to maintain that relationship stably for a considerable period of weeks or more (Maynard Smith 1982, Maynard Smith 1984, Dawkins 1995).

Thus intrinsic indeterminacy may be essential for unpredictable behaviour. We have noted the example provided by Glimcher (2005): in interactive games like matching pennies or rock–paper–scissors, any trend that deviates from random choice by an agent could be exploited to his or her opponent's advantage.

2.12.11 Dreams

Similar noise-driven processes may lead to dreams, where the content of the dream is not closely tied to the external world because the role of sensory inputs is reduced in paradoxical (desynchronized, high frequency, fast wave, dreaming) sleep, and the cortical networks, which are active in fast-wave sleep (Kandel et al. 2012, Carlson 2006, Horne 2006), may move under the influence of noise somewhat freely on from states that may have been present during the day. Thus the content of dreams may be seen as a noisy trajectory through state space, with starting point states that have been active during the day, and passing through states that will reflect the depth of the basins of attraction (which might well reflect ongoing concerns including anxieties and desires), and will be strongly influenced by noise. Moreover, the top-down attentional and monitoring control from for example the prefrontal cortex appears to be less effective during sleep than during waking, allowing the system to pass into states that may be bizarre.

I suggest that this statistical mechanics-based approach (Rolls 2008c, Rolls and Deco 2010) provides a firm theoretical foundation for understanding the interpretation of dreams, which may be contrasted with that of Freud (1900).

In this context, the following thoughts follow. Dreams, or at least the electrical activity of paradoxical sleep, may occur in cortical areas that are concerned with conscious experience, such as those involved in higher-order thoughts, and in others where processing is unconscious, and cannot be reported, such as those in the dorsal visual stream concerned with the control of actions (Rolls 2004b, Goodale 2004, Rolls 2005a, Rolls 2007a, Rolls 2008a, Rolls 2008c). Thus it may be remarked that dreams may occur in conscious and unconscious processing systems. Dreams appear to be remembered that occur just before we wake up, and consistent with this, memory storage (implemented by synaptic long-term potentiation (Rolls 2008c)) appears to be turned off during sleep. This may be adaptive, for then we do not use up memory capacity (Rolls 2008c) on noise-related representations. However, insofar as we can memorize and later remember dreams that are rehearsed while we wake up, it could be that bizarre thoughts, possibly unpleasant, could become consolidated. This consolidation could lead to the relevant attractor basins becoming deeper, and returning to the same set of memories on subsequent nights. This could be a mechanism for the formation of nightmares. A remedy is likely to be not rehearsing these unpleasant dreams while waking, and indeed to deliberately move to more pleasant thoughts, which would then be consolidated, and increase the probability of dreams on those more pleasant subjects, instead of nightmares, on later nights, given the memory attractor landscape over which noise would move one's thoughts.

In slow-wave sleep, and more generally in resting states, the activity of neurons in many cortical areas is on average low (Carlson 2006), and stochastic spiking-related noise may contribute strongly to the states that are found.

2.12.12 Multiple decision-making systems in the brain

Each cortical area can be conceived as performing a local type of decision-making using attractor dynamics of the type described (Rolls 2008c). Even memory recall is in effect the same local 'decision-making' process.

The orbitofrontal cortex for example is involved in providing evidence for decisions about which visual stimulus is currently associated with reward, in for example a visual discrimination reversal task. Its computations are about stimuli, primary reinforcers, and secondary reinforcers (Rolls 2005a). The orbitofrontal cortex appears to represent reward value on a continuous scale, with binary choice decisions being made in the immediately adjoining area anteriorly, the medial prefrontal cortex area 10, as described in Sections 3.11 and 3.12 (Rolls and Grabenhorst 2008, Grabenhorst, Rolls and Parris 2008c, Rolls, Grabenhorst and Parris 2010d, Rolls, Grabenhorst and Deco 2010b, Rolls, Grabenhorst and Deco 2010c, Grabenhorst and Rolls 2011).

The dorsolateral prefrontal cortex takes an executive role in decision-making in a working memory task, in which information must be held available across intervening stimuli (Rolls 2008c). The dorsal and posterior part of the dorsolateral prefrontal cortex may be involved in short-term memory-related decisions about where to move the eyes (Rolls 2008c).

The parietal cortex is involved in decision-making when the stimuli are for example optic flow patterns (Glimcher 2003, Gold and Shadlen 2007).

The hippocampus is involved in (providing evidence for) decision-making when the allocentric places of stimuli must be associated with rewards or objects (Rolls and Kesner 2006, Rolls 2008c).

The somatosensory cortex and ventral premotor cortex are involved in decision-making when different vibrotactile frequencies must be compared (as described in Section 2.12).

The cingulate cortex may be involved when action–outcome decisions must be taken (Rushworth et al. 2004, Rolls 2005a, Rolls 2008c, Rolls 2009a, Rushworth et al. 2011).

In each of these cases, local cortical processing that is related to the type of decision being made takes place, and all cortical areas are not involved in any one decision. The style of the decision-making-related computation in each cortical area appears to be of the form described here, in which the local recurrent collateral connections enable the decision-making process to accumulate evidence across time, falling gradually into an attractor that represents the decision made in the network. Because there is an attractor state into which the network falls, this can be described statistically as a non-linear diffusion process, the noise for the diffusion being the stochastic spiking of the neurons, and the driving force being the biasing inputs.

If decision-making in the cortex is largely local and typically specialized, it leaves open the question of how one stream for behavioural output is selected. This type of 'global decision-making' is considered in Section 5.5.1 and in *Memory, Attention, and Decision-Making* (Rolls 2008c).

2.13 Sleep

Sleep has been proposed as a state in which useful forgetting or consolidation of memories could occur. One suggestion was that if deep basins of attraction formed in a memory network, then this could impair performance, as the memories in the basins would tend to be recalled whatever the retrieval cue. If noise, present in the disorganized patterns of neural firing during sleep, caused these memories to be recalled, this would indicate that they were 'parasitic', and the suggestion was that associative synaptic weakening (LTD) of synapses of neurons with high firing during sleep would tend to decrease the depth of those basins of attraction, and improve the performance of the memory (Crick and Mitchison 1995). At least at the formal level of neural networks, the suggestion does have some merit as a possible way to 'clean up' associative networks, even if it is not a process implemented in the brain. Although the idea of some role of sleep in memory remains active, this remains to be fully established (Walker and Stickgold 2006).

The idea that sleep could be a time when memories are unloaded from the hippocampus to be consolidated in long term, possibly semantic, memories during sleep (Marr 1971) (allowing hippocampal episodic memories to then be overwritten by new episodic memories) continues to be explored. It has been shown for example that after hippocampal spatial representations have been altered by experience during the day, these changes are reflected in neuronal activity in the neocortex during sleep (Wilson and McNaughton 1994, Wilson 2002). The type of experience might involve repeated locomotion between two places, and the place fields of rat hippocampal neurons for those places may become associated with each other because of co-activity of the neurons representing the frequently visited places. The altered co-firing of the hippocampal neurons for those places may then be reflected in neocortical representations of those places. This could then result in altered representations in the neocortex, if LTP occurs during sleep in the neocortex. Of course, any change in neocortical neuronal activity might just reflect the altered representations in the hippocampus, which would be expected to influence the neocortical representations via hippocampo-neocortical backprojections, even without any neocortical learning (see Section 2.8).

2.14   Levels of explanation

We can now understand brain processing from the level of ion channels in neurons, through neuronal biophysics, to neuronal firing, through the computations performed by populations of neurons, and how their activity is reflected by functional neuroimaging, to behavioural and cognitive effects (Rolls 2008c, Rolls and Deco 2010). Activity at any one level can be used to understand activity at the next. This raises the philosophical issue of how we should consider causality with these different levels. Does the brain cause effects in the mind, or do events at the mental, mind, level influence brain activity? This issue is considered in Chapter 6.

Overall, understanding brain activity at these different levels provides a unifying approach to understanding brain function, which is proving to be so powerful that the fundamental operations involved in many aspects of brain function can be understood in principle, though with of course many details still to be discovered. These functions include many aspects of perception including visual face and object recognition, and taste, olfactory and related processing; short-term memory; long-term memory; attention; emotion; and decision-making. Predictions made at one level can be tested at another. Conceptually this is an enormous advance. But it is also of great practical importance, in medicine. For example, we now have new ways of predicting effects of possible pharmacological treatments for brain diseases by a developing understanding of how drugs affect synaptic receptors, which in turn affect neuronal activity, which in turn affect the stability of the whole network of neurons and hence cognitive symptoms such as attention vs distractibility (see Chapter 10). Perhaps the great computational unknown at present is how syntax for language is implemented in the brain.

2.15   Brain computation compared to computation on a digital computer

To highlight some of the principles of brain computation by for example the cortex described in this book, it is interesting to compare the principles of computation by the brain with those of a digital computer.

Data addressing. An item of data is retrieved from the memory of a digital computer by providing the address of the data in memory, and then the data can be manipulated (moved, compared, added to the data at another address in the computer etc.) using typically a 32 bit or 64 bit binary word of data. Pointers to memory locations are thus used extensively. In contrast, in the cortex, the data are used as the access key (in for example a pattern associator and autoassociator), and the neurons with synaptic weights that match the data respond.

Memory in the brain is thus **content-addressable**. In one time constant of the synapses/cell membranes the brain has thus found the correct output. In contrast, on a digital computer a serial search is required, in which the data at every address must be retrieved and compared in turn to the test data to discover if there is a match.

Vector similarity vs logical operations. Cortical computation including that performed by associative memories and competitive networks operates by vector similarity – the dot product of the input and of the synaptic weight vector are compared, and the neurons with the highest dot product will be most activated (Fig. 2.5). Even if an exact match is not found, some output is likely to result. In contrast, in a digital computer, logic operations (such as AND, OR, XOR) and exact mathematical operations (such as addition, subtraction, multiplication, and division) are computed. (There is no bit-wise similarity between the binary representations of 7 (0111) and 8 (1000).) The similarity computations performed by the brain may be very useful, in enabling similarities to be seen and parallels to be drawn, and this may be an interesting aspect of human creativity, realized for example in *Finnegans's Wake* by James Joyce. However, the lateral thinking must be controlled, to prevent bizarre similarities being found, and this is argued to be related to the symptoms of schizophrenia in Section 10.1.

Fault tolerance. Because exact computations are performed in a digital computer, there is no in-built fault tolerance or graceful degradation. If one bit of a memory has a fault, the whole memory chip must be discarded. In contrast, the brain is naturally fault tolerant, because it uses vector similarity (between its input firing rate vector and synaptic weight vectors) in its calculations, and linked to this, distributed representations. This makes the brain robust developmentally with respect to 'missing synapses', and robust with respect to losing some synapses or neurons later (see e.g. Section 2.5).

Word length. To enable the vector similarity comparison to have high capacity (for example memory capacity) the 'word length' in the brain is typically long, with between 10,000 and 50,000 synapses onto every neuron being common in cortical areas. (Remember that the leading term in the factor that determines the storage capacity of an associative memory is the number of synapses per neuron – see Sections 2.5.2 and 2.5.3.) In contrast, the word length in typical digital computers at 32 or 64 bits is much shorter, though with the binary and exact encoding used this allows great precision in a digital computer.

Readability of the code. To comment further on the encoding: in the cortex, the code must not be too compact, so that it can be read by neuronally plausible dot product decoding, as shown in Section 2.3. In contrast, the binary encoding used in a digital computer is optimally efficient, with one bit stored and retrievable for each binary memory location. However, the computer binary code cannot be read by neuronally plausible dot-product decoding.

Precision. The precision of the components in a digital computer is that every modifiable memory location must store one bit accurately. In contrast, it is of interest that synapses in the brain need not work with exact precision, with for example typically less that one bit per synapse being usable in associative memories (Treves and Rolls 1991, Rolls and Treves 1998). The precision of the encoding of information in the firing rate of a neuron is likely to be a few bits – perhaps 3 – as judged by the standard deviation and firing rate range of individual cortical neurons (Rolls 2008c).

The speed of computation. This brings us to the speed of computation. In the brain, considerable information can be read in 20 ms from the firing rate of an individual neuron (e.g. 0.2 bits), leading to estimates of 10–30 bits/s for primate temporal cortex visual neurons (Rolls, Treves and Tovee 1997b, Rolls and Tovee 1994), and 2–3 bits/s for rat hippocampal cells (Skaggs, McNaughton, Gothard and Markus 1993, Rolls 2008c, Rolls and Treves 2011). Though this is very slow compared to a digital computer, the brain does have the advantage that a single neuron receives spikes from thousands of individual neurons, and computes its output from all of these inputs within a period of approximately 10–20 ms (determined largely by the time constant of the synapses) (Rolls 2008c). Moreover, each neuron, up to at least the order of tens of neurons, conveys independent information, as described in Section 2.3.

Parallel vs serial processing. Computation in a conventional digital computer is inherently serial, with a single central processing unit that must fetch the data from a memory address, manipulate the word of data, and store it again at a memory address. In contrast, brain computation is parallel in at least three senses.

First, an individual neuron in performing a dot product between its input firing rate vector and its synaptic weight vector does operate in an analog way to sum all the injected currents through the thousands of synapses to calculate the activation h_i, and fire if a threshold is reached, in a time in the order of the synaptic time constant. To implement this on a digital computer would take $2C$ operations (C multiply operations, and C add operations, where C is the number of synapses per neuron – see Equation 2.1).

Second, each neuron in a single network (e.g. a small region of the cortex with of the order of hundreds of thousands of neurons) does this dot product computation in parallel, followed by interaction through the GABA inhibitory neurons, which again is fast. (It is in the order of the time constant of the synapses involved, operates in continuous time, and does not have to wait at all until the dot product operation of the pyramidal cells has been completed by all neurons given the spontaneous neuronal activity which allows some neurons to be influenced rapidly.) This interaction sets the threshold in associative and competitive networks, and helps to set the sparseness of the representation of the population of neurons.

Third, different brain areas operate in parallel. An example is that the ventral visual stream computes object representations, while simultaneously the dorsal visual stream computes (inter alia) the types of global motion described by Rolls and Stringer (2007), including for example a wheel rotating in the same direction as it traverses the visual field. Another example is that within a hierarchical system in the brain, every stage operates simultaneously, as a pipeline processor, with a good example being V1–V2–V4–IT, which can all operate simultaneously as the data are pipelined through.

We could refer to the computation that takes place in different modules, that is in networks that are relatively separate in terms of the number of connections between modules relative to those within modules, such as those in the dorsal and ventral visual streams, as being parallel computation. Within a single module or network, such as the CA3 region of the hippocampus, or inferior temporal visual cortex, we could refer to the computation as being *parallel distributed computation*, in that the closely connected neurons in the network all contribute to the result of the computation. For example, with distributed representations in an attractor network, all the neurons interact with each other directly and through the inhibitory interneurons to retrieve and then maintain a stable pattern in short-term memory (Section 2.5.3). In a competitive network involved in pattern categorization, all the neurons interact through the inhibitory interneurons to result in an active population of neurons that represents the best match between the input stimulus and what has been learned previously by the network, with neurons with a poor match being inhibited by neurons with a good match (Section 2.5.4). In a more complicated scenario with closely connected interacting modules, such as the prefrontal cortex and the inferior temporal cortex during top-down attention tasks and more generally forward and backward connections between adjacent cortical areas, we might also use the term parallel distributed computation, as the bottom-up and top-down interactions may be important in how the whole dynamical system of interconnected networks settles (see examples in Sections 2.6, 2.9 and Rolls (2008c)).

Stochastic dynamics and probabilistic computation. Digital computers do not have noise to contend with as part of the computation, as they use binary logic levels, and perform exact computation. In contrast, brain computation is inherently noisy, and this gives it a non-exact, probabilistic, character. One of the sources of noise in the brain is the spiking activity of each neuron. Each neuron must transmit information by spikes, for an all-or-none spike carried along an axon ensures that the signal arrives faithfully, and is not subject to the uncertain cable transmission line losses of analog potentials. But once a neuron needs to spike, then it turns out to be important to have spontaneous activity, so that neurons do not all have to charge up from a hyperpolarized baseline whenever a new input is received. The fact that neurons are kept near threshold, with therefore some spontaneous spiking, is inherent to the rapid operation of for example autoassociative retrieval, as described by Rolls (2008c). But keeping the neurons close to threshold, and the spiking activity received from other neurons, results in spontaneous spike trains that are approximately Poisson, that is randomly timed. The result of the interaction of all these randomly timed inputs is that in a network of finite size (i.e. with a limited number of neurons) there will be statistical fluctuations, which influence which memory is recalled, which decision is taken, etc. as described in Section 2.12. Thus brain computation is inherently noisy and probabilistic.

Syntax. Digital computers can perform arbitrary syntactical operations on operands, because they use pointers to address each of the different operands required (corresponding even for example to the subject, the verb, and the object of a sentence). In contrast, as data are not accessed in the brain by pointers that can point anywhere, but instead just having neurons firing to represent a data item, a real problem arises in specifying which neurons firing represent for example the subject, the verb, and the object, and distributed representations potentially make this even more difficult. The brain thus inherently finds syntactical operations difficult. We do not know how the brain implements the syntax required for language. But we do know that the firing of neurons conveys 'meaning' based on spatial location in the brain. For example, a neuron firing in V1 indicates that a bar or edge matching the filter characteristic of the neuron is present at a particular location in space. Another neuron in V1 encodes another feature at another position in space. A neuron in the inferior temporal visual cortex indicates (with other neurons helping to form a distributed representation) that a particular object or face is present in the visual scene. Perhaps the implementation of the syntax required for language that is implemented in the brain also utilizes the spatial location of the network in the cortex to help specify what syntactical role the representation should perform. This is a suggestion I make, as it is one way that the brain could deal with the implementation of the syntax required for language.

Modifiable connectivity. The physical architecture (what is connected to what) of a digital computer is fixed. In contrast, the connectivity of the brain alters as a result of experience and learning, and indeed it is alterations in the strength of the synapses (which implement the connectivity) that underlies learning and memory. Indeed, self-organization in for example competitive networks has a strong influence on how the brain is matched to the statistics of the incoming signals from the world, and of the architecture that develops. In a digital computer, every connection must be specified. In contrast, in the brain there are far too few genes (of order 30,000) for the synaptic connections in the brain (of order 10^{15}, given approximately 10^{11} neurons each with in the order of 10^4 synapses) for the genes to specify every connection[9]. The genes must therefore specify some much more general rules, such as that each CA3 neuron should make approximately 12,000 synapses with other CA3 neurons, and receive approximately 48 synapses from dentate granule cells (see Section 2.8). The actual connections made would then be made randomly within these constraints, and then strengthened or lost as a result of self-organization based on for example conjunctive pre- and postsynaptic activity. Some of the rules that may be specified genetically have been suggested on the basis of a comparison of the architecture of different brain areas (Rolls and Stringer 2000). Moreover, it has been shown that if these rules are selected by a genetic algorithm based on the fitness of the network that self-organizes and learns based on these rules, then architectures are built that solve different computational problems in one-layer networks, including pattern association learning, autoassociation memory, and competitive learning (Rolls and Stringer 2000). The architecture of the brain is thus interestingly adaptive, but guided in the long term by genetic selection of the building rules.

Logic. The learning rules that are implemented in the brain that are most widely accepted are associative, as exemplified by LTP and LTD. This, and the vector similarity operations implemented by neurons, set the stage for processes such as pattern association, autoassociation, and competitive learning to occur naturally, but not for logical operations such as XOR and NAND or arithmetic operations. Of course, the non-linearity inherent in the firing threshold of neurons is important in many of the properties of associative memories and competitive learning, as described by Rolls (2008c), and indeed are how some of the non-linearities that can be seen with attention can arise (Deco and Rolls 2005a).

Dynamical interaction between modules. Because the brain has populations of neurons that are simultaneously active (operating in parallel), but are interconnected, many properties arise naturally in dynamical neural systems, including the interactions that give rise to top-down attention (Section 2.9), the effects of mood on memory (Rolls 2008c) etc. Because simultaneous activity of different computational nodes does not occur in digital computers, these

---

[9]For comparison, a computer with 1 Gb of memory has approximately 10^{10} modifiable locations, and if it had a 100 Gb disk that would have approximately 10^{12} modifiable locations.

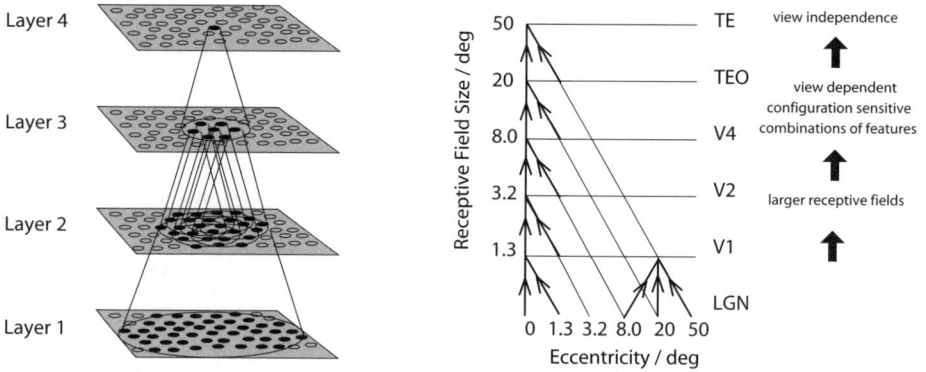

Fig. 2.31 Convergence in the visual system. Right – as it occurs in the brain. V1, visual cortex area V1; TEO, posterior inferior temporal cortex; TE, inferior temporal cortex (IT). Left – as implemented in VisNet, a model of invariant recognition (Rolls 2008c). Convergence through the network is designed to provide fourth layer neurons with information from across the entire input retina. Representations that are invariant with respect for example to the position of the object on the retina or of the view of the object are built using an associative synaptic modification rule in which the postsynaptic activity reflects a short-term memory trace of activity in the preceding few seconds. This enables neurons to learn to respond to what is seen in short periods, which is typically one object in different transforms, thus making use of the natural statistics of the visual world (Rolls 2008c, Chapter 4).

dynamical systems properties that arise from interacting subsystems do not occur naturally, though they can be simulated.

The cortex has recurrent excitatory connections within a cortical area, and reciprocal, forward and feedback, connections between adjacent cortical areas in the hierarchy. The excitatory connections enable cortical activity to be maintained over short periods, making short-term memory an inherent property of the cortex, and also autoassociative long-term memory with completion from a partial cue (given associative synaptic modifiability in these connections). Completion is a difficult and serial process to identify a possible correct partial match on a digital computer. The short-term memory property of the cortex is part of what makes the cortex a dynamical interacting system, with for example what is in short-term memory in for example the prefrontal cortex acting to influence memory recall, perception, and even what decision is taken, in other networks, by top-down biased competition (see Sections 2.6–2.12). There is a price that the brain pays for this positive feedback inherent in its recurrent cortical circuitry, which is that this circuitry is inherently unstable, and requires strong control by inhibitory interneurons to minimize the risk of epilepsy.

Modular organization. Brain organization is modular, with many relatively independent modules each performing a different function, whereas digital computers typically have a single central processing unit connected to memory. The cortex has many localized modules with dense connectivity within a module, and then connections to a few other modules. The reasons for the modularity of the brain are considered in Section 2.16.

Hierarchical organization. We can note that many brain systems are organized hierarchically. A major reason for this is that this enables the connectivity to be kept within the limits of which neurons appear capable (up to 50,000 synapses per neuron), yet for global computation (such as the presence of a particular object anywhere in the visual field) to be achieved, as exemplified by VisNet, a model of invariant visual object recognition (see Fig. 2.31) (Rolls 2008c). Another important reason is that this simplifies the learning that is required at each stage and enables it to be a local operation, in contrast to backpropagation of error networks where similar problems could in principle be solved in a two-layer network (with one hidden layer), but would require training with a non-local learning rule (Rolls 2008c) as well as potentially neurons with very large numbers of connections.

2.16 Understanding how the brain works

To conclude this chapter, I think it is very interesting to remark on how tractable the brain is to understand as a computer, relative to what it might have been, or relative to trying to reverse engineer a digital computer. This tractability applies to many brain systems, including those involved in perception, memory, attention, decision-making, and emotion (Rolls 2008c, Rolls 2005a, Rolls and Deco 2010), though not to language, where the issue of how the syntax required for language is implemented in the brain is still an enormous and fascinating mystery, or to consciousness, which is still a problematic issue (Section 6.3).

The encoding scheme. One reason for the tractability of the brain is its encoding scheme, whereby much information can be read off or decoded from the firing rates of single neurons and of populations of neurons (Section 2.3, Rolls (2008c) Appendix C, and Rolls and Treves (2011)). The deep computational reason for this appears to be that neurons decode the information by dot product decoding (Fig. 2.5 and Equation 2.1), and the consequence is that each of the independent inputs to a neuron adds information to what can be categorized by the neuron (Rolls 2008c). The brain would have been much less tractable if binary encoding of the type used in a computer was used, as this is a combinatorial code, and any single bit in the computer word, or any subset of bits, yields little evidence on its own about the particular item being represented.

Neurons have a single output. A second reason for the tractability of the vertebrate brain is that each neuron has a single output, and this enables whatever information the neuron is representing and transmitting to be read (by scientists, and by neurons) from the spikes (action potentials) emitted by the neuron. This is much simpler than many invertebrate neurons, which have multiple output connections to other neurons, each of which could convey a different signal, and each of which may be difficult to record. The reason for spiking activity in mammalian neurons is that the information may need to be transmitted long distances along the axon, and the all-or-none self-propagating action potential is an accurate way to transmit the information without signal loss due to voltage degradation along the axon. But that biological need makes the vertebrate system (relatively) easy for scientists to decode and understand, as each vertebrate neuron does have a single output signal (sent to many other neurons) (Rolls and Treves 1998).

Modular organization. A third reason is that the brain is inherently modular, so that damage to one part can be used to analyze and understand particular functions being implemented, and how each function can be dissociated from other functions. (The methodology is referred to as double dissociation.) One deep reason for this is that the cortex must operate as a set of local networks, for otherwise the total memory storage capacity of the whole cortex would be that of one of its modules or columns, given that storage capacity is determined by the number of connections C onto each neuron, not by the number of neurons in the whole system (O'Kane and Treves 1992, Rolls 2008c).

A second deep reason for this is that evolution by natural selection operates most efficiently when functions can be dissociated from each other and genes affect individual functions (implemented in the case of the brain by individual networks in different brain areas, including different cortical areas). Selection for particular characteristics or functionality can then proceed rapidly, for any change of fitness caused by gene changes related to a particular function feeds back rapidly into the genes selected for the next generation.

Another important advantage of modularity is that it helps to minimize the lengths of the connections, the axons, between the computing elements, the neurons, and thus helps to keep brain weight down, an important factor in evolution (Rolls 2008c). Minimizing connection length also simplifies the way in which brain design is specified by genes, for just connecting locally is a simple genetic specification, and reduces the need for special instructions to find the correct possibly distant neurons in the brain to which to connect (Rolls 2008c, Rolls and Stringer 2000). Having all the neurons present while the brain is developing also helps neurons to find the correct populations to which to connect during brain development. In fact, neuron numbers tend to reduce somewhat after birth, and pruning connections, and neurons, is much simpler than trying to specify which new neurons to connect to in a possibly very distant brain region in the adult brain. The only recognized case where some neurons may be added after birth is dentate granule cells of the hippocampal system, which connect only locally to nearby CA3 cells, and may help to ensure that new episodic memories that are stored are different from previous memories by helping to select new subsets of hippocampal CA3 cells (Section 2.8) (Rolls and Kesner 2006, Rolls 2008c, Clelland, Choi, Romberg, Clemenson, Fragniere, Tyers, Jessberger, Saksida, Barker, Gage and Bussey 2009, Rolls 2010c).

Conservation of architectures. A fourth reason for the tractability of the brain is that the same principles of operation implemented by similar network architectures are used in different brain areas for different purposes. An example is the use of autoassociation networks for short-term memory, episodic memory, semantic memory, and decision-making, in different brain areas. A very interesting example is that of backprojections in the neocortex, which illustrate how two simple functions, pattern association and autoassociation, can be combined (Rolls 2008c). This reflects the conservative nature of evolution: if a gene-specified implementation can be adapted for a different function, that is likely to be quicker and less risky than a completely new solution. (An example is the bones of our middle ear, the malleus, incus, and stapes, used to transmit sound to the cochlea, which were originally fish gills.) The consequence is that once we understand how the architecture is being used for one function, much of the theory can be used to understand how the same or a similar structure (network, architecture) is used elsewhere in the brain for another function.

However, we should note that the conservative nature of evolution (in a literal sense: earlier design features are often conserved and remain, even when different and better features evolve) can lead to some added complexity of the system. A good example is multiple routes to action (Fig. 2.24). In this case evolutionarily old systems (such as the amygdalar control of autonomic function, or the basal ganglia system for habits) remain and account for some behaviour, even though newer systems such as goal directed action, and rational planning, provide newer and more powerful routes to action, and enable long-term planning. In this case, to understand behaviour, we must understand that different brain systems, perhaps with different goals, may all be vying for behavioural output, and one brain system may not entirely know what another one is doing (Sections 2.11, 5.1 and 6.3.2; Figs. 2.24 and 2.25).

Ergodicity. A fifth reason for the tractability of the mammalian brain is that distributed representations are used, which involve large numbers of neurons. The result is that knowing how a subset of the neurons in a given population responds enables one to build a statistical understanding of how the whole population works computationally. Using the terminology of theoretical physics, a principle of weak *ergodicity* applies, by which studying the properties of a single element of the system for a long time (e.g. the firing rate response of a single neuron to a large set of stimuli) can yield information that can also be obtained from many elements for a short time (e.g. the firing rates of each of a large population of neurons to a single stimulus). We have shown that this principle, which holds if the elements (neurons) are statistically independent in what they respond to, does apply in the mammalian cortex (Franco, Rolls, Aggelopoulos and Jerez 2007). A condition for weak ergodicity is that the neurons are independent, and this is what the multiple cell information analyses show (Section 2.3.2). This statistical property enables a lot to be learned about the whole population if only a subset of the neurons is recorded, which is what is feasible. (Therefore the fact that we cannot record all neurons in the brain simultaneously need not limit our understanding.)

This principle also allows statistical mechanics to make quantitative predictions about for example the storage capacity of networks of neurons in the brain (Hopfield 1982, Amit 1989, Treves and Rolls 1991, Rolls 2008c). The deep reason why brain design (by Darwinian evolution) has resulted in this property is that having independent tuning by neurons to a set of stimuli (produced for example by competitive learning, Section 2.5.4) enables the information that can be represented to increase linearly with the number of neurons (and thus the number of stimuli that can be represented to increase exponentially with the number of neurons, Section 2.3), and also enables the properties of associative memories to arise that include generalization, completion, and graceful degradation so that the system can operate with diluted connectivity (Section 2.5, Rolls (2008c)).

Gene specification of rewards, and emotion. A sixth reason for the tractability of the brain is that emotion allows genes to specify behaviour in a simple way, by specifying the goals for actions (specific rewards / subjective pleasures etc.), rather than the very complex specification that would be required if actions / movements had to be gene-specified in all but the simplest of cases (Rolls 2005a), as described in the next chapter (3).

3 Neuroaffect

3.1 Introduction

Why do we have emotions? Why are emotions so important in our behaviour? What role do genes play in emotion? Is emotion a Darwinian adaption to a major problem faced by genes? Emotions are adaptive, but are they rational? Are we conscious of all the emotional influences on our behaviour? How and why are women's and men's emotions different? Why do we as individuals place different values on different goals? What motivates us? Why do some people become obese? What makes people attractive? What makes us trust, and love, someone? What is the role of hormones?

In this chapter I will show that emotions are a solution to a major problem in Darwinian evolution. I will argue that genes specify our brains so that we find some stimuli rewarding, and some punishing, in order to promote their own (the selfish genes') survival into the next generation. It is far more efficient for genes to specify what we like and dislike than to specify what behavioural responses we should perform. Once the rewards and punishers have been specified, the brain can then perform any action instrumental in obtaining the reward or avoiding the punisher. The concept goes beyond *The Selfish Gene* (Dawkins 1976, Dawkins 1989) by showing that an efficient way for 'selfish' genes to influence behaviour is by specifying what the brain decodes as rewarding and punishing. This in turn leads to a (neuro-)scientific understanding of emotion, for as I will show, emotions are the states that are associated with rewards and punishers. This fundamental understanding is a culmination of Darwinian thinking, and, as I argue in this book, provides a neuroscientific way to understand a whole host of human behaviour and thinking, with fundamental implications for understanding social behaviour, aesthetics, ethics, trust, and emotion.

Almost everyone is interested in what emotions are. There have been many answers, many of them surprisingly unclear and ill-defined. William James (1884) was at least clear about what he thought. He believed that emotional experiences were produced by sensing bodily changes, such as changes in heart rate or in skeletal muscles (the muscles involved in voluntary movements). His view was that "We feel frightened because we are running away". But he left unanswered the crucial question even for his theory, which is: Why do some events make us run away (and then feel emotional), whereas others do not?

A more modern theory is that of Frijda (1986), who argues that a change in action readiness is the central core of an emotion. Oatley and Jenkins (1996) (page 96) make this part of their definition too, stating that "the core of an emotion is readiness to act and the prompting of plans". But surely subjects in reaction time experiments in psychology who are continually for thousands of trials altering their action readiness are very far indeed from having normal or strong emotional experiences? Similarly, we can perform an action in response to a verbal request (e.g. open a door), yet may not experience great emotion when performing this action. Another example might be the actions that are performed in driving a car on a routine trip – we get ready, and many actions are performed, often quite automatically, yet little emotion occurs. So it appears that there is no necessary link between performing actions and emotion. This may not be a clear way to define emotion.

Because it is important to be able to specify what emotions are, in this chapter we consider a systematic approach to this question. Part of the approach is to ask what causes emotions. Can clear conditions be specified for the circumstances in which emotions occur? This is considered in Section 3.2. Continuing with this theme, when we have come to understand the conditions under which emotions occur, does this help us to classify and describe different emotions systematically, in terms of differences between the different conditions that cause emotions to occur. A way in which a systematic account of different emotions can be provided is described in Section 3.3.

A major help in understanding emotions would be provided by understanding what the functions of emotion are. It turns out that emotions have quite a number of different functions, each of which helps us to understand emotions a little more clearly. These different functions of emotion are described in Section 3.10. Understanding the different functions of emotion helps us to understand also the brain mechanisms of emotion, for it helps us to see that emotion can operate to affect several different output systems of the brain.

These analyses leave open though a major related question, which is why emotional states feel like something to us. This it transpires is part of the much larger, though more speculative, issue of consciousness, and why anything should feel like something to us. This aspect of emotional feelings, because it is part of the much larger issue of consciousness, is deferred until Section 6.3.

In this Chapter, in considering the function of emotions, the idea is presented that emotions are part of a system that helps to map certain classes of stimuli, broadly identified as rewarding and punishing stimuli (i.e. aversive stimuli or 'punishers'), to action systems. Part of the idea is that this enables a simple interface between such stimuli and actions. This is an important area in its own right, which goes to the heart of why animals are built to respond to rewards and punishers, and have emotions.

The suggestion made in this book is that we now have a way of systematically approaching the nature of emotions, their functions, and their brain mechanisms. Doubtless in time there will be changes and additions to the over-all picture. But the suggestion is that the ideas and theory presented here do provide a firm and systematic foundation for understanding emotions, their functions, and their brain mechanisms in a well-founded evolutionary context.

3.2   The outline of a theory of emotion

I will first introduce the essence of the definition of emotion that I propose. *The definition of emotions is that emotions are states elicited by rewards and punishers, that is, by instrumental reinforcers.* As described in more detail in Appendix A.1, a reward is anything for which an animal will work. A punisher is anything that an animal will work to escape or avoid, or that will suppress actions on which it is contingent[10]. I note that any change in the regular delivery of a reward or a punisher acts as a reinforcer. The relevant states elicited by the reinforcers are those with the particular functions described in Section 3.10.

An example of an emotion might thus be happiness produced by being given a reward, such as a hug, a pleasant touch, praise, winning a large sum of money, or being with someone whom one loves. All these things are rewards, in that we will work to obtain them. Another example of an emotion might be fear produced by the sound of a rapidly approaching bus when we are cycling, or the sight of an angry expression on someone's face. We will work to avoid such stimuli, which are punishers. Another example might be frustration, anger, or sadness produced by the omission of an expected reward such as a prize, or the termination of a reward such as the death of a loved one. Another example might be relief, produced by the omission or termination of a punishing stimulus, for example the removal of a painful stimulus, or sailing out of danger. These examples indicate how emotions can be produced by the delivery, omission, or termination of rewarding or punishing stimuli, and go some way to indicate how different emotions could be produced and classified in terms of the rewards and punishers received, omitted, or terminated.

Before accepting this proposal, we should consider whether there are any exceptions to the proposed rule. Indeed, at first this may appear to be a rather reductionist hypothesis about what produces emotions. However, one way to test the suggested definition of the events that cause emotions is to ask whether there are any rewards or punishers that do not produce emotions. Conversely, we should ask whether there are any emotions that are produced by stimuli, events, or remembered events that are not rewarding or punishing. If we cannot find exceptions, then we should accept the suggestion as a useful identification, summary, and working definition of the conditions that produce emotions. Therefore in the next few pages we consider the questions: 'Are any emotions

[10] A fuller definition in terms of reinforcement contingencies is given below.

caused by stimuli, events, or remembered events that are not rewarding or punishing? Do any rewarding or punishing stimuli not cause emotions?'

But first it is worth pointing out that in fact many approaches to or theories of emotion have in common that part of the process involves 'appraisal' (e.g. Frijda (1986); Oatley and Johnson-Laird (1987); Lazarus (1991); Izard (1993); Stein, Trabasso and Liwag (1994)). This is part, for example, of the suggestion made by Oatley and Jenkins (1996), who on page 96 write that "an emotion is usually caused by a person consciously or unconsciously evaluating an event as relevant to a concern (a goal) that is important; the emotion is felt as positive when a concern is advanced and negative when a concern is impeded". The concept of appraisal presumably involves in all these theories assessment of whether something is rewarding or punishing, that is whether it will be worked for or avoided. The description in terms of reward or punisher adopted here simply seems much more precisely and operationally specified.

The idea that rewards and punishers, that is instrumental reinforcers, are the stimuli that produce emotions has a considerable history, with origins that can be traced back to Watson (1929, 1930), Harlow and Stagner (1933), and Amsel (1958, 1962). More recently, the approach was developed by Millenson (1967), Larry Weiskrantz (1968), and Jeffrey Gray (1975, 1981). We can introduce some of the emotions that result from different reinforcement contingencies as follows. Consider the emotional effects of delivery of a 'reward' : a state such as pleasure or happiness will be produced. An example might be receiving a prize for excellent work. Now consider the emotional effects of delivery of a 'punisher' : pain or fear may be produced. For example, fear is an emotional state that might be produced by a sound that has previously been associated with a painful electrical shock. Shock in this example is the primary reinforcer, and fear is the emotional state that occurs to the tone stimulus as a result of the learning of the stimulus (i.e. tone)–reinforcer (i.e. shock) association. The tone in this example is a conditioned stimulus because of stimulus–reinforcer association learning, and has secondary reinforcing properties in that responses will be made to escape from it and thus avoid the primary reinforcer, shock.

The converse reinforcement contingencies produce the opposite effects on behaviour, and produce different emotions. The omission or termination of a reward ('extinction' and 'time out' respectively) reduce the probability of responses, and may produce the emotions of frustration, disappointment, or rage. (Imagine not receiving a prize that you deserved.) Behavioural responses followed by the omission or termination of a punisher increase in probability (this pair of reinforcement operations being termed 'active avoidance' and 'escape', respectively), and are associated with emotions such as relief.

The classification of emotions in terms of reinforcement contingencies is developed further in Section 3.3, and more formal definitions of rewards and punishers, and how they are related to learning theory concepts such as reinforcement and punishment are given in the footnote[11], and in Appendix A.1. My argument is that an affectively positive or 'appetitive' stimulus (which produces a state of pleasure) acts operationally as a **reward**, which when delivered acts instrumentally as a positive reinforcer, or when not delivered (omitted or terminated) acts to decrease the probability of responses on which it is contingent. Conversely I argue that an affectively negative or aversive stimulus (which produces an unpleasant state) acts operationally as a **punisher**, which when delivered acts instrumentally to decrease the probability of responses on which it is contingent, or when not delivered (escaped from or avoided) acts as a negative reinforcer in that it then increases the probability of the action on which its non-delivery is contingent[12].

The link between emotion and instrumental reinforcers being made is partly an operational link. Most people find that it is not easy to think of exceptions to the statements that emotions occur after rewards or punishers are given (sometimes continuing for long after the eliciting stimulus has ended, as in a mood state); or that rewards and punishers, but not other stimuli, produce emotional states. But the link is deeper than this, as we will see, in that the theory has been developed that genes specify primary reinforcers in order to encourage the animal to perform arbitrary actions to seek particular goals, thus increasing the probability of their own (the genes') survival into the next generation (Rolls 1999a, Rolls 2005a). The emotional states elicited by the reinforcers have a number of functions, described below, related to these processes.

Before considering how different emotions are related to different reinforcement contingencies in Section 3.3, I clarify a matter of terminology about

[11] Instrumental reinforcers are stimuli that, if their occurrence, termination, or omission is made contingent upon the making of an action, alter the probability of the future emission of that action (Gray 1975, Mackintosh 1983, Dickinson 1980, Lieberman 2000). Rewards and punishers are instrumental reinforcing stimuli. The notion of an action here is that an arbitrary action, e.g. turning right vs turning left, will be performed in order to obtain the reward or avoid the punisher, so that there is no pre-wired connection between the response and the reinforcer. Some stimuli are primary (unlearned) reinforcers (e.g. the taste of food if the animal is hungry, or pain); while others may become reinforcing by learning, because of their association with such primary reinforcers, thereby becoming 'secondary reinforcers'. This type of learning may thus be called 'stimulus–reinforcer association', and occurs via an associative learning process. A positive reinforcer (such as food) increases the probability of emission of a response on which it is contingent, the process is termed **positive reinforcement**, and the outcome is a reward (such as food). A negative reinforcer (such as a painful stimulus) increases the probability of emission of a response that causes the negative reinforcer to be omitted (as in active avoidance) or terminated (as in escape), and the procedure is termed **negative reinforcement**. In contrast, **punishment** refers to procedures in which the probability of an action is decreased. Punishment thus describes procedures in which an action decreases in probability if it is followed by a painful stimulus, as in passive avoidance. Punishment can also be used to refer to a procedure involving the omission or termination of a reward ('extinction' and 'time out' respectively), both of which decrease the probability of responses (Gray 1975, Mackintosh 1983, Dickinson 1980, Lieberman 2000).

[12] Note that my definition of a punisher, which is similar to that of an aversive stimulus, is of a stimulus or event that can either decrease the probability of actions on which it is contingent, or increase the probability of actions on which its non-delivery is contingent. The term punishment is restricted to situations where the probability of an action is being decreased.

moods vs emotions. A useful convention to distinguish between emotion and a *mood state* is as follows. An emotion consists of cognitive processing that results in a decoded signal that an environmental event (or remembered event) is reinforcing, together with the mood state produced as a result. If the mood state is produced in the absence of the external sensory input and the cognitive decoding (for example by direct electrical stimulation of the brain, see Rolls (2005a)), then this is described only as a mood state, and is different from an emotion in that there is no object in the environment towards which the mood state is directed. (In that emotions are produced by stimuli or objects, and thus emotions 'take or have an object', emotional states are examples of what philosophers call intentional states.) It is useful to emphasize that there is great opportunity for cognitive processing (whether conscious or not) in emotions, for cognitive processes will very often be required to determine whether an environmental stimulus or event is reinforcing (see further Section 3.4).

3.3 Different emotions

As introduced in Section 3.2, the different emotions can in part be described and classified according to whether the reinforcer is positive or negative, and by the reinforcement contingency. An outline of such a classification scheme, elaborated by Rolls (1990c), Rolls (1999a) and Rolls (2000c), is shown in Fig. 3.1. Movement away from the centre of the diagram represents increasing intensity of emotion, on a continuous scale. The diagram shows that emotions associated with the delivery of a reward (S+) include pleasure, elation and ecstasy. Of course, other emotional labels can be included along the same axis. Emotions associated with the delivery of a punisher (S−) include apprehension, fear, and terror (see Fig. 3.1). Emotions associated with the omission of a reward (S+) or the termination of a reward (S+!) include frustration, anger and rage. Emotions associated with the omission of a punisher (S−) or the termination of a punisher (S−!) include relief. Although the classification of emotions presented here (and by Rolls (1986b), Rolls (1986a), Rolls (1990c) and Rolls (1999a)) differs from earlier theories, the approach adopted here of defining and classifying emotions by reinforcing effects is one that has been developed in a number of earlier analyses (e.g. Millenson (1967), Gray (1975), Gray (1981); see Strongman (2003)).

I should make it clear that the scheme shown in Fig. 3.1 is not intended to be a dimensional scheme. [A dimensional scheme is one in which independent factors or dimensions have been identified that account for the major and independent sources of variation in a data set. Some investigators then work to show that these dimensions can be interpreted both biologically (for example as differing in autonomic, endocrine, or arousal-related ways) and psychologically (e.g. as representing anger vs fear), as described in Section 3.5.3.] However,

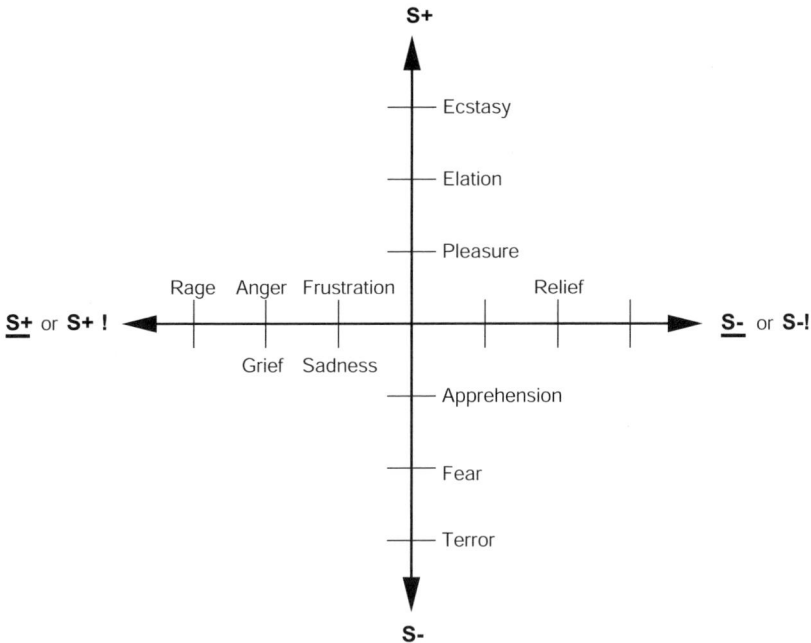

Fig. 3.1 Some of the emotions associated with different reinforcement contingencies are indicated. Intensity increases away from the centre of the diagram, on a continuous scale. The classification scheme created by the different reinforcement contingencies consists with respect to the action of (1) the delivery of a reward (S+), (2) the delivery of a punisher (S–), (3) the omission of a reward (S+) (extinction) or the termination of a reward (S+!) (time out), and (4) the omission of a punisher (S–) (avoidance) or the termination of a punisher (S–!) (escape). Note that the vertical axis describes emotions associated with the delivery of a reward (up) or punisher (down). The horizontal axis describes emotions associated with the non-delivery of an expected reward (left) or the non-delivery of an expected punisher (right). This shows emotions that might be produced by a reinforcer such as touch. There are separate systems such as this for each type of reinforcer (see text).

the import of what is shown in Fig. 3.1 is to set out a set of logical possibilities of ways in which reinforcement contingencies can vary, and to show how they may be related to some different types of emotion.

It is actually a possibility that the four directions shown in Fig. 3.1 are at least partly independent from each other, and that a four-dimensional space is spanned by what is shown in Fig. 3.1. For example, sensitivity to (that is the ability to respond to) reward (S+) could be at least partly independent from sensitivity to punishers (S–), sensitivity to non-reward (S+ and S+!), and sensitivity to non-delivery of a punisher (S– and S–!). The dimensions or independent ways in which emotions may differ from each other could thus span 4 dimensions even with what is shown in Fig. 3.1, and these ways are expanded greatly as shown by the following further effects that make different emotions different to each other.

One important point about Fig. 3.1 is that there are a large number of different primary reinforcers, and that for example the reward label S+ shows states that might be elicited by just one type of reward, such as a pleasant touch. There will be a different reward axis (S+) and non-reward axis (S+ and S+!) for each type of reward (e.g. pleasant touch vs sweet taste); and, correspondingly, a different punisher axis (S−) and non-punisher axis (S− and S−!) for each type of punisher (e.g. pain vs bitter taste).

Different reinforcement contingencies can thus be used to classify a wide range of emotions. However, some of my tutorial pupils at Oxford University sometimes expressed the view that reinforcement contingencies alone might not be able to account for the full range of human emotions. I therefore set out for them ways in which a system based on reinforcement contingencies could be developed in a number of different ways to give an account of most emotions. This extended set of ways of accounting for different emotions was published in 1986 (Rolls (1986b), Rolls (1986a)), and developed a little in later publications (e.g. Rolls (1995b), Rolls (1999a) and Rolls (2005a)). It is described, and elaborated further next. If the reader can think of any emotions that cannot be accounted for by a combination of the ways described next, then it would be interesting to consider what further extensions might be needed.

1. **Reinforcement contingency**
The first way in which different classes of emotion could arise is because of different reinforcement contingencies, as described above and indicated in Fig. 3.1.

2. **Intensity**
Second, different intensities within these classes can produce different degrees of emotion (see above and Millenson (1967)). For example, as the strength of a positive reinforcer being presented increases, emotions might be labelled as pleasure, elation, and ecstasy. Similarly, as the strength of a negative reinforcer being presented increases, emotions might be labelled as apprehension, fear, and terror (see Fig. 3.1). It may be noted here that anxiety can refer to the state produced by stimuli associated with the non-delivery of a reward or the delivery of a punisher (Gray 1987).

3. **Multiple reinforcement associations**
Third, any environmental stimulus might have a number of different reinforcement associations. For example, a stimulus might be associated both with the presentation of a reward and of a punisher, allowing states such as conflict and guilt to arise. The different possible combinations greatly increase the number of possible emotions.

4. **Different primary reinforcers**

Fourth, emotions elicited by stimuli associated with different primary reinforcers will be different even within a reinforcement category (i.e. with the same reinforcement contingency), because the original reinforcers are different. Thus, for example, the state elicited by a stimulus associated with a reward such as the taste of food will be different from that elicited by a reward such as being groomed. Indeed, it is an important feature of the association memory mechanisms described here that when a stimulus is applied, it acts as a key which 'looks up' or recalls the original primary reinforcer with which it was associated. Thus emotional stimuli will differ from each other in terms of the original primary reinforcers with which they were associated.

A summary of many different primary reinforcers is provided in Table 3.1 on page 108, and inspection of this will help to show how some different emotions are produced by different primary reinforcers. For example, from Table 3.1 it might be surmised that one of the biological origins of the emotion of jealousy might be the state elicited in a male when his partner is courted by another male, because this threatens his parental investment in the offspring he raises with his partner, as described in Chapter 4. Jealousy in females would arise in a corresponding way. Examples of how further emotions including guilt, shame, anger, forgiveness, envy and love may arise in relation to particular primary reinforcers are provided later in this section, throughout this chapter, in Chapter 4, in many other places in this book, and elsewhere (Rolls 2005a).

5. **Different secondary reinforcers**

A fifth way in which emotions can be different from each other is in terms of the particular (conditioned) stimulus that elicits the emotion, and the situation in which it occurs. Thus, even though the reinforcement contingency and even the unconditioned reinforcer may be identical, emotions will still be different cognitively, if the conditioned stimuli that give rise to the emotions are different (that is, if the objects of the emotion are different). For example, the emotional state elicited by the sight of one person may be different from that elicited by the sight of another person because the people, and thus the cognitive evaluations associated with the perception of the stimuli, are different. In another example, not obtaining a monetary reward in a gambling task might lead to frustration, but being blocked by another person from obtaining a reward might lead to anger directed at the person.

Thus evolution may have shaped different reinforcers to contribute in different ways and depending on the environmental circumstances to the exact emotion produced. For example, some emotions may be related to social reinforcers (e.g. love, anger, envy, and breaking rules of society so that shame is produced, see further Chapter 9), others to non-social reinforcers (such as fear of a painful stimulus), and others to solving difficult problems. By taking

Table 3.1 Some primary reinforcers, and the dimensions of the environment to which they are tuned

Taste

Salt taste	reward in salt deficiency
Sweet	reward in energy deficiency
Bitter	punisher, indicator of possible poison
Sour	punisher
Umami	reward, indicator of protein;
	produced by monosodium glutamate and inosine monophosphate
Tannic acid	punisher; it prevents absorption of protein; found in old leaves;
	probably somatosensory rather than strictly gustatory

Odour

Putrefying odour	punisher; hazard to health
Pheromones	reward (depending on hormonal state)

Somatosensory

Pain	punisher
Touch	reward
Grooming	reward; to give grooming may also be a primary reinforcer
Washing	reward
Temperature	reward if helps maintain normal body temperature; otherwise punisher

Visual

Snakes, etc.	punisher for, e.g. primates
Youthfulness	reward, associated with mate choice
Beauty, e.g. symmetry	reward
Secondary sexual characteristics	rewards
Face expression	reward (e.g. smile) or punisher (e.g. threat)
Blue sky, cover, open space	reward, indicator of safety
Flowers	reward (indicator of fruit later in the season?)

Auditory

Warning call	punisher
Aggressive vocalization	punisher
Soothing vocalization	reward (part of the evolutionary history of music,
	which at least in its origins taps into the channels
	used for the communication of emotions)

Table 3.1 continued **Some primary reinforcers, and the dimensions of the environment to which they are tuned**

Reproduction

Courtship	reward
Sexual behaviour	reward (different reinforcers, including a low waist-to-hip ratio,
	and attractiveness influenced by symmetry
	and being found attractive by members of the other sex
	are discussed by Rolls, 2005a)
Mate guarding	reward for a male to protect his parental investment;
	jealousy results if his mate is courted by another male,
	because this may ruin his parental investment
Nest building	reward (when expecting young)
Parental attachment	reward
Infant attachment to parents	reward
Crying of infant	punisher to parents; produced to promote successful development

Other

Novel stimuli	rewards (encourage animals to investigate the full possibilities
	of the multidimensional space in which their genes are operating)
Sleep	reward; minimizes nutritional requirements and protects from danger
Altruism to genetically related individuals	reward (kin altruism)
Altruism to other individuals	reward while the altruism is reciprocated
	in a 'tit-for-tat' reciprocation (reciprocal altruism)
	Forgiveness, honesty, and altruistic punishment are associated heuristics
	(May provide underpinning for some aspects of what is felt to be moral.)
Altruism to other individuals	punisher when the altruism is not reciprocated.
Group acceptance, reputation	reward (social greeting might indicate this)
	These goals can account for some culturally specified goals
Control over actions	reward
Play	reward
Danger, stimulation, excitement	reward if not too extreme (adaptive because of practice?)
Exercise	reward (keeps the body fit for action)
Mind reading	reward; practice in reading others' minds, which might be adaptive
Solving an intellectual problem	reward (practice in which might be adaptive)
Storing, collecting	reward (e.g. food)
Habitat preference, home, territory	reward
Some responses	reward (e.g. pecking in chickens, pigeons;
	adaptive because it is a simple way in which eating grain
	can be programmed for a fixed type of environmental stimulus)
Breathing	reward

into account the nature of the primary reinforcer, the nature of the secondary reinforcer, and the environmental circumstances in which these apply, many different emotions can thus be accounted for, and cognitive factors taken into account. The common underlying basis of emotion remains however that it is related to goals/instrumental reinforcers, and the reinforcement contingencies that operate. The variety of different goals, and the contingencies and environmental situations in which they occur, combine to contribute to the richness in the variety of emotional states.

The gene-specified reinforcer approach to emotion advocated in this book is somewhat different to the domain-specific (vs domain general) approach of some evolutionary biologists (see Nesse (2000b)). In the domain-specific approach, a modular approach to different emotions may be taken, and the temptation is to end with a large number of specialized emotional systems, each promoting particular types of action. In contrast, in the approach described here, different genes build different reinforcement systems that define the goals for actions, and arbitrary actions appropriate for reaching the goal (i.e. instrumental actions) are then performed, with action–outcome learning guiding the actions produced. This can result in a rich variety of actions being selected in different emotion-provoking situations, without a tendency to suggest that particular perhaps instinctive actions are coupled to particular emotions. Instead, 'instinct' is involved in the process whereby the *goals* for actions, which are reinforcing stimuli, are specified by genes as a result of natural selection, and the behavioural response itself is not specified or 'determined' (see further Section 3.10.4).

Further, in the approach described here, modular neural systems useful for face identification, face expression recognition, and head gesture and movement may evolve because of the different specialized computational requirements for each and the importance of minimizing wiring length in the brain (Rolls 2005a, Rolls 2008c), and because the presence of these systems helps to provide representations that are useful in defining which stimulus or object-related events in the environment are associated with primary reinforcers.

6. **The behavioural responses that are available**
A sixth possible way in which emotions can vary arises when the environment constrains the types of behavioural response that can be made. For example, if an active behavioural response can occur to the omission of an expected reward, then anger might be produced and directed at the person who prevented the reward being obtained, but if only passive behaviour is possible, then sadness, depression or grief might occur.

By realizing that these six possibilities can occur in different combinations, it can be seen that it is possible to account for a very wide range of emotions, and this is believed to be one of the strengths of the approach described here. It is also the case that the extent to which a stimulus is reinforcing on a particular occasion (and thus an emotion is produced) depends on the prior history of reinforcements (both recently through processes that include sensory-specific satiety, and in the longer term), and that the current mood state can affect the degree to which a stimulus (a term that includes cognitively decoded events and remembered events) is reinforcing (see Section 3.7).

If we wish to consider the number of independent ways in which emotions may differ from each other (for comparison with the 'dimensional' theories described in Section 3.5.3) we see immediately that a vast subtlety of emotions can be systematically described using the approach described here. For example, based on the four different reinforcement contingencies shown in Fig. 3.1 we have four at least potentially independent 'dimensions', which are combined with perhaps another 100–500 independently varying (in that they are gene-specified) primary reinforcers, some of which are included in Table 3.1. These are combined with constraints to the actions that may be possible when a reinforcer is received (the 'coping potential' of appraisal theorists), which potentially at least doubles the number of emotions that can be described. We add further combinatorial possibilities by noting (point 3 above) that a given stimulus in the world may have many different reinforcement associations producing states such as conflict. The possible number of different emotions can be further multiplied by the fact that each primary reinforcer may have associated with it almost any neutral stimulus to produce a secondary reinforcer.

The resulting number of emotional states that can be described and categorized is clearly enormous, even if we do not assume that each of the above factors operates strictly independently (factorially). For example, it is likely that if a gene were to specify a particular reward as being particularly intense in an individual, for example the pleasantness of touch, then omitting (S+) or terminating (S+!) this reward might also be expected to be particularly intense, so the contributions of reinforcement contingency and identity of the primary reinforcer might combine additively rather than multiplicatively. Even if there is only partial independence of the different processes 1–6 above, and of variation within each process, then nevertheless many different emotions can be systematically classified and described. It does of course remain an interesting issue of how the processes described above do combine, and of the extent to which a few factors actually do account for a great deal in the variation between different emotions. For example, if an individual's sensitivity to non-reward is generally much more intense than the individual's sensitivity to reward, then this will shape the emotions in that individual, and account for quite a deal of the variance between that individual's emotional states. Such a factor might also account for quite an amount of the variation in emotions and personality between individuals (see Section 3.6).

Some examples of how different emotions might be classified using the above criteria now follow. Fear is a state that might be produced by a stimulus that has become a secondary reinforcer by virtue of its learned association with a primary negative reinforcer such as pain (see Fig. 3.1). Anger is a state that might be produced by the omission of an expected reward, frustrative non-reward, when an active behavioural response is possible (see Fig. 3.1). (In particular, anger may occur if another individual prevents an expected reward from being obtained.) Guilt may arise when there is a conflict between an available reward and a rule or law of society. Jealousy is an emotion that might be aroused in a male if the faithfulness of his partner seems to be threatened by her liaison (e.g. flirting) with another male. In this case the reinforcement contingency that is operating is produced by a punisher, and it may be that males are specified genetically to find this punishing because it indicates a potential threat to their paternity and paternal investment, as described in Chapter 4 and Section 3.10. Similarly, a female may become jealous if her partner has a liaison with another female, because the resources available to the 'wife' useful to bring up her children are threatened. Again, the punisher here may be gene-specified, as described in Section 3.10. Envy or disappointment might be produced if a prize is obtained by a competitor. In this case, part of the way in which the frustrative non-reward is produced is by the cognitive understanding that this is a competition in which there will be a winner, and that the person has set himself or herself the goal of obtaining it.

The partial list of primary reinforcers provided in Table 3.1 should provide readers with a foundation for starting to understand the rich classification scheme for different types of emotion that can be classified in this way.

Many other similar examples can be surmised from the area of evolutionary psychology (see e.g. Ridley (1993b), Buss (2008) and Barrett, Dunbar and Lycett (2002)). For example, there may be a set of reinforcers that are genetically specified to help promote social cooperation and even reciprocal altruism. Such genes might specify that emotion should be elicited, and behavioural changes should occur, if a cooperating partner defects or 'cheats' (Cosmides and Tooby 1999). Moreover, the genes may build brains with genetically specified rules that are useful heuristics for social cooperation, such as acting with a strategy of 'generous tit-for tat', which can be more adaptive than strict 'tit-for-tat', in that being generous occasionally is a good strategy to help promote further cooperation that has failed when both partners defect in a strict 'tit-for-tat' scenario (Ridley 1996). Genes that specify good heuristics to promote social cooperation may thus underlie such complex emotional states as feeling forgiving.

It is suggested that many apparently complex emotional states have their origins in designing animals to perform well in such sociobiological and socioeconomic situations (Ridley 1996, Glimcher 2003, Glimcher 2004). Indeed, many principles that humans accept as ethical may be closely related to strategies that are useful heuristics for promoting social cooperation, and emotional feelings associated with ethical behaviour may be at least partly related to the adaptive value of such gene-specified strategies. These ideas are developed in Chapter 9.

These examples indicate that an emotional state can be systematically specified and classified using the six principles described above in this section. The similarity between particular emotions will depend on how close they are in the space defined by the above principles.

3.4 Refinements of the theory of emotion

The definition of emotions given above, that they are states produced by instrumental reinforcing stimuli, and have particular functions, is refined now.

First, when positively reinforcing (rewarding) stimuli (such as the taste of food or water) are relevant to a drive state produced by a change in the *internal milieu* (such as hunger and thirst), then we do not normally classify these stimuli as emotional, though they do produce pleasure, and indeed we describe the state they produce as affective (see Section 3.13). In contrast, emotional states are normally initiated by reinforcing stimuli that have their origin in the external environment, such as an (external) noise associated with pain (delivered by an external stimulus). We may then have identified a class of reinforcers (in our example, food) that we do not want to say cause emotions. This then is a refinement of the definition of emotions given above. Fortunately, we can encapsulate the set of reinforcing stimuli that we wish to exclude from our definition of stimuli that produce emotion. They are the set of external reinforcers (such as the sight of food) that are relevant to motivational states such as hunger and thirst, which are controlled by internal need-related (i.e. homeostatic) signals such as the concentration of glucose in the plasma (see Section 3.13). However, there is room for plenty of further discussion and refinement here. Perhaps some people (especially French people?) might say that they do experience emotion when they savour a wonderful food. There may well be cultural differences here in the semantics of whether such reinforcing stimuli should be included within the category that produce emotions.

Another area for discussion is how we wish to categorize the reinforcers associated with sexual behaviour. Such stimuli may be made to be rewarding, and to feel pleasurable, partly because of the internal hormonal state. Does this mean that we wish to exclude such stimuli from the class that we call emotion-provoking, in the same way that we might exclude food reward from the class of stimuli that are said to cause emotion, because the reward value of food depends on an internal controlling signal? I am not sure that there is a perfectly clear answer to this. But this may not matter, as long as we understand that there are some rewarding stimuli that some may wish to exclude from those that cause emotional states.

Second, emotional states can be produced by *remembered reinforcing stimuli*. (Indeed, when we remember stimuli or events, many of the cortical areas activated by the original sensory stimulus are also activated by the remembered stimuli or events. This is the case for most of the cortical areas in each sensory system, apart perhaps from the first (Rolls 1989b, Rolls 2008c). Thus if we recall a particular event, and this leads to reinstatement of activity in the higher parts of the visual system, this activity will provide inputs to the later parts of the brain involved in emotion, so that emotional states may then be produced.

Third, the stimulus that produces the emotional state does not have to be shown to be a reinforcer when producing the emotional state – it simply has to be *capable of being shown to have instrumental reinforcing properties*. An emotion-provoking stimulus can act as a reward or punisher, and is a goal for possible action.

Fourth, the definition given provides great opportunity for *cognitive processing* (whether conscious or not) in emotions, for cognitive processes will very often be required to determine whether an environmental stimulus or event is a reward or punisher. Normally an emotion consists of this cognitive processing that results in a decoded signal that the environmental event is reinforcing, together with the mood state produced as a result. If the mood state is produced in the absence of the external sensory input and the cognitive decoding (for example by direct electrical stimulation of the amygdala, see Rolls (1975) and Rolls (2005a)), then this is described only as a mood state, and is different from an emotion in that there is no object in the environment towards which the mood state is directed. The external reinforcing stimulus may alter the mood state very rapidly, and then the firing of the neurons that represent the mood state may gradually return back to their baseline firing rate, depending on the time course of the emotional state that is produced by the external reinforcing stimulus.

While discussing *mood*, it is worth pointing out that mood may be a particularly difficult state for the brain to maintain at a relatively constant level. In sensory systems the situation is different, for most sensory systems work by contrast, rather than absolute level. For example, early in the visual system it is the difference in brightness levels present at an edge, rather than the absolute brightness that is signaled. This is achieved by a process of lateral inhibition, which means that neighbouring neurons effectively inhibit each other. The result is that it is only at a dark–light boundary, where there is contrast, that neurons are firing. (In fact the firing will be fast on the bright side of the edge, and low, below a spontaneous level of firing, on the dark side of the edge. In the middle of a large bright area few neurons will be active, because the nearby neurons will be inhibiting each other.) However, for mood, the situation may be different. Here, the absolute firing rates of the neurons that represent mood state must be set to fire at the appropriate rate for long periods. Any drift in firing rates would represent a change of mood level. The situation for a brain system that represents mood may thus be different from that involved in most sensory and motor processing in the brain, both because in sensory systems it is the local contrast of firing rate that is important, not the absolute level, and because in sensory systems the inputs keep changing, so that it is not necessary to maintain an absolute value for long. The difficulty of maintaining a constant absolute level of firing in neurons may contribute to 'spontaneous' mood swings, depression that occurs without a clear external cause, and the multiplicity of hormonal and transmitter systems that seem to be involved in the control of mood (Rolls 2005a).

Having said this, it also seems to be the case that there is some '*regression to a constant value*' for emotional states. What I mean by this is that we are sensitive to some extent not just to the absolute level of reinforcers being received, but also to the change in the rate, probability, or magnitude of reinforcers being received. This is well shown by the phenomenon of *positive and negative contrast* effects with rewards. Imagine that an animal is working at a moderate rate for a moderate reward. If the reward is suddenly increased, the animal will work very much harder for a period (perhaps lasting for minutes or longer), but will then gradually revert back to work at a rate close to that at which the animal was working for the moderate reinforcement. This is called positive contrast (see Fig. 3.2). A comparable contrast effect is seen when the reward magnitude (or rate at which rewards are obtained, or the probability of obtaining rewards) is reduced – there is a negative overshoot in the rate of working for a time, but then the rate reverts back to a value close to that at which the animal worked for the moderate reward. This phenomenon is adaptive. It is evidence that animals are in part sensitive to a change in reinforcement, and this helps them to 'climb reward gradients' to obtain better rewards. In effect, regardless of the absolute level of reinforcement being received, it is adaptive to be sensitive to a change in reinforcement. If this were not true, an animal receiving very little reinforcement but then obtaining a small increase in positive reinforcement might still be working very little for the reward. But it is much more adaptive to work hard in this situation, as the extra small amount of reward might make the difference between survival or not, and might lead the animal in the direction of even better rewards if what has just been done leads to an improvement in rewards. A similar phenomenon may be evident in humans. People who have very little in the way of rewards, who may be poor, have a

Positive and Negative Contrast

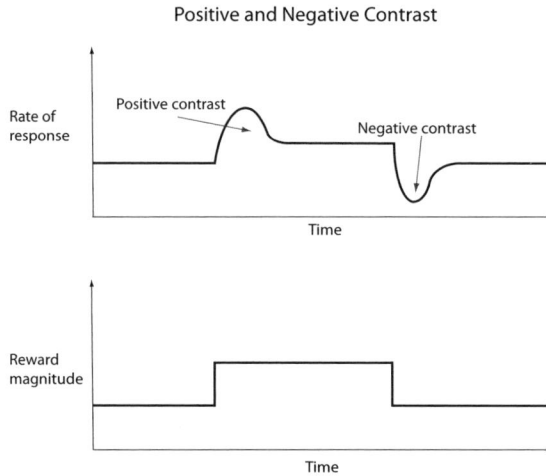

Fig. 3.2 Positive and negative contrast. If the magnitude of a moderate reward (positive reinforcer) is increased, then the rate of working increases markedly, but then drops back to a rate just greater than that to the moderate reinforcer. The positive overshoot is positive contrast. The converse happens if the magnitude of the reward is decreased.

poor diet, and may suffer from disease, may nevertheless not have a baseline level of happiness that is necessarily very different from that of a person in an affluent society who in absolute terms apparently has many more rewards. This may be due in part to resetting of the baseline of expected rewards to a constant value, so that we are especially sensitive to changes in rewards (or punishers).

Fifth, in a case where the sight of a stimulus associated with pain produces fear, some philosophers categorize fear as an emotion, but not pain. The distinction they make may be that primary (unlearned) reinforcers do not produce emotions, whereas secondary reinforcers (stimuli associated by stimulus–reinforcement learning with primary reinforcers) do. They describe the pain as a sensation. But neutral stimuli (such as a table) can produce sensations when touched. It accordingly seems to be much more useful to categorize stimuli according to whether they are reinforcing (in which case they produce emotions), or are not reinforcing (in which case they do not produce emotions). Clearly there is a difference between primary reinforcers and learned reinforcers; but this is most precisely caught by noting that this is the difference, and that it is whether a stimulus is reinforcing that determines whether it is related to emotion. Primary and secondary reinforcers have in common that they produce affective states, whereas neutral, non-reinforcing, stimuli do not produce affective states. The major division thus seems to be between stimuli that produce affective states and those that do not; and it is reinforcing stimuli that produce affective states.

Sixth, as we are about to see, emotional states (i.e. those elicited by instrumental reinforcers) have many functions, and the implementations of only some of these functions by the brain are associated with emotional feelings (see Section 6.3). Indeed there is evidence for interesting dissociations in some patients with brain damage between actions performed to reinforcing stimuli and what is subjectively reported. In this sense it is biologically and psychologically useful to consider emotional states to include more than those states associated with feelings of emotion.

Seventh, the role of learning in many emotions should be emphasized. The approach described above shows that the learning of stimulus–reinforcer (i.e. stimulus–reward and stimulus–punisher) associations is the learning involved when emotional responses are learned. In so far as the majority of stimuli that produce our emotional responses do so as a result of learning, this type of learning, and the brain mechanisms that underlie it, are crucial to the majority of our emotions. This, then, provides a theoretical basis for understanding the functions of some brain systems such as the amygdala in emotion, as described in Sections 2.11 and 3.11 (Rolls 2005a).

It also follows from this approach towards a theory of emotion that brain systems involved in disconnecting stimulus–reinforcer associations when they are no longer appropriate will also be very important in emotion. Failure of this function would be expected to lead, for example, in frustrating situations to inappropriate perseveration of behaviour to stimuli no longer associated with rewards. The inability to correct behaviour when reinforcement contingencies change would be evident in a number of emotion-provoking situations, such as frustration (i.e. non-reward), and the punishment of previously rewarded behaviour. This approach, which emphasizes the necessity, in for example social situations, to update and correct the decoded reinforcement value of stimuli continually and rapidly, helps to provide a basis for understanding the functions of some other brain regions such as the orbitofrontal cortex in emotion (Rolls 2005a).

Eighth, understanding the functions of emotion is also important for understanding the nature of emotions, and for understanding the brain systems involved in the different types of response that are produced by emotional states. Emotion appears to have many functions, which are not necessarily mutually exclusive. Some of these functions are described in Section 3.10.

However, a fundamentally important function of emotion that I will propose in Section 3.10 draws out a close link with the definition given here of emotions as states elicited by instrumental reinforcing stimuli, which are the goals for action. I show in Section 3.10 that genes define the goals for (instrumental) actions, and that this is an important Darwinian, adaptive, aspect of brain design.

These goals for action are instrumental reinforcers, and this thus helps us to see that by understanding emotions as states elicited by reinforcers, we gain important insight into the nature of emotions. The treatment of the nature of emotion given in this chapter is thus seen to be directly relevant to understanding this fundamentally important role of emotion in brain design which is related to the role that reinforcers have in guiding actions.

3.5 Other theories of emotion

In the following subsections, I outline some other theories of emotion, and compare them with the above (Rolls') theory of emotion. Surveys of some of the approaches to emotion that have been taken in the past are provided by Strongman (2003) and Oatley and Jenkins (1996).

3.5.1 The James–Lange and other bodily theories of emotion including Damasio's theory

James (1884) believed that emotional experiences were produced by sensing bodily changes, such as changes in heart rate or in skeletal muscles. Lange (1885) had a similar view, although he emphasized the role of autonomic feedback (for example from the heart) in producing the experience of emotion. The theory, which became known as the James–Lange theory, suggested that there are three steps in producing emotional feelings (see Fig. 3.3). The first step is elicitation by the emotion-provoking stimulus of peripheral changes, such as skeleto-muscular activity to produce running away, and autonomic changes, such as alteration of heart rate. But, as pointed out above, the theory leaves unanswered perhaps the most important issue in any theory of emotion: Why do some events make us run away (and then feel emotional), whereas others do not? This is a major weakness of this type of theory. The second step is the sensing of the peripheral responses (e.g. running away, and altered heart rate). The third step is elicitation of the emotional feeling in response to the sensed feedback from the periphery.

The history of research into peripheral theories of emotion starts with the fatal flaw that step one (the question of which stimuli elicit emotion-related responses in the first place) leaves unanswered this most important question. The history continues with the accumulation of empirical evidence that has gradually weakened more and more the hypothesis that peripheral responses made during emotional behaviour have anything to do with producing the emotional behaviour (which has largely already been produced anyway according to the James–Lange theory), or the emotional feeling. Some of the landmarks in this history are as follows.

First, the peripheral changes produced during emotion are not sufficiently distinct to be able to carry the information that would enable one to have

James-Lange theory of emotion

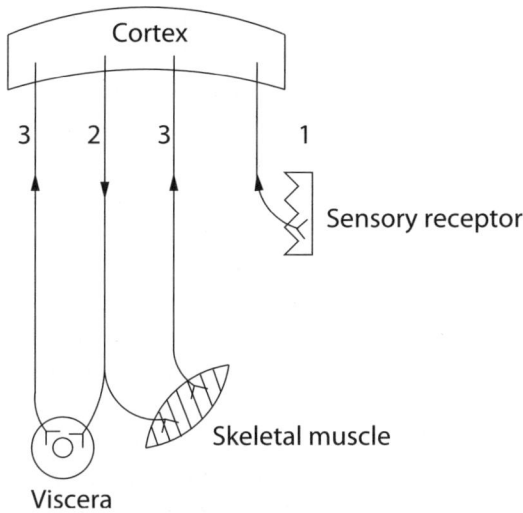

Fig. 3.3 The James–Lange theory of emotion proposes that there are three steps in producing emotional feelings. The first step is elicitation by the emotion-provoking stimulus (received by the cortex via pathway 1 in the Figure) of peripheral changes, such as skeleto-muscular activity to run away, and autonomic changes, such as alteration of heart rate (via pathways labelled 2 in the Figure). The second step is the sensing of the peripheral responses (e.g. altered heart rate, and somatosensory effects produced by running away) (via pathways labelled 3 in the Figure). The third step is elicitation of the emotional feeling in response to the sensed feedback from the periphery.

subtly different emotional feelings to the vast range of different stimuli that can produce different emotions. The evidence suggests that by measuring many peripheral changes in emotion, such as heart rate, skin conductance, breathing rate, and hormones such as adrenaline and noradrenaline (known in the United States by their Greek names epinephrine and norepinephrine), it may be possible to make coarse distinctions between, for example, anger and fear, but not much finer distinctions (Wagner 1989, Cacioppo, Klein, Berntson and Hatfield 1993, Oatley and Jenkins 1996).

Second, when emotions are evoked by imagery, then the peripheral responses are much less marked and distinctive than during emotions produced by external stimuli (Ekman, Levenson and Friesen 1983, Stemmler 1989, Levenson, Ekman and Friesen 1990). This makes sense in that although an emotion evoked by imagery may be strong, there is no need to produce strong peripheral responses, because no behavioural responses are required.

Third, disruption of peripheral responses and feedback from them either surgically (for example in dogs, (Cannon 1927, Cannon 1929, Cannon 1931), or as a result of spinal cord injury in humans (Hohmann 1966, Bermond, Fasotti, Niewenhuyse and Schuerman 1991)), does not abolish emotional responses.

What was found was that in some patients there was apparently some reduction in emotions in some situations (Hohmann 1966), but this could be related to the fact that some of the patients were severely disabled (which could have produced its own consequences for emotionality), and that in many cases the patients were considerably older than before the spinal cord damage, and this could have been a factor. What was common to both studies was that emotions could be felt by all the patients; and that in some cases, emotions resulting from mental events were even reported as being stronger (Hohmann 1966, Bermond, Fasotti, Niewenhuyse and Schuerman 1991).

Fourth, when autonomic changes are elicited by injections of, for example, adrenaline or noradrenaline, particular emotions are not produced. Instead, the emotion that is produced depends on the cognitive decoding of the reinforcers present in the situation, for example an actor who insults your parents to make you angry, or an actor who plays a game of hula hoop to make you feel happy (Schachter and Singer 1962). In this situation, the hormone adrenaline or noradrenaline can alter the magnitude of the emotion, but not which emotion is felt. This is further evidence that it is the decoded reinforcement value of the input stimulus or events that determines which emotion is felt. The fact that the hormone injections produced some change in the magnitude of an emotion is not very surprising. If you felt your heart pounding for no explicable reason, you might wonder what was happening, and therefore react more or abnormally.

Fifth, if the peripheral changes associated with emotion are blocked with drugs, then this does not block the perception of emotion (Reisenzein 1983).

Sixth, it is found that in normal life, behavioural expressions of emotion (for example smiling when at a bowling alley) do not usually occur when one might be expected to feel happy because of a success, but instead occur when one is looking at one's friends (Kraut and Johnson 1979). These body responses, which can be very brief, thus often serve the needs of communication, or of action, not of producing emotional feelings.

Despite this rather overwhelming evidence against an important role for body responses in producing emotions or emotional feelings, Damasio (1994) has effectively tried to resurrect a weakened version of the James–Lange theory of emotion from the 19th century, by arguing with his somatic marker hypothesis that after reinforcers have been evaluated, a bodily response ('somatic marker') normally occurs, then this leads to a bodily feeling, which in turn is appreciated by the organism to then make a contribution to the decision-making process[13]. The James–Lange theory has a number of major weaknesses just outlined that apply also to the somatic marker hypothesis.

The somatic marker hypothesis postulates that emotional decision-making is facilitated by peripheral feedback from for example the autonomic nervous system. In a direct test of this, Heims, Critchley, Dolan, Mathias and Cipolotti (2004) measured emotional decision-making using the Iowa Gambling Task (Bechara, Damasio, Damasio and Anderson 1994, Bechara, Tranel, Damasio and Damasio 1996, Bechara, Damasio, Tranel and Damasio 1997, Damasio 1994) in patients with pure autonomic failure. In this condition, there is degeneration of the peripheral autonomic system, and thus autonomic responses are severely impaired, and there can be no resulting feedback to the brain. (In the Iowa Gambling Task, patients may choose cards from a deck that provides some large gains, but in the long term more losses than the other deck. The patients seem to not learn correctly from their losses (Rolls 2005a, Rolls 2008c).) It was found that performance in the Iowa Gambling Task was not impaired, and nor were many other tests of emotion and emotional performance, including face expression identification, theory of mind tasks of social situations, and social cognition tasks. Thus emotional decision-making does not depend on the ongoing feedback from somatic markers related to autonomic function. Damasio might argue that feedback from the autonomic system is not actually important, and that it is feedback from skeletomotor responses such as arm movements or muscle tension that is important. He might also argue that the autonomic feedback is not usually necessary for emotional decision-making, because it can be 'simulated' by the rest of the brain. However, the study by Heims et al. (2004) does show that ongoing autonomic feedback is not necessary for normal emotional decision-making, and this leaves the somatic marker hypothesis more precarious.

Part of the evidence for the somatic marker hypothesis was that normal participants in the Iowa Gambling Task were described as deciding advantageously before knowing the advantageous strategy (Bechara, Damasio, Tranel and Damasio 1997). The interpretation was that they had implicit (unconscious) knowledge implemented via a somatic marker process that was used in the task, which was not being solved by explicit (conscious) knowledge. Maia and Mc-Clelland (2004) (see also Maia and McClelland (2005)) however showed that

---

[13]In the James–Lange theory, it was emotional feelings that depend on peripheral feedback; for Damasio, it is the decision of which behavioural response to make that is normally influenced by the peripheral feedback. A quotation from Damasio (1994, p190) follows: "The squirrel did not really think about his various options and calculate the costs and benefits of each. He saw the cat, was jolted by the body state, and ran." Here it is clear that the pathway to action uses the body state as part of the route. Damasio would also like decisions to be implemented using the peripheral changes elicited by emotional stimuli. Given all the different reinforcers that may influence behaviour, Damasio (1994) even suggests that the net result of them all is reflected in the net peripheral outcome, and then the brain can sense this net peripheral result, and thus know what decision to take.

with more sensitive questioning, normal participants at least had available to them explicit knowledge about the outcomes of the different decks that was as good as or better than the choices made, weakening the arguments of Bechara et al. (1997) and Bechara, Damasio, Tranel and Damasio (2005) that the task was being solved implicitly and using somatic markers. Further evidence on factors that contribute to the effects found in the Iowa Gambling Task are described by Rolls (2005a) and Rolls (2008c).

Another argument against the somatic marker hypothesis is that there can be dissociations between autonomic and other indices of emotion, thus providing evidence that behaviour may not follow from autonomic and other effects. For example, lesions of different parts of the amygdala influence autonomic responses and instrumental behaviour differently, as shown in Section 2.11 and Fig. 3.6.

Another major weakness, which applies to both the James–Lange and to Damasio's somatic marker hypothesis, is that they do not take account of the fact that once an information processor has determined that a response should be made or inhibited based on reinforcement association, a function attributed to the orbitofrontal cortex in the theory proposed in this chapter and by Rolls (1986b, 1986a, 1990c, 1999a, 2005a), it would be very inefficient and noisy to place in the execution route a peripheral response, and transducers to attempt to measure that peripheral response, itself a notoriously difficult procedure (see, e.g. Grossman (1967)). Even for the cases when Damasio (1994) might argue that the peripheral somatic marker and its feedback can be by-passed using conditioning of a representation in, e.g. the somatosensory cortex to a command signal (which might originate in the orbitofrontal cortex), he apparently would still wish to argue that the activity in the somatosensory cortex is important for the emotion to be appreciated or to influence behaviour. (Without this, the somatic marker hypothesis would vanish.) The prediction would apparently be that if an emotional response were produced to a visual stimulus, then this would necessarily involve activity in the somatosensory cortex or other brain region in which the 'somatic marker' would be represented. This prediction could be tested (for example in patients with somatosensory cortex damage), but it seems most unlikely that an emotion produced by a visual reinforcer would require activity in the somatosensory cortex to feel emotional or to elicit emotional decisions. However, Adolphs, Tranel and Denburg (2000) have pursued this general line of enquiry, and report that the more damage there is to somatosensory cortex, the greater the impairment in the emotional state reported by patients. However, the parts of the somatosensory system that appear to be damaged most frequently in the patients with emotional change are often in the anterior and ventral extensions of the somatosensory cortex in insular and nearby areas, and it would be useful to know whether this damage interrupted some of the connections or functions of the orbitofrontal cortex areas just anterior.

More recently, Damasio has stated the somatic marker hypothesis in a weak form, suggesting that somatic markers do not even reflect the valence of the reinforcer, but just provide a signal that depends on the intensity of the emotion, independently of the type of emotion. On this view, the role of somatic markers in decision-making would be very general, providing, as Damasio says, just a jolt to spur the system on (A.R.Damasio, paper delivered at the 6th Annual Wisconsin Symposium on Emotion, April 2000).

The alternative view proposed here and elsewhere (Rolls 1986b, Rolls 1986a, Rolls 1990c, Rolls 1999a, Rolls 2000c, Rolls 2005a) is that where the reinforcement value of the visual stimulus is decoded, namely in the orbito-frontal cortex and the amygdala, is the appropriate part of the brain for outputs to influence behaviour (via, e.g. the orbitofrontal-to-striatal connections), and that the orbitofrontal cortex and amygdala, and brain structures that receive connections from them, are the likely places where neuronal activity is directly related to emotional states and to felt emotions (see further Section 6.3 and Rolls (2005a)).

3.5.2 Appraisal theory

Appraisal theory, developed and described by Frijda (1986), Oatley and Johnson-Laird (1987), Lazarus (1991), Izard (1993), Stein, Trabasso and Liwag (1994), Oatley and Jenkins (1996), and Scherer (1999) (see also Scherer (2001) and Scherer, Schorr and Johnstone (2001)) generally holds that two types of appraisal are involved in emotion. Primary appraisal holds that "an emotion is usually caused by a person consciously or unconsciously evaluating an event as relevant to a concern (a goal) that is important; the emotion is felt as positive when a concern is advanced and negative when a concern is impeded" (from Oatley and Jenkins (1996), p. 96). As noted above, the concept of appraisal presumably involves assessment of whether something is a reward or punisher, that is whether it will be worked for or avoided. The description in terms of rewards and punishers adopted here simply seems much more precisely and operationally specified. If primary appraisal is defined with respect to goals, it might be helpful to note that goals may just be the reinforcers specified in Rolls' theory, and if so the reinforcer/punisher approach provides clear definitions of goals (as reinforcers, see Appendix A.1), which is helpful, precise, and makes a link to what may be specified by genes.

Secondary appraisal is concerned with coping potential, that is with whether for example a plan can be constructed, and how successful it is likely to be.

I note that appraisal theory is in many ways quite close to the theory that I outline here and in *Emotion Explained* (Rolls 2005a), and I do not see them as rivals. Instead, I hope that those who have an appraisal theory of emotion will consider whether much of what is encompassed by primary appraisal is not actually rather close to assessing whether stimuli or events are reinforcers; and whether much of what is encompassed by secondary appraisal is rather close to taking into account the actions that are possible in particular circumstances, as described above in Section 3.2.

An aspect of some flavours of appraisal theory with which I do not agree is that emotions have as one of their functions releasing particular actions, which seems to make a link with species-specific action tendencies or responses (Tomkins 1995, Panksepp 1998) or more 'open motor programs' (Ekman 2003). I argue in Section 3.10 that rarely are behavioural responses programmed by genes (see Table 3.1), but instead genes optimise their effects on behaviour if they specify the goals for (flexible) actions, that is if they specify rewards and punishers. The difference is quite considerable, in that specifying goals is much more economical in terms of the information that must be encoded in the genome; and in that specifying goals for actions allows much more flexibility to the actual actions that are produced. Of course I acknowledge that there is some preparedness to learn associations between particular types of secondary and primary reinforcers, and see this just as an economy of sensory–sensory convergence in the brain, whereby for example it does not convey much advantage to be able to learn that flashing lights (as contrasted with the taste of a food just eaten) are followed by sickness.

3.5.3   Dimensional and categorical theories of emotion

These theories suggest that there are a number of fundamental or basic emotions. Charles Darwin for example in his book *The Expression of the Emotions in Man and Animals* (1872) showed that some basic expressions of emotion are similar in animals and humans. Some of the examples he gave are shown in Table 3.1. His focus was on the continuity between animals and humans of how emotion is expressed.

In a development of this approach, Ekman and colleagues (Ekman 1982, Ekman 1992, Ekman 1993, Ekman, Friesen and Ellsworth 1972, Ekman, Levenson and Friesen 1983) have suggested that humans categorize face expressions into a number of basic categories that are similar across cultures. These face expression categories include happy, fear, anger, surprise, grief and sadness.

A related approach is to identify a few clusters of variables or factors that result from multidimensional analysis of questionnaires, and to identify these factors as basic emotions. (Multidimensional analyses such as factor analysis seek to identify a few underlying sources of variance to which a large number of data values such as answers to questions are related.) The categories of emotions identified in these ways may be supported by correlating them with autonomic measures (e.g. Ekman et al. (1983)).

One potential problem with some of these approaches is that they risk finding seven plus or minus two categories, which is the normal maximal number of categories with which humans normally operate, as described in a famous paper by George Miller (1956). A second problem is that there is no special reason why the first few factors (which account for most of the variance) in a factor analysis should provide a complete or principled classification of different emotions, or of their functions. In contrast, the theory described here does produce a principled classification of different emotions based on reinforcement contingencies, the nature of the primary and secondary reinforcers, etc., as set out in Sections 3.2 and 3.3. Moreover, the present theory links the functions of emotions to the classification produced, by showing how the functions of emotion can be understood in terms of the gene-specified reinforcers that produce different emotions (see Section 3.10).

An opposite approach to the dimensional or categorical approach is to attempt to describe the richness of every emotion (e.g. Ben-Ze'ev (2000)). Although it is important to understand the richness of every emotion, I believe that this is better performed with a set of underlying principles of the type set out above (in Section 3.2), rather than without any obvious principles to approach the subtlety of emotions.

3.5.4   Other approaches to emotion

LeDoux (1992, 1995, 1996) has described a theory of the neural basis of emotion that is probably conceptually similar to that of Rolls (1975, 1986b, 1986a, 1990c, 1995b, 1999a, 2000c, 2005a) (and this book), except that he focuses mostly on the role of the amygdala in emotion (and not on other brain regions such as the orbitofrontal cortex, which are poorly developed in the rat); except that he focuses mainly on fear (based on his studies of the role of the amygdala and related structures in fear conditioning in the rat); and except that he suggests from his neurophysiological findings that an important route for conditioned emotional stimuli to influence behaviour is via the subcortical inputs (especially auditory from the medial part of the medial geniculate nucleus of the thalamus) to the amygdala. This theory is discussed further elsewhere (Rolls 2005a, Rolls 2008c, Rolls 2008a).

Panksepp's (1998) approach to emotion has its origins in neuroethological investigations of brainstem systems that when activated lead to behaviours like fixed action patterns, including escape, flight and fear behaviour. His views about consciousness include the postulate that "feelings may emerge when endogenous sensory and emotional systems within the brain that receive direct inputs from the outside world as well as the neurodynamics of the SELF (a Simple Ego-type Life Form) begin to reverberate with each other's changing neuronal firing rhythms" (Panksepp 1998) (p. 309).

Other approaches to emotion are summarized by Strongman (2003).

3.6 Individual differences in emotion, personality, and emotional intelligence

Hans J. Eysenck developed the theory that personality might be related to different aspects of conditioning. He analysed the factors that accounted for the variance in the differences between the personality of different humans (using, for example, questionnaires), and suggested that the first two factors in personality (those which accounted for most of the variance) were introversion vs extraversion, and neuroticism (related to a tendency to be anxious). He performed studies of classical conditioning on groups of subjects, and also obtained measures of what he termed arousal. Based on the correlations of these measures with the dimensions identified in the factor analysis, he suggested that introverts showed greater conditionability (to weak stimuli) and are more readily aroused by external stimulation than extraverts; and that neuroticism raises the general intensity of emotional reactions (see Eysenck and Eysenck (1968) and Eysenck and Eysenck (1985)).

Jeffrey A. Gray (1970) reinterpreted the findings, suggesting that introverts are more sensitive to punishment and frustrative non-reward than are extraverts; and that neuroticism reflects the extent of sensitivity to both reward and punishment (see Matthews and Gilliland (1999)). A related hypothesis is that extraverts may show enhanced learning in reward conditions, and may show enhanced processing of positively valent stimuli (Rusting and Larsen 1998). Matthews and Gilliland (1999), reviewing the evidence, show that there is some support for both hypotheses about introversion vs extraversion, namely that introverts may in general condition readily, and that extraverts may be relatively more responsive to reward stimuli (and correspondingly, introverts to punishers). However, Matthews and Gilliland (1999) go on to show that extraverts may perform less well at vigilance tasks (in which the subject must detect stimuli that occur with low probability); may tend to be more impulsive; and perform better when arousal is high (e.g. later in the day), and when rapid responses rather than reflective thought is needed (see also Matthews, Zeidner and Roberts (2002)). With respect to neuroticism and trait anxiety, anxious individuals tend to focus attention on potentially threatening information (punishers) at the cost of neglecting neutral or positive information sources; and may make more negative judgements, especially in evaluating self-worth and personal competence (Matthews, Zeidner and Roberts 2002).

More recent evidence comes from functional neuroimaging studies. For example, Canli, Sivers, Whitfield, Gotlib and Gabrieli (2002) have found that happy face expressions are more likely to activate the human amygdala in extraverts than in introverts. In addition, positively affective pictures interact with extraversion, and negatively affective pictures with neuroticism to produce activation of the amygdala (Canli, Zhao, Desmond, Kang, Gross and Gabrieli 2001, Hamann and Canli 2004). This supports the conceptually important point made above that part of the basis of personality may be differential sensitivity to different rewards and punishers, and omission and termination of different rewards and punishers.

The observations just described are consistent with the hypothesis that part of the basis of extraversion is increased reactivity to positively affective (as compared to negatively affective) face expressions and other positively affective stimuli including pictures.

Another example is the impulsive behaviour that is a part of Borderline Personality Disorder (BPD), which could reflect factors such as less sensitivity to the punishers associated with waiting for rational processing to lead to a satisfactory solution, or changes in internal timing processes that lead to a faster perception of time (Berlin, Rolls and Kischka 2004, Berlin and Rolls 2004). It was of considerable interest that the BPD group (mainly self-harming patients), as well as a group of patients with damage to the orbitofrontal cortex, scored highly on a Frontal Behaviour Questionnaire that assessed inappropriate behaviours typical of orbitofrontal cortex patients including disinhibition, social inappropriateness, perseveration, and uncooperativeness. In terms of measures of personality, using the Big Five personality measure, both groups were also less open to experience (i.e. less open-minded). In terms of other personality measures and characteristics, the orbitofrontal and BPD patients performed differently: BPD patients were less extraverted and conscientious and more neurotic and emotional than the orbitofrontal group (Berlin, Rolls and Kischka 2004, Berlin and Rolls 2004, Berlin, Rolls and Iversen 2005). Thus some aspects of personality, such as impulsiveness and being less open to experience, but not other aspects, such as extraversion, neuroticism and conscientiousness, were differentially related to orbitofrontal cortex function.

Daniel Goleman (1995) has popularized the concept of *emotional intelligence*. The rather sweeping definition given was "Emotional intelligence [includes] abilities such as being able to motivate oneself and persist in the face of frustrations, to control impulse and delay gratification; to regulate one's moods and keep distress from swamping the ability to think; to empathize and to hope" (Goleman (1995), p. 34).

One potential problem with this definition of emotional intelligence as an ability is that different aspects within this definition (such as impulse control and hope) may be unrelated, so a unitary ability described in this way seems unlikely. An excellent critical evaluation of the concept has been produced by Matthews, Zeidner and Roberts (2002). They note (p. 368) that in a rough and ready way, one might identify personality traits of emotional stability (low neuroticism), extraversion, agreeableness, and conscientiousness/self-control as dispositions that tend to facilitate everyday social interaction and to promote more positive emotion. (Indeed, one measure of emotional intelligence, the EQ-i (Bar-On 1997), has high correlations with some of the Big Five personality traits, especially, negatively, with neuroticism, and the EQ-i may reflect three constructs, self-esteem, empathy, and impulse control (Matthews et al. 2002).) But these personality traits are supposed to be independent, so linking them to a single ability of emotional intelligence is inconsistent. Moreover, this combination of personality traits might well not be adaptive in many circumstances, so the concept of this combination as an 'ability' is inappropriate (pp. 368–370).

However, the concept of emotional intelligence does appear to be related in a general way to the usage of the (mainly clinical) term 'alexithymia', in a sense the opposite, which includes the following components: (a) difficulty in identifying and describing emotions and distinguishing between feelings and the bodily sensations of arousal, (b) difficulty in describing feelings to other people, (c) constricted imaginal processes, as evidenced by a paucity of fantasies, and (d) a stimulus-bound externally oriented cognitive style, as evidenced by preoccupation with the details of external events rather than inner emotional experiences (Matthews et al. 2002). In terms of personality, alexithymia converges with the first three dimensions of the Five Factor Model of personality (FFM, the Big Five model), with high N (vulnerability to emotional distress), low E (low positive emotionality), and a limited range of imagination (low O) (Matthews et al. 2002). Indeed, alexithymia is strongly inversely correlated with measures of emotional intelligence, suggesting that emotional intelligence may be a new term that encompasses much of the opposite of what has been the important concept of alexithymia in the clinical literature for more than 20 years (Matthews et al. 2002). Alexithymics have difficulties in identifying face expressions (Lane, Sechrest, Reidel, Weldon, Kaszniak and Schwartz 1996), suggesting some impairments in the fundamental processing of emotion-related information, in particular capacities known to require the orbitofrontal and anterior cingulate cortices (Hornak, Bramham, Rolls, Morris, O'Doherty, Bullock and Polkey 2003). Consistently, it has been found that anterior cingulate cortex activation is correlated across individuals with their ability to recognize and describe emotions induced either by films or by the recall of personal experiences (Lane, Reiman, Axelrod, Yun, Holmes and Schwartz 1998).

Rolls (2007c) has argued in a neurobiological approach to emotional intelligence, and what is largely its opposite, alexithymia, that emotional intelligence need not be viewed as a particular ability, and is not independent of existing personality measures, but does encompass a number of probably different ways in which individuals may differ in their emotion-related processing.

I do not consider this research area in much more detail. However, I do point out that insofar as sensitivity to rewards and punishers, and the ability to learn and be influenced by rewards and punishers, may be important in personality, and are closely involved in emotion according to the theory developed here, there may be close links between the neural bases of emotion, to be described in Section 3.11, and personality. An extreme example might be that if humans were insensitive to social punishers following orbitofrontal cortex damage, we might expect social problems, and indeed Tranel, Bechara and Denburg (2002) have used the term 'acquired sociopathy' to describe some of these patients.

More generally, we might expect sensitivity to different types of reinforcer (including social reinforcers) to vary between individuals both as a result of gene variation and as a result of learning, and this, operating over a large number of different social reinforcers, might produce many different variations of personality based on the sensitivity to a large number of different reinforcers. Further, insofar as the functions of particular brain regions may be related to particular processes involved in emotion (with evidence for example that the human orbitofrontal cortex is involved in face expression decoding, and in impulsiveness, but not in some other aspects of personality), then it may be possible in future to understand different particular modules for inter-relations between reward/punishment and personality systems.

The concept of the relation between differential sensitivity to different types of reward and punisher might produce individuals showing many types of conditional evolutionarily stable strategies, where the conditionality of the strategy might be influenced in different individuals by differential sensitivity to different rewards and punishers (Rolls 2005a). Examples of behaviours that might be produced in this way are included in Section 4.4.

3.7 Cognition and Emotion

It may be noted that while the definition of emotions as states elicited by reinforcers (with particular functions) is operational, it should not be criticized as behaviourist (Katz 2000). For example, the definition has nothing to do with stimulus–response (habit) associations, but instead with a two-stage type of learning, in which a first stage is learning which environmental stimuli or events are associated with reinforcers, which potentially is a very rapid and flexible process; and a second stage produces appropriate instrumental and arbitrary actions performed in order to achieve the goal (which might be to obtain a reward or avoid a punisher). In the instrumental stage, animals learn about the outcomes of their actions (see Dickinson (1994), Pearce (1997)).

To determine what is a goal for an action, every type of cognitive operation may be involved. The proposal is that whatever cognitive operations are involved, then if the outcome is that a certain event, stimulus, thought (or any one of these remembered) leads to the evaluation that the event is a reward or punisher, then an emotion will be produced. So cognition is far from excluded.

Indeed, cognitive operations may produce emotions when operating at three levels of the architecture, as described more fully in Section 3.10. The first is the implicit level (see Fig. 5.1), where a primary reinforcer, or a stimulus or event associated with a primary reinforcer, may lead to emotions. The second level is where a (first order) syntactic symbol processing system performing "what ... if" computations to implement planning results in identification of a rewarding or punishing outcome. The third level is the higher-order linguistic thought level described in Section 6.3, where thinking about and evaluating the operations of a first order linguistic processor may result in a reinforcing outcome such as "I should not spend further time thinking about that set of plans, as it would be better now to devote my linguistic resources (which are limited and serial) to this other set of plans".

Another way in which cognition influences emotion is that cognitive states, even at the level of language, can modulate subjective and brain responses to affective stimuli, as analysed in Section 3.11 (Rolls and Grabenhorst 2008, Grabenhorst and Rolls 2011). There an experiment is described in which a word label ('cheese' vs 'body odour') influences the pleasantness ratings, and the activations in olfactory stages at least as early as the secondary olfactory cortex in the orbitofrontal cortex, to a standard test odour (De Araujo et al. 2005). An implication of these findings is that language-based cognitive states can influence even relatively early cortical representations of rewards and punishers, and thus potentially modulate how much emotion is felt subjectively to an emotion-provoking stimulus.

I suggest that this top-down modulation occurs in a way that is exactly analogous to top-down attentional effects, which are believed to be implemented by a top-down biased competition mechanism (Rolls and Deco 2002, Deco and Rolls 2003, Deco and Rolls 2005a, Rolls and Stringer 2001b). In this case, the semantic, language-based, representation is the source of the biased competition, and the effect could be not only to bias the early cortical representation of a reward or punisher in one direction or another, but also by providing much or little top-down modulation, to influence how much emotion is felt (see Section 6.3), by modulating the processing of emotion-related stimuli (including remembered stimuli or events) at relatively early processing stages. This could be a mechanism by which cognition can influence how much emotion is felt under conditions in which emotions such as empathy and pity may occur, and when for example reading a novel, attending a play, listening to music, etc. (see Chapter 7). Analysis of the mechanisms by which the top-down biased competition operates are becoming detailed (Desimone and Duncan 1995, Rolls and Deco 2002, Deco and Rolls 2003, Deco and Rolls 2004, Deco and Rolls 2005a, Rolls 2008c), and are included in a model in which a rule module exerts a top-down influence on neurons that represent stimulus–reward and stimulus–punisher combinations to influence which stimulus should currently be interpreted as reward-related (see also Deco and Rolls (2005c)).

Another way in which cognitive factors are related to emotion is that mood can affect cognitive processing, and one of the effects of this is to promote continuity of behaviour (see Section 3.10). One of the mechanisms utilizes backprojections to cortical areas from the amygdala and orbitofrontal cortex, so that reciprocal interactions between cognition and emotion are made possible (Rolls 2005a, Rolls 2008c).

3.8   Emotion, motivation, reward, and mood

It is useful to be clear about the difference between motivation, emotion, reward, and mood (Rolls 2000c, Rolls 2005a). **Motivation** makes one work to obtain a reward, or work to escape from or avoid a punisher. One example of motivation is hunger, and another thirst, which in these cases are states set largely by internal homeostatically-related variables such as plasma glucose concentration and plasma osmolality. A reward is a stimulus or event that one works to obtain, such as food, and a punisher is what one works to escape from or avoid (or which suppresses an action on which its delivery is contingent), such as a painful stimulus or the sight of an object associated with a painful stimulus. Obtaining the reward or avoiding the punisher is the goal for the action. A motivational state is one in which a goal is *desired*, i.e. will be worked for. An **emotion** is a state elicited when a goal is obtained, that is by an instrumental reinforcer (i.e. a reward or punisher, or omission or termination of a reward or punisher), for example fear produced by the sight of the object associated with pain. This makes it clear that emotions are states elicited by rewards or punishers that have particular functions.

Of course, one of the functions of emotions is that they are motivating, as exemplified by the case of the fear produced by the sight of the object that can produce pain, which motivates one to avoid receiving the painful stimulus, which is the goal for the action. In that emotion-provoking stimuli or events produce motivation, then arousal is likely to occur, especially for reinforcers that lead to the active initiation of actions. However, arousal alone is not sufficient to define motivation or emotion, in that the motivational state must specify the particular type of goal that is the object of the motivational state, such as water if we are thirsty, food if we are hungry, and avoidance of the painful unconditioned stimulus signalled by a fear-inducing conditioned stimulus.

A **mood** is a continuing state normally elicited by a reinforcer, and is thus part of what is an emotion. The other part of an emotion is the decoding of the stimulus in terms of whether it is a reward or punisher, that is, of what causes the emotion, or in philosophical terminology of what the emotion is about or the object of the emotion. Mood states help to implement some of the persistence-related functions of emotion, can continue when the originating stimulus may be forgotten (by the explicit system described in Section 6.3), and may occur spontaneously not because such spontaneous mood swings may have been selected for, but because of the difficulty of maintaining stability of the neuronal firing that implements mood (or affective) state (see *Emotion Explained*, Rolls (2005a)). Mood states are thus not necessarily about an object.

Thus, motivation may be seen as a state in which one is working for a goal, and emotion as a state that occurs when the goal, a reinforcer, is obtained, and that may persist afterwards.

The concept of gene-defined reinforcers providing the goals for action helps to understand the relation between motivational states (or desires) and emotion, as the organism must be built to be motivated to obtain the goals, and to be placed in a different state (emotion) when the goal is or is not achieved by the action. Emotional states may be motivating, as in frustrative non-reward. The close but clear relation between motivation and emotion is that both involve what humans describe as affective states (e.g. feeling hungry, liking the taste of a food, feeling happy because of a social reinforcer), and both are about goals. The Darwinian theory of the functions of emotion developed in Section 3.10 which shows how emotion is adaptive because it reflects the operation of a process by which genes define goals for action applies just as much to motivation (see further Section 3.10.5), in that *emotion can be thought of as states elicited by goals (rewards and punishers); and motivation can be thought of as states present when goals are being sought.* By specifying goals the genes must specify both that we must be motivated to obtain those goals, and that when the goals are obtained, further states, emotional states with further functions, are produced.

3.9 Is the concept of emotion still useful when we understand its mechanisms?

Kralik and Hauser (2000) ask whether it is helpful to maintain the concept of an emotional state when one starts to understand the mechanisms of reward and punisher decoding, the selection of actions, etc. My view is that emotion is a helpful concept, for a number of reasons.

First, the state is produced by clearly defined stimuli (see above).

Second, the state has many different functions, summarized in Section 3.10, so that a model in which a stimulus is connected to a single output is inappropriate. In these circumstances, an intervening state that implements many functions is useful.

Third, one of the functions of emotion is to support the selection of any appropriate action to a reward or punisher, or its omission or termination, as in two-process learning. In the first stage, an emotional state is produced, and in the second stage, any action is selected that is appropriate given the emotional state. For example, if fear is the emotional state produced by a pain-associated stimulus, an action will be selected to escape from or avoid the emotion-provoking stimulus. In that emotion is a state that guides the elicitation of an action to a stimulus, the emotional state is not itself a behavioural response.

Fourth, other functions of emotional states include the biasing of cognitive function to influence the interpretation of future events, which is clearly not a response.

Fifth, emotional states have the important properties that they persist for times in the order of minutes or hours, thus maintaining persistence of behaviour and consistency of action even after the emotion-provoking stimulus has disappeared.

Sixth, the concept of emotional states just described maps neatly onto folk-psychological concepts of emotions, and provides a convenient conceptual level that bridges to the low-level description of exactly how the stimuli are decoded to elicit the state, how the state is maintained, and how it performs its many functions.

The concept of an emotional state is thus clearly defined in terms of how stimuli elicit the state, and of the many functions of the state including the selection of action. Emotional states are not the stimuli themselves, nor the stimulus decoding, nor the responses finally selected, but consist of on-going states elicited by stimuli in the way described, and performing the functions described. We are indeed starting to understand how the different types of processing involved are implemented in the brain, and these are some of the types of advance described in *Emotion Explained* (Rolls 2005a). But understanding the implementation of the processes involved in emotion does not mean that emotion itself as a useful concept at its own level will disappear.

In addition, understanding 'how' emotion works will not address a number of important questions about emotion, including the 'why' questions about for example the evolutionary adaptive value of emotions (see Section 3.10).

Advantages of Rolls' theory of emotion are described in *Emotion Explained* (Rolls 2005a), and neuroscience developments based on it are described by Rolls (2008c), Rolls and Grabenhorst (2008), and Grabenhorst and Rolls (2011).

3.10 The functions of emotion: reward, punishment, and emotion in brain design

We now confront the fundamental issue of why we, and other animals, are built to have emotions, as well as motivational states. I propose that it is because we (and many other animals) use rewards and punishers to guide or determine our behaviour, and that this is a good design for a system that is built by genes where some of the genes are increasing their survival by specifying the goals for behaviour. The emotions arise and are an inherent part of such a system because they are the states, typically persisting, that are elicited by the rewards and punishers. I will show that this is a very adaptive way for evolution to design complex animals without having to specify the details of the behavioural responses (Rolls 2005a).

What results from this analysis is thus a thoroughly Darwinian theory (though not anticipated by Darwin, and operating at the level of individual genes) that places emotion at the heart of brain design because it reflects the way in which genes build our brains in such a way that our genes can specify the goals of our actions, and thus what we do. There is thus a close conceptual link between instrumental learning and emotion, for primary reinforcers (primary rewards and punishers) are the gene-specified goals for our actions, and we use instrumental learning to learn any actions during our life-times that will lead to the gene-specified goals. The approach goes beyond the concepts in the *The Selfish Gene* by Richard Dawkins (1976) by arguing that a very efficient way for selfish genes to operate is by specifying the goals for action, and then by going on to argue that this is why we and other animals have emotions.

In Section 3.10.1, I outline several types of brain design, with differing degrees of complexity, and suggest that evolution can operate much better with only some of these types of design.

Understanding the functions of emotion is important not only for understanding the nature of emotions, but also for understanding the different brain systems involved in the different types of response that are produced by emotional states. Indeed, answers to 'why' questions in nature (for example, 'Why do we have emotions? What are the functions of emotion?') are important, as are answers to 'how' questions (for example, 'How is emotion implemented in the brain? How do disorders of emotion arise, and how can they be understood and treated?').

In this book and elsewhere (Rolls 1999a, Rolls 2005a), the question of why we have emotions is a fundamental issue that I answer in terms of a Darwinian, functional, approach, producing the answer that emotions are states elicited by goals (rewards and punishers), and that this is part of an adaptive process by which genes can specify the behaviour of the animal by specifying goals for behaviour rather than fixed responses. I believe that this approach leads to a fundamental understanding of why we have emotions which is likely to stand the test of time, in the same way that Darwinian thinking itself provides a fundamental way of understanding biology and many 'why' questions about life.

While considering 'why' (or 'ultimate') questions (which are important in their own right, and will not be answered by answers to 'how' or 'proximate' questions), it may be helpful to place into perspective the approaches taken to understanding the adaptive value of behaviour (Tinbergen 1963) that have led to sociobiology (Wilson 1975) and evolutionary psychology (see Buss (2008)). These approaches are relevant to understanding why we have emotions, including many of the issues discussed in Section 4.4 on sexual behaviour. 'Adaptation' refers to characteristics of living organisms – such as their colour, shape, physiology, and behaviour – that enable them to survive and reproduce successfully in the environments in which they live M.S.Dawkins (1995).

Sociobiology and evolutionary psychology have sometimes been criticized as producing 'just-so' stories in which the purported adaptive explanation for a behaviour seems too facile and untestable (Gould and Lewontin 1979), but we should note that there are rigorous approaches to testing evolutionary hypotheses for the adaptive value of a behaviour or other characteristic (Dawkins 1995, Rolls 2005a). The tests include the following:

1. *Making use of existing genetic variation.* A famous example is that of the peppered moth, *Biston betularia*. In its dark form it is found to survive better than the light form when both are released into industrial areas in which the genetically specified black form is better camouflaged (Kettlewell 1955).

2. *Using artificially produced variation.* The variation could be genetically produced, or variation produced in the testing conditions in an experiment. Using the latter approach, Tinbergen, Broekhuysen, Feekes, Houghton, Kruuk and Szule (1967) showed that black-headed gulls' newly born chicks survive better if the parents remove broken egg shells to a good distance from the nest. This supports the hypothesis that the behaviour is adaptive because removing the shells, which are white inside, makes the nest less conspicuous to predators.

3. *The comparative method.* Comparing species in which a trait has evolved genetically and independently can provide good evidence at the correlative level. For example, kittiwakes do not remove hatched egg shells from their nest, and in contrast with black-headed gulls, this is related to the fact that kittiwakes nest in places on cliffs that are inaccessible even to other birds (Cullen 1957).

4. *Adaptation through design features.* If it can be shown that design features such as the bat sonar echolocation system are very well suited to detecting small animals of prey such as moths, then that provides some evidence that the features are genetically selected because of their adaptive value. Even more telling are examples where the behaviour is not as adaptive as it might be – for example, in some schooling fish, the spacing during swimming is not hydrodynamically optimal, and this implies that the details of the behaviour are under selective pressure other than only swimming efficiency (Dawkins 1995).

Thus adaptive accounts of behaviour can be tested, and need not be 'just-so' stories.

We should also note that by no means all behaviour reflects optimal adaptation (Dawkins 1982).

3.10.1 Brain design and the functions of emotion

3.10.1.1 Taxes, rewards, and punishers: gene-specified goals for actions, and the flexibility of actions

Taxes

A simple design principle is to incorporate mechanisms for taxes into the design of organisms. Taxes consist at their simplest of orientation towards stimuli in the environment, for example the bending of a plant towards light that results in maximum light collection by its photosynthetic surfaces. (When just turning rather than locomotion is possible, such responses are called tropisms.) With locomotion possible, as in animals, taxes include movements towards sources of nutrient, and movements away from hazards such as very high temperatures. The design principle here is that animals have, through a process of natural selection, built receptors for certain dimensions of the wide range of stimuli in the environment, and have linked these receptors to response mechanisms in such a way that the stimuli are approached or escaped from.

Rewards and punishers

As soon as we have approach to stimuli at one end of a dimension (e.g. a source of nutrient) and away from stimuli at the other end of the dimension (in this case lack of nutrient), we can start to wonder when it is appropriate to introduce the terms 'rewards' and 'punishers' for the stimuli at the different ends of the dimension. By convention, if the response consists of a fixed response to obtain the stimulus (e.g. locomotion up a chemical gradient), we shall call this a taxis not a reward. On the other hand, if an arbitrary operant response can be performed by the animal in order to approach the stimulus, then we will call this rewarded behaviour, and the stimulus that the animal works to obtain a reward. (The arbitrary operant response can be thought of as any arbitrary response the animal will perform to obtain the stimulus. It can be thought of as an action.) This criterion, of an arbitrary operant response, is often tested by bidirectionality. For example, if a rat can be trained to either raise its tail, or lower its tail, in order to obtain a piece of food, then we can be sure that there is no fixed relationship between the stimulus (e.g. the sight of food) and the response, as there is in a taxis. Some authors reserve the term 'motivated behaviour' for that in which an arbitrary operant response will be performed to obtain a reward or to escape from or avoid a punisher. If this criterion is not met, and only a fixed response can be performed, then the term 'drive' can be used to describe the state of the animal when it will work to obtain or escape from the stimulus.

We can thus distinguish a first level of approach/avoidance mechanism complexity in a taxis, with a fixed response available for the stimulus, from a second level of complexity in which any arbitrary response (or action) can be performed, in which case we use the term reward when a stimulus is being approached, and punisher when the action is to escape from or avoid the stimulus.

The role of natural selection in this process is to guide animals to build sensory systems that will respond to dimensions of stimuli in the natural environment along which actions of the animals can lead to better survival to enable genes to be passed on to the next generation, which is what we mean by fitness[14]. The animals must be built by such natural selection to perform actions that will enable them to obtain more rewards, that is to work to obtain stimuli that will increase their fitness. Correspondingly, animals must be built to perform actions that will enable them to escape from, or avoid when learning mechanisms are introduced, stimuli that will reduce their fitness. There are likely to be many dimensions of environmental stimuli along which actions of the animal can alter fitness. Each of these dimensions may be a separate reward–punisher dimension. An example of one of these dimensions might be food reward. It increases fitness to be able to sense nutrient need, to have sensors that respond to the taste of food, and to perform behavioural responses to obtain such reward stimuli when in that need or motivational state. Similarly, another dimension is water reward, in which the taste of water becomes rewarding when there is body-fluid depletion (Rolls and Rolls 1982, Rolls 2005a).

One aspect of the operation of these reward–punisher systems that these examples illustrate is that with very many reward–punisher dimensions for which actions may be performed, there is a need for a selection mechanism for actions performed to these different dimensions. In this sense, rewards and punishers provide a common currency that provides one set of inputs to action selection mechanisms. Evolution must set the magnitudes of each of the different reward systems so that each will be chosen for action in such a way as to maximize overall fitness. Food reward must be chosen as the aim for action if some nutrient depletion is present, but water reward as a target for action must be selected if current water depletion poses a greater threat to fitness than does the current degree of food depletion. This indicates that for a competitive selection process for rewards, each reward must be carefully calibrated in evolution to have the right value in the common currency in the selection process (Rolls 2005a, Grabenhorst and Rolls 2011). Other types of behaviour, such as sexual behaviour, must be performed sometimes, but probably less frequently, in order to maximize fitness (as measured by gene transmission into the next generation).

There are many processes that contribute to increasing the chances that a wide set of different environmental rewards will be chosen over a period of time, including not only need-related satiety mechanisms that reduce the rewards within a dimension, but also sensory-specific satiety mechanisms, which

---

[14]Fitness refers to the fitness of genes, but this must be measured by the effects that the genes have on the organism.

facilitate switching to another reward stimulus (sometimes within and some-times outside the same main dimension), and attraction to novel stimuli. (As noted in Section 3.10.3.5, attraction to novel stimuli, i.e. finding them reward-ing, is one way that organisms are encouraged to explore the multidimensional space within which their genes are operating. The suggestion is that animals should be built to find somewhat novel stimuli rewarding, for this encourages them to explore new parts of the environment in which their genes might do better than others' genes. Unless animals are built to find novelty somewhat rewarding, the multidimensional genetic space being explored by genes in the course of evolution might not find the appropriate environment in which they might do better than others' genes.)

Stimulus–response learning reinforced by rewards and punishers

In this second level of complexity, involving reward or punishment, learning may occur. If an organism performs trial-and-error responses, and as the result of performing one particular response is more likely to obtain a reward, then the response may become linked by a learning process to that stimulus as a result of the reward received. The reward is said to reinforce the response to that stimulus, and we have what is described as stimulus–response or habit learning. The reward acts as a positive reinforcer in that it increases the probability of a response on which it is made contingent. A punisher reduces the probability of a response on which it is made contingent. (It should be noted that this is an operational definition, and that there is no implication that the punisher feels like anything – the punisher just has in the learning mechanism to reduce the probability of responses followed by the punisher.) Stimulus–response or habit learning is typically evident after over-training, and once habits are being executed, the behaviour becomes somewhat independent of the reward value of the goal, as shown in experiments in which the reward is devalued (Section 2.11) (Rolls 2005a).

Stimulus–reinforcer association learning, and two-factor learning theory

Two-process learning introduces a third level of complexity and capability into the ways in which behaviour can be guided. Rewards and punishers still provide the basis for guiding behaviour within a dimension, and for selecting the dimension towards which action should be directed.

The first stage of the learning is stimulus–reinforcer association learning, in which the reinforcing value of a previously neutral, e.g. visual or auditory, stimulus is learned because of its association with a primary reinforcer, such as a sweet taste or a painful touch. This learning is of an association between one stimulus, the conditioned or secondary reinforcer, and the primary reinforcer, and is thus stimulus–stimulus association learning. This stimulus–reinforcer learning can be very fast, in as little as one trial. For example, if a new visual stimulus is placed in the mouth and a sweet taste is obtained, a simple approach response such as reaching for the object will be made on the next trial. Moreover, this stimulus–reinforcer association learning can be reversed very rapidly. For example, if subsequently the object is made to taste of salt, then approach no longer occurs to the stimulus, and the stimulus is even likely to be actively pushed away.

The second process or stage in this type of learning is instrumental learning of an operant response made in order to obtain the stimulus now associated with reward (or avoid a stimulus associated by learning with the punisher). This is action–outcome learning. The outcome could be a primary reinforcer, but often involves a secondary reinforcer learned by stimulus–reinforcer association learning. The action–outcome learning may be much slower, for it may involve trial-and-error learning of which action is successful in enabling the animal to obtain the stimulus now associated with reward or avoid the stimulus now associated with a punisher. However, this second stage may be greatly speeded if an operant response or strategy that has been learned previously to obtain a different type of reward (or avoid a different punisher) can be used to obtain (or avoid) the new stimulus now known to be associated with reinforcement. It is in this flexibility of the response that two-factor learning has a great advantage over stimulus–response learning. The advantage is that any response (even, at its simplest, approach or withdrawal) can be performed once an association has been learned between a stimulus and a primary reinforcer. This flexibility in the response is much more adaptive (and could provide the difference between survival or not) than no learning, as in taxes, or stimulus–response learning. The different processes that are involved in instrumental learning are described in more detail in Section 2.11.

Another key advantage of this type of two-stage learning is that after the first stage the different rewards and punishers available in an environment can be compared in a selection mechanism, using the common currency of rewards and punishers for the comparison and selection process. In this type of system, the many dimensions of rewards and punishers are again the basis on which the selection of a behaviour to perform is made.

Part of the process of evolution can be seen as identifying the factors or dimensions that affect the fitness of an animal, and providing the animal with sensors that lead to rewards and punishers that are tuned to the environmental dimensions that influence fitness. The example of sweet taste receptors being set up by evolution to provide reward when physiological nutrient need is present has been given above.

To help specify the way in which stimulus–reinforcer association learning operates, a list of what may be in at least some species primary reinforcers is provided in Table 3.1 on page 108. The reader will doubtless be able to add to this list, and it may be that some of the reinforcers in the list are actually secondary reinforcers. The reinforcers are categorized where possible by modality, to help the list to be systematic. Possible dimensions to which each reinforcer is tuned are suggested.

3.10.1.2 Explicit systems, language, and reinforcement

A fourth level of complexity to the way in which behaviour is guided is by processing that includes syntactic operations on semantically grounded symbols (see Section 6.3.2). This allows multistep one-off plans to be formulated. Such a plan might be: if I do this, then B is likely to do this, C will probably do this, and then X will be the outcome. Such a process cannot be performed by an animal that works just to obtain a reward, or secondary reinforcers. The process may enable an available reward to be deferred for another reward that a particular multistep strategy could lead to. What are the roles of rewards and punishers in such a system?

The language system can still be considered to operate to obtain rewards and avoid punishers. This is not merely a matter of definition, for many of the rewards and punishers will be the same as those described above, those which have been tuned by evolution to the dimensions of the environment that can enable an animal to increase fitness. The processing afforded by language can be seen as providing a new type of strategy to obtain such gene-specified rewards or avoid such punishers. If this were not generally the case, then the use of the language system would not be adaptive: it would not increase fitness.

However, once a language system has evolved, a consequence may be that certain new types of reward become possible. These may be related to primary reinforcers already present, but may develop beyond them. For example, music may have evolved from the system of non-verbal communication that enables emotional states to be communicated to others. An example might be that lullabies could be related to emotional messages that can be sent from parents to offspring to soothe them. Music with a more military character might be related to the sounds given as social signals to each other in situations in which fighting (or co-operation in fighting) might occur. The prosodic quality of voice expression may be part of the same emotion communication system, and brain systems that are activated by prosody may be strongly engaged in women even in tasks that do not require prosody to be analysed (Schirmer, Zysset, Kotz and von Cramon 2004). Then on top of this, the intellectualization afforded by linguistic (syntactic) processing would contribute further aspects to music.

Another example here is that solving problems by intellectual means should itself be a primary reinforcer as a result of evolution, for this would encourage the use of intellectual abilities that have potential advantage if used. A further set of examples of how, when a language system is present, there is the possibility for further types of reinforcer, comes from the possibility that the evolution of some mental abilities may have been influenced by sexual selection (see Section 7.9).

3.10.1.3   Special-purpose design by an external agent vs evolution by natural selection

The above mechanisms, which operate in an evolutionary context to enable animals' behaviour to be tuned to increase fitness by evolving reward–punisher systems tuned to dimensions in the environment that increase fitness, may be contrasted with typical engineering design. In the latter, we may want to design a robot to work on an assembly line. Here there is an external designer, the engineer, who defines the function to be performed by the robot (e.g. picking a nut from a box, and attaching it to a particular bolt in the object being assembled). The engineer then produces special-purpose design features that enable the robot to perform this task, by for example providing it with sensors and an arm to enable it to select a nut, and to place the nut in the correct position in the 3D space of the object to enable the nut to be placed on the bolt and tightened. This contrast with a real animal allows us to see important differences between these types of control for the behaviour of the system.

In the case of the animal, there is a multidimensional space within which many optimizations to increase fitness must be performed. The solution to this is to evolve multiple reward–punisher systems tuned to each dimension in the environment that can lead to an increased fitness if the animal performs the appropriate actions. Natural selection guides evolution to find these dimensions. In contrast, in the robot arm, there is an externally defined behaviour to be performed, of placing the nut on the bolt, and the robot does not need to tune itself to find the goal to be performed. The contrast is between design by evolution that is 'blind' to the purpose of the animal, and design by a designer who specifies the job to be performed (cf. Dawkins (1986)).

Another contrast is that for the animal the space will be high-dimensional, so that selection of the most appropriate reward for current behaviour (taking into account the costs of obtaining each reward) is needed, whereas for the robot arm, the function to perform at any one time is specified by the designer. Another contrast is that the behaviour, that is the instrumental action, that is most appropriate to obtain the reward must be selected by the animal, whereas the movement to be made by the robot arm is specified by the design engineer.

The implication of this comparison is that operation by animals using reward and punisher systems tuned to dimensions of the environment that increase fitness provides a mode of operation that can work in organisms that evolve by natural selection. It is clearly a natural outcome of Darwinian evolution to operate using reward and punisher systems tuned to fitness-related dimensions of the environment, if arbitrary responses are to be made by the animals rather than just preprogrammed movements such as are involved in tropisms and taxes. Is there any alternative to such a reward–punisher-based system in this evolution by natural selection situation? I am not clear that there is. This may be the reason why we are built to work for rewards, avoid punishers, have emotions, and feel needs (motivational states). These concepts start to bear on developments in the field of artificial life (see, e.g. Boden (1996)).

The sort of question that some philosophers might ponder is whether if life evolved on Mars it would have emotions. My answer to this is 'Yes', if the organisms have evolved genetically by natural selection, and the genes have elaborated behavioural mechanisms to maximize their fitness in a flexible way, which as I have just argued would imply that they have evolved reward and punisher systems that guide behaviour. They would have emotions in the sense that they would have states that would be produced by rewards or punishers, or by stimuli associated with rewards and punishers (Rolls 2002, Rolls 2005b, Rolls 2005a).

It is of course a rather larger question to ask whether our extraterrestrial organisms would have emotional feelings. My answer to this arises out of the theory of consciousness introduced in Section 6.3, and would be 'Only if the organisms have a linguistic system that can think about and correct their first order linguistic thoughts'. However, even if such higher-order thought processes are present, it is worth considering whether emotional feelings might despite these higher-order thoughts not be present. After all, we know that much behaviour can be guided unconsciously, implicitly, by rewards and punishers. My answer to this issue is that the organisms would have emotional feelings; for as suggested above, the explicit system has to work in general for rewards of the type that are rewarding to the implicit system, and for the explicit system to be guided towards solutions that increase fitness, it should feel good when the explicit system works to a correct solution. Otherwise, it is difficult to explain how the explicit system is guided towards solutions that are not only solutions to problems, but that are also solutions that tend to have adaptive value. If the system has evolved so that it feels like something when it is performing higher-order thought processing, then it seems likely that it would feel like something when it obtained a reward or punisher, for this is the way that the explicit, conscious, thoughts would be guided (see Section 6.3 for further explanation).

3.10.2 Selection of behaviour: cost–benefit 'analysis'

One advantage of a design based on rewards and punishers is that the decoding of stimuli to a reward or punisher value provides a common currency, that is on a *common scale*, for the mechanism that selects which behavioural action should be performed (Rolls 2005a, Rolls and Grabenhorst 2008). Thus, for example, a moderately sweet taste when little hunger is present would have a smaller reward value than the taste of water when thirst is present. An action-selection mechanism could thus include in its specification competition between the different rewards, all represented in a common currency, with the most rewarding stimulus being that most likely to be selected for action. As described above, to make sure that different types of reward are selected when appropriate, natural selection would need to ensure that different types of reward would operate on similar scales (from minimum to maximum), so that each type of reward would be selected if it reaches a high value on its scale. Evidence that this common scale is implemented in the orbitofrontal cortex has been found in an fMRI study in humans (Grabenhorst, D'Souza, Parris, Rolls and Passingham 2010, Grabenhorst and Rolls 2011). Mechanisms such as sensory-specific satiety can be seen as contributing usefully to this mechanism which ensures that different types of reward will be selected for action.

However, the action selection mechanisms must take into account not only the relative value of each type of reward, but also the cost of obtaining each type of reward (Rolls 2005a, Rolls and Grabenhorst 2008, Grabenhorst and Rolls 2011). If there is a very high cost of obtaining a particular reward, it may be better, at least temporarily, until the situation changes, to select an action that leads to a smaller reward, but is less costly. It appears that animals do operate according to such a cost–benefit analysis, in that if there is a high cost for an action, that action is less likely to be performed. One example of this comes from the fighting of deer. A male deer is less likely to fight another if he is clearly inferior in size or signalled prowess (Dawkins 1995).

There may also be a cost to switching behaviour. If the sources of food and water are very distant, it would be costly to switch behaviour (and perhaps walk a mile) every time a mouthful of food or a mouthful of water was swallowed. This may be part of the adaptive value of incentive motivation or the 'salted nut' phenomenon – that after one reward is given early on in working for that reward the incentive value of that reward may increase. This may be expressed in the gradually increasing rate of working for food early on in a meal. By increasing the reward value of a stimulus for the first minute or two of working for it, hysteresis may be built into the behaviour selection mechanism, to make behaviour 'stick' to one reward for at least a short time once it is started.

When the mechanisms for cost–benefit analysis are analysed, it is useful to distinguish two types of cost (Grabenhorst and Rolls 2011). One type is *intrinsic costs*, which describe that fact that a stimulus may have both rewarding proper-ties (e.g. a sweet taste) and aversive properties (e.g. a bitter taste). Intrinsic costs, properties of the stimuli, appear to be represented in the human orbitofrontal cortex, where the reward and the punisher value of stimuli are represented simultaneously (Grabenhorst, Rolls, Margot, da Silva and Velazco 2007, Rolls, Grabenhorst and Parris 2008a, Grabenhorst and Rolls 2011). A second type is *extrinsic costs*, the costs of actions required to obtain a rewarding goal stimulus. This action-related cost appears to involve the cingulate cortex (Rushworth, Buckley, Behrens, Walton and Bannerman 2007b, Grabenhorst and Rolls 2011), which receives its intrinsic cost information from the orbitofrontal cortex (Rolls 2005a, Rolls 2009a, Grabenhorst and Rolls 2011).

When one refers to a 'cost–benefit analysis', one does not necessarily mean at all that the animal thinks about it and plans with 'if ... then' multistep linguistic processing the benefits and costs of each possible course of action. Instead, in many cases 'cost–benefit analysis' is likely to be built into animals to be performed with simple implicit processing or rules. One example would be the incentive motivation just described, which provides a mechanism for an animal to persist for at least a short time in one behaviour without having to explicitly plan to do this by performing as an individual a cost–benefit analysis of the relative advantages and costs of continuing or switching. Another example might be the way in which the decision to fight is made by male deer: the decision may be based on simple processes such as reducing the probability of fighting if the other individual is larger, rather than thinking through the consequences of fighting or not on this occasion. Thus, many of the costs and benefits or rewards that are taken into account in the action selection process may in many animals operate according to simply evaluated rewards and costs built in by natural selection during evolution (Krebs and Kacelnik 1991, Dawkins 1995). Animals may take into account, for example, quite complex information, such as the mean and variance of the rewards available from different sources, in making their selection of behaviour, yet the actual selection may then be based on quite simple rules of thumb, such as, 'if resources are very low, choose a reliable source of reward'. It may only be in some animals, for example humans, that explicit, linguistically based multistep cost–benefit analysis can be performed. It is important when interpreting animal behaviour to bear these arguments in mind, and to be aware that quite complex behaviour can result from very simple mechanisms. It is important not to over-interpret the factors that underlie any particular example of behaviour.

Reward and punisher signals provide a common currency for different sensory inputs, and can be seen as important in the selection of which actions are performed. Evolution ensures that the different reward and punisher signals are made potent to the extent that each will be chosen when appropriate. For example, food will be rewarding when hungry, but as hunger falls, the current level of thirst may soon become sufficient to make the reward produced by the taste of water greater than that produced by food, so that water is ingested. If however a painful input occurs or is signalled at any time during the feeding or drinking, this may be a stronger signal in the common currency, so that behaviour switches to that appropriate to reduce or avoid the pain. After the painful stimulus or threat is removed, the next most rewarding stimulus in the common currency might be the taste of water, and drinking would therefore be selected. The way in which a part of the brain such as the striatum and rest of the basal ganglia may contribute to the selection of behaviour by implementing competition between the different rewards and punishers available, all expressed in a common currency, is described elsewhere (Rolls 2005a, Rolls and Grabenhorst 2008, Grabenhorst and Rolls 2011). The functions of the orbitofrontal cortex and cingulate cortex in behavioural switching are considered in Section 3.11.

Many of the rewards for behaviour consist of stimuli, or remembered stimuli. Examples of some primary reinforcers are included in Table 3.1. However, for some animals, evolution has built-in a reward value for certain types of response. For example, it may be reinforcing to a pigeon or chicken to peck. The adaptive value of this is that for these animals, simply pecking at their environment may lead to the discovery of rewards. Another example might be exercise, which may have been selected to be rewarding in evolution because it keeps the body physically fit, which could be adaptive. While for some animals making certain responses may thus act as primary reinforcers (see Glickman and Schiff (1967)), this is likely to be adaptive only if animals operate in limited environmental niches. If one is built to find pecking very rewarding, this may imply that other types of response are less able to be made, and this tends to restrict the animal to an environmental niche. In general, animals with a wide range of behavioural responses and strategies available, such as primates, are able to operate in a wider range of environments, are in this sense more general-purpose, are less likely to find that particular responses are rewarding per se, and are more likely to be able to select behaviour based on which of a wide range of stimuli is most rewarding, rather than based on which response-type they are pre-adapted to select.

The overall aim of the cost–benefit analysis in animals is to maximize fitness. By fitness we mean the probability that an animal's genes will be passed on into the next generation. To maximize fitness, there may be many ways in which different stimuli have been selected during evolution to be rewards and punishers (and among punishers we could include costs). All these rewards and punishers should operate together to ensure that over the lifetime of the animal there is a high probability of passing on genes to the next generation; but in doing this, and maximizing fitness in a complex and changing environment, all these rewards and punishers may be expected to lead to a wide variety of behaviour.

Once language enables rewards and punishers to be intellectualized, so that, for example, solving complex problems in language, mathematics, or music becomes rewarding, behaviour might be less obviously seen as adapted for fitness. However, it was suggested above that the ability to solve complex problems may be one way in which fitness, especially in a changing environment, can be maximized. Thus we should not be surprised that working at the level of ideas, to increase understanding, should in itself be rewarding. These circumstances, that humans have developed language and other complex intellectual abilities, and that natural selection in evolution has led problem-solving to be rewarding, may lead to the very rapid evolution of ideas (see also Section 7.9).

3.10.3 Further functions of emotion

The fundamental function of emotion, to enable an efficient way for the goals for actions to be defined by genes during evolution and to be implemented in the brain, has been described in Sections 3.10.1 and 3.10.2. The simple brain implementation provided as a result of evolution allows the different goals to be selected and compared by using reward and punisher evaluation or appraisal of stimuli, and of the stimuli that may be obtained by different courses of action. This function allows flexibility of the behavioural responses that will be performed to obtain gene-specified goals. Next we consider some further functions and properties of emotion, and also highlight some particularly interesting examples of the types of emotional behaviour that result from the fundamental operation of emotional systems described above.

3.10.3.1 Autonomic and endocrine responses

An additional function of emotion is the elicitation of autonomic responses (e.g. a change in heart rate) and endocrine responses (e.g. the release of adrenaline). It is of clear survival value to prepare the body, for example by increasing the heart rate, so that actions such as running which may be performed as a consequence of the reinforcing stimulus can be performed more efficiently. The neural connections from the amygdala and orbitofrontal cortex via the hypothalamus as well as directly towards the brainstem autonomic motor nuclei may be particularly involved in this function (see Section 3.11). I have presented arguments that feedback of effects from the periphery as postulated in the James–Lange theory and by Damasio are not at all necessary for emotion (Section 3.5.1).

3.10.3.2 Emotional states are motivating

Another function of emotion is that it is motivating. For example, fear learned by stimulus–reinforcer association formation provides the motivation for actions performed to avoid noxious stimuli. Similarly, positive reinforcers elicit motivation, so that we will work to obtain the rewards. Another example where emotion affects motivation is when a reward becomes no longer available, that is frustrative non-reward (see Fig. 3.1). If an action is possible, then increased motivation facilitates behaviour to produce harder working to obtain that reinforcer again or another reinforcer. If no action is possible to obtain again that reward (e.g. after a death in the family), then grief or sadness may result. This may be adaptive, by preventing continuing motivated attempts to regain the positive reinforcer that is no longer available, and helping the animal in due course to therefore be sensitive to other potential reinforcers to which it might be adaptive to switch. If such frustrative non-reward occurs in humans when no action is possible, depression may occur (Clark and Beck 2010).

A depressed state that lasts for a short time may be seen as being adaptive for the reason just given. However, the depression may last for a very long time perhaps because long-term explicit (conscious) knowledge in humans enables the long-term consequences of loss of the positive reinforcer to be evaluated and repeatedly brought to mind as described in Section 6.3, and this may make long-term (psychological) depression maladaptive. Thus a discrepancy between the evolutionary and current environment caused by the rapid development of an explicit system may contribute to some emotional states that are no longer adaptive.

In an interesting evolutionary approach to depression, Nesse (2000a) has argued that humans may set long-term goals for themselves that are difficult to attain, and may spend years trying to attain these goals. An example of such a goal might be obtaining a particular position in one's career, or professional qualification, which may take years of a person's life. If the goal is not attained, then the lack of the reinforcer may lead to prolonged depression. Humans may find it difficult to reorganize their long-term aims to identify other, replacement and more attainable, goals, and without facility at this reorganization of long-term goals, the depression may be prolonged. The evolutionary aspect of this is that with our long-term explicit planning system (described in Section 6.3) and the value that society places on long-term goals and the status that attaining these may confer, humans find themselves in an environmental situation in which their explicit long-term planning system did not evolve, so that it is not well adapted to identifying goals that are realistic. The explicit system then provides a long-lasting non-reward signal to the emotion system (which in addition did not evolve to deal with such long-lasting non-reward inputs), and this contributes to long-lasting depression. A therapeutic solution would be to help depressed people identify possible precipitating factors such as unachieved long-term goals, and readjust their life aims so that positive reinforcers start to be obtained again, helping to lift the person out of the depression.

As described in Section 3.10.5, motivation is an important aspect of an emotional state, but does not require an inherently different mechanism, in that if genes are specifying the reward value of some stimuli ('primary reinforcers'), then the behavioural system must be built to seek to obtain these stimuli (i.e. be motivated to treat them as goals), for otherwise the stimuli would be operationally describable as rewards.

3.10.3.3 Communication

Because of its survival value, the ability to decode signals from other animals as being rewarding or punishing is important. The reward or punisher value may in some cases be innate, and in other cases learned. It may also be adaptive to send such signals, and in some cases the sending of such signals may be 'honest' and in other cases 'deceptive'. These communicated signals may indicate for example the extent to which animals are willing to compete for resources, and they may influence the behaviour of other animals (Hauser 1996). Communicating emotional states may have survival value, for example by reducing fighting.

Darwin (1872), in his book entitled *The Expression of the Emotions in Man and Animals* had as a goal emphasizing the similarity, and therefore the possible phylogenetic closeness, of the expressions of man and his closest living relatives, but nevertheless noted the communicative value of such expressions.

The observation that expressions can evoke a response in the receiver underlies the idea that expressions can also be communicative rather than simply outward signs of affect (Chevalier-Skolnikoff 1973). Expression can be seen as a way of inviting/inducing certain responses in the receiver, much as a smile can appease, a laugh can invite participation, and fear could enlist assistance. In non-human primates, expression is used as a tool with which to regulate and maintain social relations. For example, in the macaque if a subordinate grimaces in an aggressive encounter with a dominant, this signals submission. However, the use of expression is not necessarily so straightforward, for if the same expression were to be given by a dominant individual approaching a subordinate, it no longer signals submission but the positive intention of the dominant. In this way, the communicative effect of expression can be said to be context-dependent, and it depends on the age, sex, dominance and kinship of the senders and receivers (Chevalier-Skolnikoff 1973).

Zeller (1987) also describes what may be said to be the manipulation of social relations through facial expression in non-human primates, that is threat faces are given to coerce another into a desired activity, and friendly expressions are given to enlist co-operation from another.

To argue that the expression of emotion is utilized in social communication, then the ability to decode these signals must be demonstrated. That is, are others able to perceive the content of an individual's expression? Humans have been shown to be remarkable in this ability, even cross-culturally (Ekman 1998), and many would say the happiness in a smile and the anger in furrowed brows are intuitively easy signals to understand.

As social groups are at once both competitive and cooperative, one of the adaptive functions of emotion, and its display, could be to signal, whether honestly or dishonestly, the shifting intentions and dispositions of an individual to another in the social group. Overall, it would be expected that the emphasis on the communicative value of emotion would be greater in species with a socially complex organization, and the evidence described elsewhere (Rolls 2005a) seems to bear this out.

3.10.3.4 Social attachment

Another area in which emotion is important is in social bonding. Examples of this are the emotions associated with the attachment of the parents to their young, with the attachment of the young to their parents, and with the attachment of the parents to each other (see Section 4.4). In the theory of the ways in which the genes affect behaviour ('selfish gene' theory, see Dawkins (1989), Ridley (2003)), it is held that (because, e.g. of the advantages of parental care) all these forms of emotional attachment have the effect that genes for such attachment are more likely to survive into the next generation. Kin-altruism can also be considered in these terms (see e.g. Dawkins (1989) and Section 4.2). In these examples, social bonding is related to primary (gene-specified) reinforcers. In other cases, the emotions involved in social interactions may arise from reinforcers involved in reciprocal altruism, utilizing for example 'tit-for-tat' strategies. In these cases it is crucial to remember which reinforcers are exchanged with particular individuals, so that cheating does not lead to disadvantages for some of those involved. This type of social bonding can be stable when there is a net advantage to both parties in cooperating (see Chapter 9).

Some investigators have argued that the main functions of emotion are in social situations (see Strongman (2003)). While it is certainly the case that many emotions are related to social situations (as can be inferred from Table 3.1), many are not, including for example the fear of snakes by primates, or the fear that is produced by the sight of an object that has produced pain previously.

3.10.3.5 Separate functions for each different primary reinforcer

It is useful to highlight that each primary (gene-specified) reinforcer (of which a large number are suggested in Table 3.1) not only leads to a different set of emotions, but also implements a different function, as described elsewhere (Rolls 2005a).

3.10.3.6 The mood state can influence the cognitive evaluation of moods or memories

Another property of emotion is that the current mood state can affect the cognitive evaluation of events or memories (see Blaney (1986)), and this may have the function of facilitating continuity in the interpretation of the reinforcing value of events in the environment. A theory of how this occurs is presented by Rolls (2005a).

3.10.3.7 Facilitation of memory storage

An eighth function of emotion is that it may facilitate the storage of memories. One way in which this occurs is that episodic memory (i.e. one's memory of particular episodes) is facilitated by emotional states. This may be advantageous in that storage of as many details as possible of the prevailing situation when a strong reinforcer is delivered may be useful in generating appropriate behaviour in situations with some similarities in the future. This function may be implemented in the brain by the relatively non-specific projecting systems to the cerebral cortex and hippocampus, including the cholinergic pathways in the basal forebrain and medial septum (Rolls and Treves 1998, Rolls 1999a, Wilson and Rolls 1990c, Wilson and Rolls 1990b, Wilson and Rolls 1990a, Rolls 2005a), and the ascending noradrenergic pathways (Rolls 2005a).

A second way in which emotion may affect the storage of memories is that the current emotional state may be stored with episodic memories, providing a mechanism for the current emotional state to affect which memories are recalled. In this sense, emotion acts as a contextual retrieval cue, that as with other contextual effects influences the retrieval of episodic memories (Rolls 2008c, Rolls 2010c).

A third way in which emotion may affect the storage of memories is by guiding the cerebral cortex in the representations of the world that are set up. For example, in the visual system, it may be useful to build perceptual representations or analysers that are different from each other if they are associated with different reinforcers, and to be less likely to build them if they have no association with reinforcers. Ways in which backprojections from parts of the brain important in emotion (such as the amygdala) to parts of the cerebral cortex could perform this function are discussed by Rolls (2005a).

3.10.3.8 Emotional and mood states are persistent, and help to produce persistent motivation

A ninth function of emotion is that by enduring for minutes or longer after a reinforcing stimulus has occurred, it may help to produce persistent motivation and direction of behaviour. For example, if an expected reward is not obtained, the persisting state of frustrative non-reward may usefully keep behaviour directed for some time at trying to obtain the reward again.

3.10.3.9 Emotions may trigger memory recall and influence cognitive processing

A tenth function of emotion is that it may trigger recall of memories stored in neocortical representations. Amygdala and orbitofrontal cortex backprojections to cortical areas could perform this for emotion in a way analogous to that in which the hippocampus could implement the retrieval in the neocortex of recent memories of particular events or episodes (Rolls 2008c, Rolls 2011b, Rolls and Kesner 2006). This is thought to operate as follows. When a memory is stored in a neocortical area or hippocampus, any mood state that is present and reflected in the firing of neurons in the orbitofrontal cortex or amygdala will become associated with that memory by virtue of the associatively modifiable synaptic connections from the backprojecting neurons onto the neocortical or hippocampal system neurons. Then later, a particular mood state represented by the firing of neurons in the amygdala or orbitofrontal cortex will by the associatively modified backprojection connections enhance or produce the recall of memories stored when that mood state was present. These effects have been formally modelled by Rolls and Stringer (2001b), and are described further in Section 3.7. One consequence of these effects is that once in a particular mood state, memories associated with that mood state will tend to be recalled and incoming stimuli will be interpreted in the light of the current mood state. The result may be some continuity of emotional state and thus of behaviour. This continuity may sometimes be advantageous, by keeping behaviour directed towards a goal, and making behaviour interpretable by others, but it may become useful in human psychiatric conditions to break this self-perpetuating tendency.

It is useful to have these functions of emotion in mind when considering the neural basis of emotion, for each function is likely to activate particular output pathways from emotional systems associated with it.

3.10.4   The functions of emotion in an evolutionary, Darwinian, context

In this book, the question of why we have emotions is a fundamental issue that I answer in terms of a Darwinian, functional, approach, producing the answer that emotions are states elicited by goals (rewards and punishers), and that this is part of an adaptive process by which genes can specify the behaviour of the animal by specifying goals for behaviour rather than fixed responses. The emotional states elicited with respect to the goals depend on the reinforcement contingencies, as illustrated in Fig. 3.1. The states themselves may be the goals for action, such as reducing fear, and additionally maintain behaviour by being persistent, and act in other ways as described in Section 3.10.3.

This theory of emotion provides I believe a powerful approach to understanding how genes influence behaviour. In much thinking in zoology, an approach has been to understand how genes may determine particular behaviours. For example, Tinbergen (1951, 1963) considered that innate releasing stimuli might elicit fixed action patterns. An example is the herring gull chick's pecking response elicited as a fixed action pattern by the innate releasing stimulus of a red spot on its parent's bill. A successor in this approach in the context of emotion is Panksepp (1998). The instinctive pecking response may be improved by learning (Hailman 1967), but is not an arbitrary, flexible, response as in instrumental, action–outcome, learning (see Section 2.11). The details of the stimulus, or the context in which it occurs, may be taken into account, to influence the instinctive response (Dawkins 1995), but there is still no arbitrary relation between the stimulus and the action, as in instrumental learning.

In contrast, the most important function of emotion that I propose is for genes to specify the stimuli that are the goals for behaviour. This means that the genetic specification can be kept relatively simple, in that it is stimuli that are specified by the genes, such as a taste or touch, and this is generally simpler than specifying the details of a response (such as climbing a tree, running along a branch, picking an apple, and placing it into the mouth). It also means that relatively few genetic specifications are needed, for instead of having to encode many relations between particular stimuli and particular behavioural responses, the genes need to span the dimensionality of the stimulus space of primary reinforcers. Examples of some of these primary, gene-encoded, reinforcers are shown in Table 3.1.

Another way in which the genetic specification required can be kept low is that stimulus–reinforcer association learning can then be used to enable quite arbitrary stimuli occurring in the lifetime of an animal to become associated with primary reinforcers by stimulus–reinforcer association learning, and thus to lead to actions.

But *the most important advantage conferred by emotion* is that the behaviour, in particular the actions, required need not be genetically specified, for arbitrary actions can be learned in the lifetime of the animal by instrumental, action–outcome, learning to obtain or avoid the goals specified by the genes. The actions are arbitrary operants, in that any action may be made to obtain the goal. Thus the genetic specification of the behaviour that emotion allows is one in which the behaviour is not pre-programmed with respect to the stimulus (as in instinctive behaviour such as fixed action patterns), but instead the action is not specified by the genes, and the goals to which actions are directed are specified by the genes. Of course this does not deny that some behavioural responses are genetically specified as responses, and examples might include pecking to particular stimuli in birds, orientation to and suckling of the nipple in mammals, and some examples of preparedness to learn (Rolls 2005a).

Darwinian natural selection of genes that encode the goals for action (i.e. encode reinforcers) rather than the actions themselves, and thus allows great flexibility of the resulting behaviour, can be thought of as liberating 'The Selfish Gene' (Dawkins 1976, Dawkins 1989). When Richard Dawkins wrote *The Selfish Gene* (Dawkins 1976), he was careful to make it clear that the concept that selection and competition operate at the level of genes (Hamilton 1964, Hamilton 1996) does not lead inevitably to genetic determinism of behaviour. Nevertheless the concept was criticised on these grounds, and Dawkins devoted a whole chapter of *The Extended Phenotype* to addressing this further (Dawkins 1982). The concepts developed in *Emotion Explained* help to resolve this further, for I argue that an important way in which genes influence hehaviour (and in doing so produce emotion), is by specifying the reinforcers, the goals for actions, rather than particular behaviours. This helps to avoid the charge that selfish genes 'determine' the behaviour. Instead, many of the genes that influence behaviour operate by competing with each other in a world of reinforcing stimuli or goals for actions, and thus there is great flexibility in the behaviour that results. We are led to think not of behaviours being inherited or 'determined by selfish genes', but instead of genes exploring by natural selection reinforcers that may guide behaviour successfully so that the fitness of the genes is increased. In this sense, the selfish gene (in particular, those involved in specifying reinforcers) is liberated from directly 'determining' behaviour, to providing goals for (instrumental) actions that can involve completely flexible behaviour made to obtain the goal. In these cases, the heritability of behaviour is best understood as the heritability of reinforcers in a stimulus space not in a behavioural or response space.

An interesting consequence of this fundamental adaptive value of emotion that I propose is that the genetic specification does need to include specification for several synapses through the nervous system from the sensory input to the brain region where the reward or punishment value of the goal stimulus is made explicit in the representation. It is thus a prediction that genes specify the connectivity to the stage of processing in the brain where goals are specified, so that appropriate actions can be learned to the goals. Evidence is described in Sections 3.11 and 3.13 that the goals may not be made explicit, that is related to neuronal firing, until stages of information processing such as the orbitofrontal cortex and amygdala. An example of this specification is that sweet taste receptors on the tongue must be connected to neurons that specify food reward, and whose responses are modulated by hunger signals (see Section 3.13).

The definition I provide of emotions, that they are states (with particular functions) elicited by reinforcers, thus is consistent with what I see as the most important function of emotion, that of being part of a design by which genes can specify (some) goals or reinforcers of our actions. This means that the theory of emotion that I propose should not be seen as behaviourist, but instead as part of a much broader theory that takes an adaptive, Darwinian, approach to the functions of emotion, and how they are important in brain design. Further, the theory shows how cognitive states can produce and modulate emotion, and in turn how emotional states can influence cognition (Section 3.7).

I believe that this approach leads to a fundamental understanding of why we have emotions that is likely to stand the test of time, in the same way that Darwinian thinking itself provides a fundamental way of understanding biology and many 'why' questions about life. This is thus intended to be a thoroughly Darwinian theory of the adaptive value of emotion in the design of organisms.

3.10.5 The functions of motivation in an evolutionary, Darwinian, context

Motivation may be seen as a state in which one is working for a goal, and emotion as a state that occurs when the goal, a reinforcer, is obtained, and that may persist afterwards. The concept of gene-specified reinforcers providing the goals for action helps to understand the relation between motivational states and emotion, as the organism must be built to be motivated to obtain the goals, and to be placed in a different state (emotion) when the goal is or is not achieved by the action. The close but clear relation between motivation and emotion is that both involve what humans describe as affective states (e.g. feeling hungry, liking the taste of a food, feeling happy because of a social reinforcer), and both are about goals. The Darwinian theory of the functions of emotion developed in this chapter, which shows how emotion is adaptive because it reflects the operation of a process by which genes define goals for action, applies just as much to motivation. By specifying goals, the genes must specify both that we must be motivated to obtain those goals, and that when the goals are obtained, further states, emotional states with further functions, are produced.

3.10.6 Are all goals for action gene-specified?

Finally in this section, we can ask whether all goals are gene-specified. An important concept of this chapter has been that part of the adaptive value of emotion is that it is part of the process that results from the way in which genes specify reinforcers, that is the goals for action. Emotions may thus be elicited by primary reinforcers, or by stimuli that become associated by learning with primary reinforcers, i.e. secondary reinforcers. But are there goals related to emotional and motivational states that are not related to goals defined in this way by gene-specified reinforcers?

I think it is likely that most reinforcers can be traced back to a gene-specified goal, even if they are in some cases rather general goals. Some examples of these types of reinforcer (and there are likely to be many others) are included in Table 3.1, such as goals for social cooperation and group acceptance, mind reading, and solving an intellectual problem. However, when an explicit, rational, reasoning system capable of syntactic operations on symbols (as described in Section 6.3) evolves, it is possible that goals that are not very directly related to gene specifications become accepted. This may be seen in some of the effects of culture. Indeed, some goals are defined within a culture, for example writing a novel. But it is argued that it is primary reinforcers specified by genes of the general type shown in Table 3.1 that make us want to be recognised in society because of the advantages this can bring, to solve difficult problems, etc., and therefore to perform actions such as writing novels (see further Sections 4.4 and 6.3, Ridley (2003) Chapter 8, Ridley (1993a) pp. 310 ff, Laland and Brown (2002) pp. 271 ff, and Dawkins (1982)).

Indeed, culture is influenced by human genetic propensities, and it follows that human cognitive, affective, and moral capacities are the product of a unique dynamic known as *gene-culture coevolution* (Gintis 2007, Bowles and Gintis 2005, Gintis 2003, Boyd, Gintis, Bowles and Richerson 2003). Nevertheless, there may be cases where the explicit, reasoning, system might specify a goal, and thus lead to emotions, that could not be related to any genetic adaptive value, whether current or specified in evolutionary history. In these cases I would argue that although emotion has evolved and is generally adaptive in relation to gene-specified reinforcers, when the explicit, reasoning, system evolves, this can set up alternative goals that tap into and utilize the existing emotional system for facilitating actions, but with respect to which the goals might be genetically unspecified and even non-adaptive. Examples of such goals are described in Sections 5.4, 9.2 and 13.2 where I discuss the concept of a *selfish phenotype* (Rolls 2011b).

3.11 Brain mechanisms of emotion

Some of the main brain regions implicated in reward and punishment representations and learning, and thus in emotion, will now be considered. These brain regions include the amygdala, orbitofrontal cortex, cingulate cortex, and basal forebrain areas including the hypothalamus, which are shown in Figs. 3.4 and 2.25. Particular emphasis is placed on investigations of the functions of these regions in primates (in practice monkeys or humans), for in primates many areas of the neocortex undergo great development and provide major inputs to these regions, in some cases to parts of these structures thought not to be present in non-primates. Key principles are described here, with more detailed evidence presented elsewhere (Rolls 2005a, Rolls 2008c, Rolls and Grabenhorst 2008, Grabenhorst and Rolls 2011).

A first principle is that the set of structures shown as *Tier 1* in Fig. 2.25 on page 74 *represents the identity of stimuli* and objects in each sensory modality independently of reward value. This enables us to perceive and represent objects and stimuli independently of how much we like them at any particular time. This is very important, for then we can learn about the properties of objects, for example where they are in the world, independently of whether for example they are rewarding or not at the time, which might be influenced by many factors, including motivational states such as hunger. Outputs can be taken from the end of Tier 1 to other brain regions involved in these other processes, including language areas. In the case of vision, it is a feature of brain design that invariant representations for objects are present at the end of Tier 1 (in the inferior temporal visual cortex, IT), for then if we learn that an object or person is rewarding by associations made at the next stage of processing, that learning automatically generalizes to other views, etc. of the object or person.

A second principle is that *the reward (or punishment) value of representations is made explicit in the firing of neurons in Tier 2* in Fig. 2.25. For example, in the primary taste and olfactory cortex in Tier 1 the identity and subjective

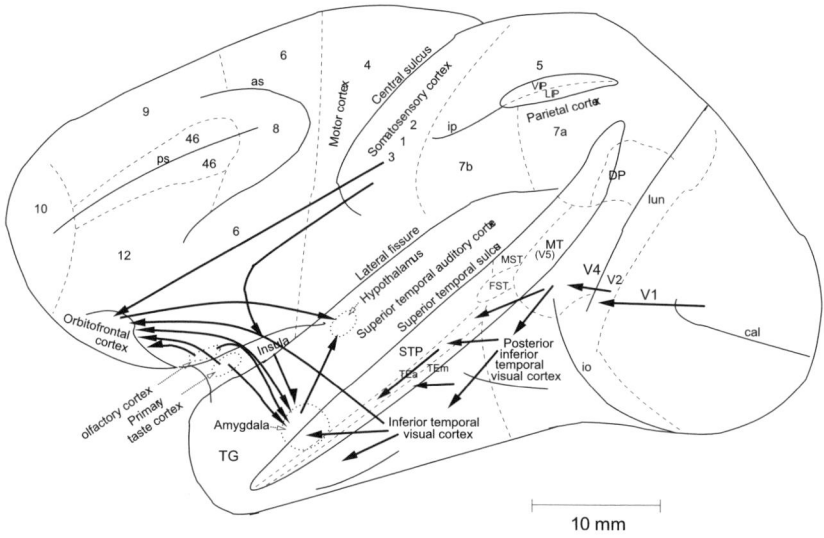

Fig. 3.4 Some of the pathways involved in emotion described in the text are shown on this lateral view of the brain of the macaque monkey. Connections from the primary taste and olfactory cortices to the orbitofrontal cortex and amygdala are shown. Connections are also shown in the 'ventral visual system' from V1 to V2, V4, the inferior temporal visual cortex, etc., with some connections reaching the amygdala and orbitofrontal cortex. In addition, connections from the somatosensory cortical areas 1, 2, and 3 that reach the orbitofrontal cortex directly and via the insular cortex, and that reach the amygdala via the insular cortex, are shown. as, arcuate sulcus; cal, calcarine sulcus; cs, central sulcus; lf, lateral (or Sylvian) fissure; lun, lunate sulcus; ps, principal sulcus; io, inferior occipital sulcus; ip, intraparietal sulcus (which has been opened to reveal some of the areas it contains); sts, superior temporal sulcus (which has been opened to reveal some of the areas it contains). AIT, anterior inferior temporal cortex; FST, visual motion processing area; LIP, lateral intraparietal area; MST, visual motion processing area; MT, visual motion processing area (also called V5); PIT, posterior inferior temporal cortex; STP, superior temporal plane; TA, architectonic area including auditory association cortex; TE, architectonic area including high order visual association cortex, and some of its subareas TEa and TEm; TG, architectonic area in the temporal pole; V1–V4, visual areas V1–V4; VIP, ventral intraparietal area; TEO, architectonic area including posterior visual association cortex. The numbers refer to architectonic areas, and have the following approximate functional equivalence: 1, 2, 3, somatosensory cortex (posterior to the central sulcus); 4, motor cortex; 5, superior parietal lobule; 7a, inferior parietal lobule, visual part; 7b, inferior parietal lobule, somatosensory part; 6, lateral premotor cortex; 8, frontal eye field; 12, part of orbitofrontal cortex; 46, dorsolateral prefrontal cortex.

intensity of the taste and smell are represented, but in the orbitofrontal cortex, a key Tier 2 structure, the reward value of the stimuli, and the subjective pleasantness of the taste and smell, are represented. This is illustrated in Fig. 2.4 on page 15, which shows a taste neuron in the orbitofrontal cortex which responds by increasing its firing rate to the taste of glucose when hunger is present, and which gradually stops responding to the taste of a food (glucose) as its reward value decreases to zero while food is fed to satiety. To emphasize

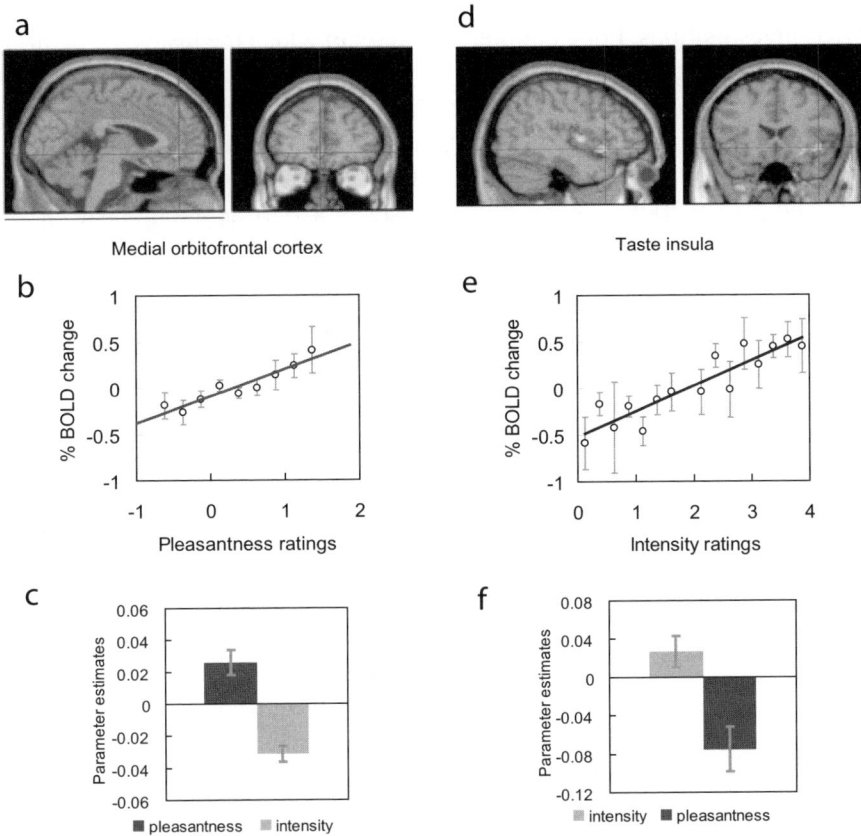

Fig. 3.5 (a-c). Human subjective pleasure related to activations in the medial orbitofrontal cortex and pregenual cingulate cortex. (a). Brain scans with a midline view (left) and coronal (cross-section) view on the right showing the activations in colour. The cursor is located in the pregenual cingulate cortex, and in the midline view the activation can be seen extending back to the medial orbitofrontal cortex. (b). The fMRI BOLD signal is directly proportional in both these brain regions to the subjective pleasantness ratings of the taste. (c). When the humans pay attention to the pleasantness of the taste, there are larger activations in these regions (reflected in the parameter estimates) than when the participants pay attention to the intensity of the identical taste. (d-f). Human subjective intensity ratings are related to activations in Tier 1 structures such as the primary taste cortex in the insula. (d). Brain scans with a parasaggittal (parallel to the midline) view (left) and coronal (cross-section) view on the right showing the activations in colour. The cursor is located in the primary taste cortex in the anterior insula. (e). The fMRI BOLD signal is directly proportional in the taste insula to the subjective intensity ratings of the taste. (f). When the humans pay attention to the intensity of the taste, there are larger activations in the primary taste cortex (reflected in the parameter estimates) than when the participants pay attention to the pleasantness of the identical taste. (After Grabenhorst, Rolls and Parris 2008b.) (See colour plates Appendix B.)

the difference, in Tier 1, neurons in the primary taste cortex are not affected when the reward value of the taste is decreased to zero by feeding to satiety.

This principle is also reflected in human functional magnetic resonance (fMRI) investigations. For example, in a study in which the taste of umami was being delivered into the mouth, it was found that the conscious subjective pleasantness ratings of the taste were directly (linearly) related to the activations, measured by the BOLD (blood oxygenation level dependent) signal, in the orbitofrontal cortex and pregenual cingulate cortex (Fig. 3.5a,b) (Grabenhorst, Rolls and Parris 2008b). (The umami taste is a rich delicious taste of protein that can be produced by monosodium glutamate present in many natural foods including human mothers' milk, tomatoes, mushrooms, meat, and fish (Rolls 2009c).) In contrast, the conscious subjective pleasantness ratings of the intensity of the same taste stimuli were related to the activations in the primary taste cortex in the insula, a Tier 1 brain region (Fig. 3.5d,e) (Grabenhorst et al. 2008b).

Other neurons in the orbitofrontal cortex learn, by association with a primary reinforcer, the reward (or punishment) value of for example a visual stimulus, when it is paired with a primary reward, or is no longer paired with a primary reward. This can be very fast, occurring in one trial (Rolls 2005a, Rolls 2008c). Thus emotions can be learned in one trial to a stimulus, and this is highly adaptive.

A third principle is that *the representation of particular rewards can be highly specific in Tier 2*. For example, the neuron shown in Fig. 2.4 stopped firing to glucose but still responded to fruit juice (blackcurrant juice, BJ) after feeding to satiety with glucose. This type of investigation is how we discovered *sensory-specific satiety* (Rolls 1981, Rolls 2005a), the most important principle that controls food intake in a meal, and which I believe is a property of all reward systems, for it encourages exploration of new rewards after one type of reward has been received for some time. This enables us to obtain for example different nutrients, and more generally a wide range of rewards, which is essential for genes to progress to the next generation. To procreate, we are encouraged by this biological process to eat, drink, sleep, compete to become attractive by demonstrating that we have resources that will provide for our mates, have status that means we can provide protection, and are likely to have similarly attractive children with our mates. This is a key biological process in all reward systems that contributes greatly to culture, by encouraging continuing exploration of new directions of choice that might provide thereby more reward. This process may also be called **reward-specific satiety**. This is a key process underlying fashion and creativity in culture, and is a good example of neuroculture: how understanding neural processes helps us to understand our behaviour, wants, desires, emotions, fashion consciousness, aesthetics, etc. (Chapter 7).

Many examples of how specific the reward representations can be are provided in the sources cited, with different neurons receiving different combinations of the sensory properties of stimuli, and often responding to only particular combinations of the sensory inputs (Rolls 2005a, Rolls 2008c, Rolls and Grabenhorst 2008, Grabenhorst and Rolls 2011). It is the specificity of reward representations that allows sensory-specific satiety to be computed by a computationally simple mechanism: reduction of the sensitivity of the reward-specific neurons by a process such as adaptation over a time period measured typically over minutes, but which doubtless is different for different types of reward, and may vary with their intensity (Rolls 2005a). Rewards must also be highly specific, for we may need to perform one action to obtain one reward, and a different action for another, and the action selection depending on current reward value could only work if each reward is represented in a specific way by different neurons firing. Love, affection, caring, etc. are all stimulus-specific because of this specificity of reward representations in brain structures such as the orbitofrontal cortex (Rolls 2005a).

Drugs of addiction that operate through the dopamine system in the brain, including the psychomotor stimulants amphetamine and cocaine, may be addictive in part because they bypass the stage of processing at which sensory-specific satiety is computed in the brain, and act to influence the firing of reward neurons in brain regions such as the orbitofrontal cortex and a structure to which it projects, the ventral striatum (Rolls 2005a, Rolls 2008c, Voellm, De Araujo, Cowen, Rolls, Kringelbach, Smith, Jezzard, Heal and Matthews 2004, Phillips, Mora and Rolls 1981).

A fourth principle is that *multimodal representations that may be required to represent particular reinforcers are formed in the orbitofrontal cortex*, by associative learning to primary reinforcers if necessary. Different single neurons respond to different combinations of taste, somatosensory including food texture signals from the mouth, olfactory, and visual stimuli.

A fifth principle is that a very *wide range of reinforcers is represented in the orbitofrontal cortex*, so that it is involved in a very wide range of emotions. For example, not only are taste, touch, olfactory and visual stimuli represented in terms of their reinforcer value, but there are representations of actual monetary reward and loss, and expected monetary reward and loss (a type of Expected Value, see Chapter 8) (Rolls 2005a, Rolls and Grabenhorst 2008, Grabenhorst and Rolls 2011). There are also neurons that represent face identity, and others that represent face expression (Rolls et al. 2006a), and the combination may be important in order to determine an appropriate social response to a particular face expression from a particular person. Consistent with this principle, we have shown that patients with damage to the orbitofrontal cortex and behavioural have subjective changes in a wide range of emotions, and are impaired in categorising face and voice expressions (Hornak, Rolls and Wade 1996, Hornak et al. 2003, Berlin et al. 2005).

A sixth principle is that there are *representations of negative reinforcers, experienced as subjectively unpleasant, in the orbitofrontal cortex,* typically more laterally in humans than rewards (Grabenhorst and Rolls 2011). Examples are unpleasant odours, losing money, and unpleasant touch. Not only are inherently aversive stimuli represented in the orbitofrontal cortex, but there is a special system for detecting when an expected reward is not obtained, so that behaviour can be changed. For example, some orbitofrontal cortex neurons respond only when an expected reward is not obtained, and are called error neurons, or *negative reward prediction error neurons* (Thorpe, Rolls and Maddison 1983, Rolls 2009b, Rolls 2011e). Consistently, parts of the lateral orbitofrontal cortex respond in a simple social task when one of two people is selected, but an expected smile is not obtained (Kringelbach and Rolls 2003). Further, patients with damage to the orbitofrontal cortex are especially impaired when they must change their behaviour when an expected reward is not obtained (Rolls, Hornak, Wade and McGrath 1994a, Hornak, O'Doherty, Bramham, Rolls, Morris, Bullock and Polkey 2004, Berlin et al. 2004). The mechanism involved is one which allows behaviour to be switched very quickly, within one trial, and it is very important to survival to have a fast mechanism of this type. To bring the point home: we have an understanding of some of the reasons why people have altered social and emotional behaviour after damage to the orbitofrontal cortex: it is because they may be insensitive to some rewards, but especially that they cannot alter their social and emotional behaviour rapidly if expected rewards are not obtained (Rolls 2005a).

This raises an issue that is very relevant to neuroculture: to unravel the implications for society and culture of our rapidly developing understanding of brain function. Damage to the orbitofrontal cortex is quite common after head injury in road traffic accidents. Our modern understanding helps us with potential therapeutic approaches to the treatment of such patients, including tests that can diagnose any changes that may be present, and suggestions for behavioural strategies that may enable such patients to cope better (Rolls 2005a).

The understanding also raises important questions for society. Should individuals with such brain damage be held fully responsible for their actions? It might seem reasonable that any such brain damage might be taken into account (Chapter 9). But it would be very difficult to extend this to individuals who as a result of inherited genes might have differences from the average on particular capacities in these functions implemented in the brain regions that implement social and emotional behaviour.

A seventh principle is that *cognition, even from the linguistic level, operates by top-down effects to modulate the responsiveness of the orbitofrontal cortex reward system* to incoming rewards. We showed this in a task in which a standard brie-like odour (isovaleric acid) was being delivered, but was accompanied on some trials by a word label describing it as cheddar cheese, and on other trials as body odour. The orbitofrontal cortex reward-related activations to the identical odour were higher on trials on which the positive rather than the negative word label was applied, showing that cognitive factors modulate the reward value representations in the orbitofrontal cortex (De Araujo, Rolls, Velazco, Margot and Cayeux 2005). Cognition descends down into the orbitofrontal cortex to modulate processing at the first brain stage where reward value is represented. In a sense, the representations in the orbitofrontal cortex to which subjective pleasure and reward value are related can be modulated by cognition: the represented reward value changes. We have shown similar effects for taste, flavour, and tactile stimuli (Grabenhorst, Rolls and Bilderbeck 2008a, McCabe, Rolls, Bilderbeck and McGlone 2008, Rolls 2011a).

The implication for neuroculture is that we may be able to enhance subjective pleasure to rewarding, including aesthetic, stimuli by the appropriate top-down cognitive modulation. For example, if we appreciate the full perhaps symbolic and compositional aspects of a work of art, this may be a way for cognition to directly influence the pleasure produced by the art (Chapter 7).

Cognition can of course enhance pleasure in other ways, for example by developing an understanding of what aspects of music can be used to enhance the properties of the reward stimuli. This occurs in Richard Wagner's cognitive understanding of how to build up an auditory sound world that produces almost supernormally intensive emotional stimuli in *Tristan and Isolde*. The more we understand the operation of our brains, the better we may be able to produce enhanced emotional feelings of pleasure, as in opera.

An eighth principle is that we can use *selective attention to enhance (or not) processing in the reward systems.* For example, if one pays attention to the pleasantness of an odour or taste, brain activations in the orbitofrontal and pregenual cingulate cortex are larger than if selective attention is directed to the intensity of the stimulus (Rolls, Grabenhorst, Margot, da Silva and Velazco 2008b, Grabenhorst et al. 2008b). Conversely, if selective attention is paid to the intensity instead of the pleasantness, the brain activations in the primary olfactory or taste cortex to the stimuli are increased relative to the pleasantness condition. This is illustrated for taste in Fig. 3.5c,f, and the mechanisms for this top-down biased activation mechanism of attention are illustrated in Fig. 2.19.

This ability to select attention to affective brain systems, or not, again has important implications for understanding, and adjusting, how we respond to rewarding stimuli, and implies that by selective attention we can enhance the raw emotional effects produced in us by aesthetic and emotional stimuli; or vice versa where this may be important. This is a way that the rational system (which can control top-down attention) can exert some control over the emotional system, and this in turn has interesting implications for ethical and legal issues involving for example responsibility (Chapter 9).

A ninth principle or property of the value system is that there appears to be a *common neural scale in the orbitofrontal cortex for the subjective value of different rewards* (Grabenhorst et al. 2010). The importance of this is that when reward value representations are sent as the inputs to a choice, decision, system, it is easier if the scale for the two rewards is the same, otherwise without special adjustments the decision mechanism might always choose one of the rewards (Rolls and Grabenhorst 2008, Grabenhorst and Rolls 2011).

For a similar reason, we argue (Rolls and Grabenhorst 2008, Grabenhorst and Rolls 2011) that the costs associated with obtaining a particular reward should be subtracted from the reward value to produce a net reward value, for then the net value of a choice is the most straightforward input to a decision-making network of the type described in Section 2.12. The costs in this case refer to the intrinsic costs of a particular rewarding stimulus, for example the fact that a pleasant touch might also if too intense be slightly painful, or a sculpture although inherently beautiful might be somewhat damaged (Rolls and Grabenhorst 2008, Grabenhorst and Rolls 2011).

A 10th important property of the value representation system in the orbitofrontal cortex is that both *the absolute subjective value of a reward, and the relative subjective value of rewards, are separately represented* (Grabenhorst and Rolls 2009). The utility of a representation of absolute value is that this provides a basis for long-term economic choice (Glimcher, Camerer, Fehr and Poldrack 2009). The utility of the representation of relative value is that this is a useful first step towards a decision between two rewards on an individual trial.

An 11th very interesting principle of how emotions are implemented in the brain is that *the representation of value (in the orbitofrontal cortex in Tier 2 in Fig. 2.25) is kept separate from the computation of the choice* between a set of values on a particular trial which takes place more anteriorly in the medial prefrontal cortex area 10, in Tier 3 in Fig. 2.25 on page 74. The utility of this is that a stable representation of value itself is provided even when on a particular trial we may choose one of two rewarding stimuli, and on another trial we may choose the other. This separation means that we can do long-term planning based on the absolute value of a reward on a continuous scale at the same time as on a particular trial we have to make a binary choice between a pair of rewards. Consistent with this, humans can rate the absolute value

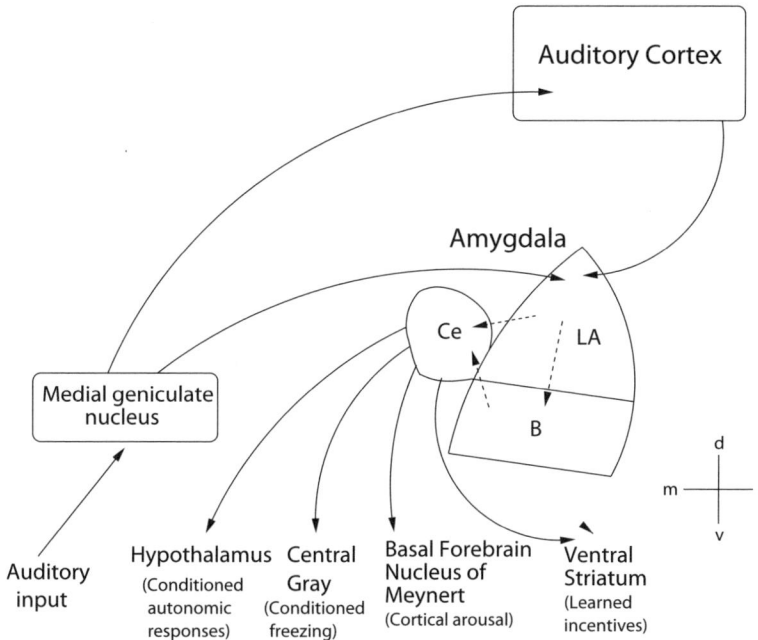

Fig. 3.6 The pathways for fear-conditioning to pure-tone auditory stimuli associated with footshock in the rat (after Quirk, Armony, Repa, Li and LeDoux 1996). The lateral amygdala (LA) receives auditory information directly from the medial part of the medial geniculate nucleus (the auditory thalamic nucleus), and from the auditory cortex. Intra-amygdala projections (directly and via the basal and basal accessory nuclei, B) end in the central nucleus (Ce) of the amygdala. Different output pathways from the central nucleus and the basal nucleus mediate different conditioned fear-related effects. d, dorsal; v, ventral; m, medial; l, lateral.

of a reward when a few seconds later they may or may not be selecting it when faced with a choice between it and another reward (Grabenhorst, Rolls and Parris 2008c, Rolls, Grabenhorst and Parris 2010d, Rolls, Grabenhorst and Deco 2010b, Rolls, Grabenhorst and Deco 2010c). The actual mechanisms for the choice are described in Section 2.12.

Another reason for separating choice from value representations is that value representations are on a continuous scale, whereas choice involves categorical decisions (this one or that one) (Rolls and Deco 2010). The choice mechanism is influenced by (neuronal spiking) noise in the brain, and this has many advantages, including the fact that probabilistic choice results in the less favoured stimulus being chosen sometimes, which can enable changes in what is actually available in the world to be revealed in an uncertain world (Section 2.12).

A 12th principle is that evolutionarily older structures than the orbitofrontal and pregenual cingulate cortex are also involved in emotion (illustrating the general principle of brain design that new systems are often added on top of older systems, which do not totally disappear, though their functions may change). An example is provided by *the amygdala, which is also involved in emotion* (Fig. 3.4 and Fig. 2.25). The amygdala is much older evolutionarily than the orbitofrontal cortex (the amygdala is present for example in reptiles), and is a useful model system for studying emotion and especially fear (Fig. 3.6) (Quirk, Armony, Repa, Li and LeDoux 1996, Pare, Quirk and LeDoux 2004, Davis 2006, Antoniadis, Winslow, Davis and Amaral 2009, Johansen, Tarpley, LeDoux and Blair 2010). However with the great development of the neocortex including the orbitofrontal and cingulate cortex in primates including humans, the amygdala is much less important in humans than the cortical structures such as the orbitofrontal and anterior cingulate cortex in emotion (Rolls 2005a, Grabenhorst and Rolls 2011). Indeed, in humans the amygdala appears to be important mainly for some fear responses to some stimuli, such as whether an individual backs off in a social encounter (Adolphs 2003, Adolphs, Gosselin, Buchanan, Tranel, Schyns and Damasio 2005, Phelps 2004, Schiller et al. 2010, Feinstein, Adolphs, Damasio and Tranel 2011).

3.12 Brain mechanisms of value-based decision-making

The mechanisms that are being unravelled for the mechanisms by which the brain makes decisions are described in Section 2.12. When the choices are between the reward value of stimuli, the relevant brain area appears to be the medial prefrontal cortex area 10 in Tier 3, just in front of the value representations in Tier 2 in the orbitofrontal cortex (Fig. 2.25) (Grabenhorst, Rolls and Parris 2008c, Rolls, Grabenhorst and Parris 2010d, Rolls, Grabenhorst and Deco 2010b, Rolls, Grabenhorst and Deco 2010c, Grabenhorst and Rolls 2011). As noted above, the intrinsic costs of a particular rewarding stimulus need to be evaluated (by subtracting the intrinsic cost from the reward value) in an intrinsic cost–benefit analysis before the net reward value enters the value-based decision process (Rolls and Grabenhorst 2008, Grabenhorst and Rolls 2011). I emphasize that the choice described above is between stimuli of different value, and this is what is represented in the orbitofrontal cortex, not actions or behavioural responses (Rolls 2005a, Rolls and Grabenhorst 2008).

When the action must be selected to obtain a reward, then a different brain area, in which actions are represented, is involved, the cingulate cortex (Rushworth et al. 2007b, Rushworth et al. 2011). There are projections from the orbitofrontal cortex to the anterior cingulate cortex, which contains representations of rewards and (intrinsic) punishers or intrinsic costs in its anterior part (Rolls 2005a, Rolls 2009a, Grabenhorst and Rolls 2011). The evidence

Obesity: sensory and cognitive factors
that make food increasingly palatable
may over-ride existing satiety signals

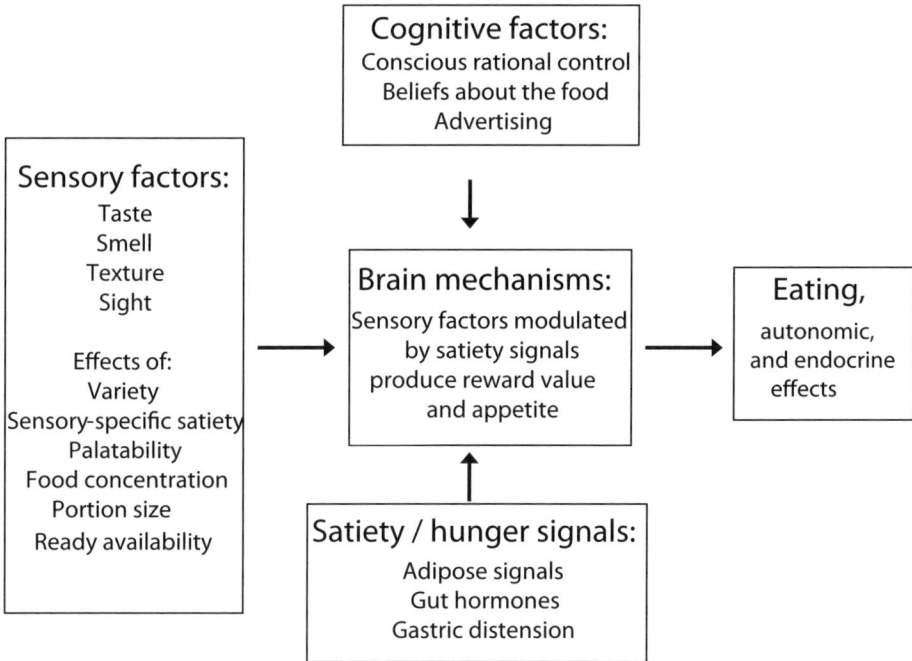

Cognitive factors:
Conscious rational control
Beliefs about the food
Advertising

Sensory factors:
Taste
Smell
Texture
Sight

Effects of:
Variety
Sensory-specific satiety
Palatability
Food concentration
Portion size
Ready availability

Brain mechanisms:
Sensory factors modulated
by satiety signals
produce reward value
and appetite

Eating,
autonomic,
and endocrine
effects

Satiety / hunger signals:
Adipose signals
Gut hormones
Gastric distension

Fig. 3.7 Schematic diagram to show how sensory factors interact in the orbitofrontal cortex with satiety signals to produce the hedonic, rewarding value of food, which leads to appetite and eating. Cognitive and attentional factors directly modulate the reward system in the brain.

indicates that this value information is then projected backwards to the mid-cingulate cortex where action-outcome learning is implemented to select the correct action (Rushworth et al. 2007b, Rolls 2008c, Rolls 2009a, Grabenhorst and Rolls 2011, Rushworth et al. 2011).

3.13   Hunger, the desire to eat, and obesity

3.13.1   Food reward and appetite

The overall brain circuitry involved in the control of food intake is shown in Fig. 2.25 (Rolls 2005a). In Tier 1 food-related stimuli are represented in terms of their identity in the different sensory modalities. Then in Tier 2 the representation is converted into one of the food reward value of the stimuli. For example, neurons in the orbitofrontal cortex represent food reward value in that they respond to the taste, smell, and sight of food only if hunger is present (see example in Fig. 2.4) (Rolls, Sienkiewicz and Yaxley 1989,

Critchley and Rolls 1996a, Kringelbach, O'Doherty, Rolls and Andrews 2003). Indeed, sensory-specific satiety, which evolved partly to promote variety in nutrition, is implemented in the orbitofrontal cortex. Satiety signals, which arise from gastric distension, gut stimulation by food, the absorbed food, and related changes in hormones, reach the hypothalamus and orbitofrontal cortex to help implement these effects. Once food reward has been computed in the orbitofrontal cortex, these signals are sent to the hypothalamus, and thereby produce peripheral autonomic and endocrine changes including an increase in insulin release which promotes the storage of energy from the absorbed food (Fig. 2.25). The food reward signals can also influence action by the habit system involving the basal ganglia, and the goal-directed action system involving action-outcome learning via the cingulate cortex (Fig. 2.25).

A fundamental concept towards which this neuroscience leads about some of the major causes of obesity is that, over the last 30 years, sensory stimulation produced by the taste, smell, texture and appearance of food, as well as its availability, have increased dramatically, yet the satiety signals produced by stomach distension, satiety hormones, etc. have remained essentially unchanged. The resulting effect on the brain's control system for appetite (shown in Fig. 3.7) is to lead to a net average increase in the reward value and palatability of food which over-rides the satiety signals, and contributes to the tendency to be overstimulated by food and to overeat.

In this scenario, it is important to understand much better the rules used by the brain to produce the representation of the pleasantness of food and how the system is modulated by eating and satiety. This understanding, and how the sensory factors can be designed and controlled so as not to override satiety signals, are important research areas in the understanding, prevention, and treatment of obesity.

3.13.2 Food reward and obesity

These concepts and related research are leading towards the following understanding of the factors that contribute to obesity (Rolls 2005a, Rolls 2011i). An important point is that as any one of the following factors may tend to lead to overeating and obesity, it may be important to correct behaviour, at least in some individuals, so that all factors are addressed in order to successfully avoid overeating and obesity.

Understanding the mechanisms that control appetite is becoming an increasingly important issue, given the increasing incidence of obesity (a three-fold increase in the UK since 1980 to a figure of 20% defined by a Body Mass Index > 30) and the realization that it is associated with major health risks (with 1000 deaths each week in the UK alone attributable to obesity). It is important to understand and thereby be able to minimize and treat obesity because many diseases are associated with a body weight that is much above normal. These diseases include hypertension, cardiovascular disease, hypercholesterolaemia, and gall bladder disease; and in addition obesity is associated with some deficits in reproductive function (e.g. ovulatory failure), and with an excess mortality from certain types of cancer (Rolls 2011i).

1. Genetic factors.

These are of some importance, with some of the variance in weight and resting metabolic rate in a population of humans attributable to inheritance (Morton, Cummings, Baskin, Barsh and Schwartz 2006, O'Rahilly 2009). However the 'obesity epidemic' that has occurred since 1990 cannot be attributed to genetic changes, for which the time scale is far too short, but instead to factors such as the increased palatability, variety, and availability of food (as well as less exercise) which are some of the crucial drivers of food intake and the amount of food that is eaten and that are described below.

2. Endocrine factors, and their interaction with brain systems.

A small proportion of cases of obesity can be related to gene-related dysfunctions of the peptide systems in the hypothalamus, with for example 4% of obese people having deficient (MC4) receptors for melanocyte stimulating hormone (Morton et al. 2006, O'Rahilly 2009). Cases of obesity that can be related to changes in the leptin hormone satiety system are very rare. Further, obese people generally have high levels of leptin, so leptin production is not the problem, and instead leptin resistance (i.e. insensitivity) may be somewhat related to obesity, with the resistance perhaps related in part to smaller effects of leptin on hypothalamic arcuate nucleus neurons. However, although there are similarities in fatness within families, these are as strong between spouses as they are between parents and children, so that these similarities cannot be attributed to genetic influences, but presumably reflect the effect of family attitudes to food and weight.

3. Brain processing of the sensory properties and pleasantness of food.

The way in which the sensory factors produced by the taste, smell, texture and sight of food interact in the brain with satiety signals (such as gastric distension and satiety-related hormones) to determine the pleasantness and palatability of food, and therefore whether and how much food will be eaten, is described above and shown in Fig. 3.7. The concept is that convergence of sensory inputs produced by the taste, smell, texture and sight of food occurs in the orbitofrontal cortex to build a representation of food flavour. The orbitofrontal cortex is where the pleasantness and palatability of food are represented, as shown by the discoveries that these representations of food are only activated if hunger is present, and correlate with the subjective pleasantness of the food flavour (see above). The orbitofrontal cortex representation of whether food is pleasant (which takes into account any satiety signals present) then drives brain areas such as the striatum and cingulate cortex that then lead to eating behaviour.

The fundamental concept this leads to about some of the major causes of obesity is that, over the last 30 years, sensory stimulation produced by the taste, smell, texture and appearance of food, as well as its availability, have increased dramatically, yet the satiety signals produced by stomach distension, satiety hormones, etc. have remained essentially unchanged, so that the effect on the brain's control system for appetite (shown in Figs. 3.7 and 2.25) is to lead to a net average increase in the reward value and palatability of food which overrides the satiety signals, and contributes to the tendency to be overstimulated by food and to overeat.

In this scenario, it is important to understand much better the rules used by the brain to produce the representation of the pleasantness of food and how the system is modulated by eating and satiety. This understanding, and how the sensory factors can be designed and controlled so as not to override satiety signals, are important research areas in the understanding, prevention, and treatment of obesity. Advances in understanding the receptors that encode the taste and olfactory properties of food (Buck 2000, Zhao, Zhang, Hoon, Chandrashekar, Erlenbach, Ryba and Zucker 2003), and the processing in the brain of these and related properties (Rolls 2005a, Rolls 2011i, Rolls 2011g, Rolls 2012a), are also important in providing the potential to produce highly palatable food that is at the same time nutritious and healthy.

An important aspect of this hypothesis is that different humans may have reward systems that are especially strongly driven by the sensory and cognitive factors that make food highly palatable. In a test of this, we showed that activation to the sight and flavour of chocolate in the orbitofrontal and pregenual cingulate cortex were much higher in chocolate cravers than non-cravers (Rolls and McCabe 2007). This concept that individual differences in responsiveness to food reward are reflected in brain activations in regions related to the control food intake (Rolls and McCabe 2007) may provide a way for understanding and helping to control food intake.

4. Food palatability.

A factor in obesity (as shown under point 3) is food palatability, which with modern methods of food production can now be greater than would have been the case during the evolution of our feeding control systems. These brain systems evolved so that internal signals from for example gastric distension and glucose utilization could act to decrease the pleasantness of the sensory sensations produced by feeding sufficiently by the end of a meal to stop further eating. However, the greater palatability of modern food may mean that this balance is altered, so that there is a tendency for the greater palatability of food to be insufficiently decreased by a standard amount of food eaten, so that extra food is eaten in a meal (see Fig. 3.7).

5. Sensory-specific satiety, and the effects of variety on food intake.

Sensory-specific satiety is the decrease in the appetite for a particular food as it is eaten in a meal, without a decrease in the appetite for different foods (Rolls 2005a), as shown above. It is an important factor influencing how much of each food is eaten in a meal, and its evolutionary significance may be to encourage eating of a range of different foods, and thus obtaining a range of nutrients. As a result of sensory-specific satiety, if a wide variety of foods is available, overeating in a meal can occur. Given that it is now possible to make available a very wide range of food flavours, textures, and appearances, and that such foods are readily available, this variety effect may be a factor in promoting excess food intake.

6. Fixed meal times, and the availability of food.

Another factor that could contribute to obesity is fixed meal times, in that the normal control of food intake by alterations in inter-meal interval is not readily available in humans, and food may be eaten at a meal-time even if hunger is not present (Rolls 2005a). Even more than this, because of the high and easy availability of food (in the home and workplace) and stimulation by advertising, there is a tendency to start eating again when satiety signals after a previous meal have decreased only a little, and the consequence is that the system again becomes overloaded.

7. Food saliency, and portion size.

Making food salient, for example by placing it on display, may increase food selection particularly in the obese (Schachter 1971, Rodin 1976), and portion size is a factor, with more being eaten if a large portion of food is presented (Kral and Rolls 2004), though whether this is a factor that can lead to obesity and not just alter meal size is not yet clear. The driving effects of visual and other stimuli, including the effects of advertising, on the brain systems that are activated by food reward may be different in different individuals, and may contribute to obesity.

8. Energy density of food.

Although gastric emptying rate is slower for high energy density foods, this does not fully compensate for the energy density of the food. The implication is that eating energy dense foods (e.g. high fat foods) may not allow gastric distension to contribute sufficiently to satiety. Because of this, the energy density of foods may be an important factor that influences how much energy is consumed in a meal (Kral and Rolls 2004). Indeed, it is notable that obese people tend to eat foods with high energy density, and to visit restaurants with high energy density (e.g. high fat) foods. It is also a matter of clinical experience that gastric emptying is faster in obese than in thin individuals, so that gastric distension may play a less effective role in contributing to satiety in the obese. It is also important to remember that the flavour of a food can be conditioned to its energy density, leading over a few days to more eating of low than high energy dense foods, in the phenomenon known as conditioned appetite and satiety (Rolls 2005a).

9. Eating rate.

A factor related to the effects described above in point 8 is eating rate, which is typically fast in the obese, and may provide insufficient time for the full effect of satiety signals as food reaches the intestine to operate.

10. Stress.

Another potential factor in obesity is stress, which can induce eating and could contribute to a tendency to obesity. (In a rat model of this, mild stress in the presence of food can lead to overeating and obesity. This overeating is reduced by antianxiety drugs.)

11. Food craving.

Binge eating has some parallels to addiction. In one rodent model of binge eating, access to sucrose for several hours each day can lead to binge-like consumption of the sucrose over a period of days (Rolls 2010b). The binge-eating is associated with the release of dopamine. This model brings binge eating close to an addictive process, at least in this model, in that after the binge-eating has become a habit, sucrose withdrawal decreases dopamine release in the ventral striatum (a part of the brain involved in addiction to drugs such as amphetamine), altered binding of dopamine to its receptors in the ventral striatum is produced, and signs of withdrawal from an addiction occur including teeth chattering. In withdrawal, the animals are also hypersensitive to the effects of amphetamine.

12. Energy output.

If energy intake is greater than energy output, body weight increases. Energy output is thus an important factor in the equation. A lack of exercise, or the presence of high room temperatures, may tend to limit energy output, and thus contribute to obesity. It should be noted though that obese people do not generally suffer from a very low metabolic rate: in fact, as a population, in line with their elevated body weight, obese people have higher metabolic rates than normal weight humans (at least at their obese body weights: it might be interesting to investigate this further).

13. Cognitive factors, and attention.

As shown above, cognitive factors, such as preconceptions about the nature of a particular food or odour, can reach down into the olfactory and taste system in the orbitofrontal cortex which controls the palatability of food to influence how pleasant an olfactory, taste, or flavour stimulus is (De Araujo, Rolls, Velazco, Margot and Cayeux 2005, Grabenhorst, Rolls and Bilderbeck 2008a). This has implications for further ways in which food intake can be controlled by cognitive factors, and this needs further investigation.

For example, the cognitive factors that have been investigated in these studies are descriptors of the reward value of the food, such as 'rich and delicious'. But it could be that cognitive descriptions of the consequences of eating a particular food, such as 'this food tends to increase body weight', 'this food tends to alter your body shape towards fatness', 'this food tends to make you less attractive', 'this food will reduce the risk of a particular disease', etc., could also modulate the reward value of the food as it is represented in the orbitofrontal cortex. If so, these further types of cognitive modulation could be emphasized in the prevention and treatment of obesity.

Further, attention to the affective properties of food modulates processing of the reward value of food in the orbitofrontal cortex (Rolls, Grabenhorst, Margot, da Silva and Velazco 2008b, Grabenhorst, Rolls and Parris 2008b), and this again suggests that how attention is directed may be important in the extent to which food over-stimulates food intake. Not drawing attention to the reward properties of food, or drawing attention to other properties such as its nutritional value and energy content, could reduce the activation of the brain's reward system by the food, and could be another useful way to help prevent and treat obesity.

Thus research in modern neuroscience and related areas is leading to indications of how the tendency in Western culture to overeat and become obese may be controlled (Rolls 2011i).

Overall, in this chapter we have seen not only how to define emotion and understand its adaptive value in a Darwinian, evolutionary, context. We have also built a foundation for understanding how reward-based systems in the brain help us to understand the cultural phenomena of aesthetics and ethics, and issues such as free will, which are considered in later chapters.

4 Neurosociality

4.1 Introduction

What are the fundamental bases of social behaviour? How is social behaviour adaptive, and is it adaptive in different ways in women and men? What are the roles of kin altruism and reciprocal altruism in promoting cooperation within a society, and what are the factors that promote competition between individuals with a society, and between societies? We consider these issues, and then move on to consider the neurobiological adaptations that promote pair-bonding, love for one's partner, reproduction, and love for one's children.

4.2 Altruism

An important process in social behaviour is altruism, and we consider now how this leads to neurobiologically selected mechanisms for behavioural heuristics including love of one's partner and children, reciprocation, trust, honesty, fairness-detection, defence of reputation, and even altruistic punishment.

There are neurobiological forces that can promote altruism because altruism can be to the advantage of an individual's genes, and they may operate unconsciously. Those mechanisms are described here. In addition, the reasoning system (Chapter 5) might produce behaviour that is altruistic with no advantage to the individual, or the individual's genes. However, even in these cases the altruistic behaviour might be shaped by neurobiological factors that encourage membership of a group and sociality. In addition, unless the altruism is anonymous, one's reputation may be enhanced through processes such as the attractiveness of wealth being made evident by conspicuous consumption.

4.2.1 Kin altruism

Kin altruism is altruism to genetically related individuals (Dawkins 1989). Kin altruism promotes the likelihood that the kin (who contain some of one's genes, and are likely to share the genes for kin altruism) survive and reproduce. This tends to produce supportive behaviour towards individuals likely to be related, especially towards children, grandchildren, siblings, etc., depending on how closely they are genetically related[15]. Richard Dawkins (1976, 1989) used the term 'Selfish Gene' to refer to this behaviour which although apparently altruistic is performed for the advantage of some of one's own genes, which can thus be considered as 'selfish' (in a non-anthropomorphic sense). A good example is the behaviour of worker bees, which work selfishly for their queen bee, because they cannot reproduce, but she has very similar genes (Dawkins 1976, Dawkins 1995). The altruism implied by love must have its origins in selfish genes, which shape human behaviour to in this case produce a state that promotes the production of and survival of offspring. Kindness may also be related to kin altruism (Hamilton 1964), and the etymology of 'kindness' (Middle English kin, relation) is telling.

Kin altruism is a powerful force. There is typically almost no critical thought given to the idea that the 'right' thing to do with at least part of one's wealth is to leave it to one's children.

Male lions use infanticide to remove offspring in a newly acquired pride that are not genetically related to the male coalition. Infanticide is advantageous to the incoming males in that the females become fertile sooner and have cubs by the incoming males more quickly (Packer and Pusey 1983). While there may be no direct parallels in human behaviour, this does illustrate the potential importance of kinship and kin altruism which has shaped behaviour in the past, and having understood this potential effect that can operate in some species, it can be taken into account by the reasoning system.

4.2.2 Reciprocal altruism

Some animals, including primates, co-operate with others in order to achieve ends that turn out to be on average to their advantage, including genetic advantage (Trivers 1971). There is sometimes a 'tit-for-tat' reciprocation. One example includes the coalitions formed by male baboons in order to obtain a female, followed by reciprocation of the good turn later (Ridley 1996, Buss 2008). This is an example of reciprocal altruism, in this case by non-human primates, which is to the advantage of both individuals provided that neither individual cheats, in which case the rules for social interaction must change to keep the strategy stable. For reciprocal altruism, the individuals must recognise each other so that the reciprocation can be returned. Although reciprocal altruism is not very common in non-humans with not many well-established examples (Ridley 1996, Buss 2008, Silk 2009), it does appear to be common in humans (Barkow, Cosmides and Tooby 1992, Cosmides and Tooby 1999).

[15] Kin selection genes spread because of kin altruism. Such genes direct their bodies to aid relatives because those relatives have a high chance of having the same relative-helping gene. This is a specific mechanism, and it happens to be incorrect to think that genes direct their bodies to aid relatives because those bodies 'share genes' in general (see Hamilton (1964); and the chapter on inclusive fitness in M.S.Dawkins (1995)).

The ability to be able to detect that the other partner is still reciprocating is important for stability of the reciprocation, and the ability to detect a *cheater or defector* and to be sensitiveness to *fairness* may be a biological heuristic that has evolved genetically to help the stability of social interactions including reciprocal altruism (Cosmides and Tooby 1999, Fehr 2009).

Similarly, *honesty* may be a behavioural heuristic that has evolved because of the advantages that can be provided (in terms of eventual reproductive success) by reciprocal altruism. In these situations, a single cheating incident, slip, or lie can be very costly to the reputation of the individual, and this is part of the biological underpinning of honesty. It also accounts for the importance to an individual of protecting their *reputation*, and the individual's sensitivity to reputation, both of which may be biologically in-built (Barrett et al. 2002).

Similarly, *forgiveness* may be seen as an adaptive behavioural characteristic that can help the stale-mate in a tit-for tat reciprocation that has failed when one then the other partner cheats or defects, for occasional forgiveness may help to restart the tit-for tat reciprocation, to the advantage of both individuals (Ridley 1996).

Trust may have evolved to help with stable reciprocal interaction (Fehr 2009). Very interestingly trusting bonds between individuals in economic interactions (Fehr 2009) may be promoted by the release of the hormone oxytocin produced when some rewards are provided, including some social and sexual rewards (Section 4.5) (Carter 1998, Insel and Young 2001, Winslow and Insel 2004, Kosfeld, Heinrichs, Zak, Fischbacher and Fehr 2005).

Another example is *altruistic punishment*, in which an individual will suffer personal loss in order to punish a cheater or defector, but which may be a stable strategy that has evolved because of the importance of removing cheaters and defectors from a society where reciprocal altruism is a possibility (Fehr and Gächter 2002, Fehr 2009).

In all these cases, neural mechanisms have evolved to provide behavioural heuristics to promote reciprocal altruism. In some of these examples, a key component is the ability to detect when an expected reward is not provided or returned, and this type of mechanism is present in the (especially lateral) orbitofrontal cortex (Rolls 2005a). This system for non-reward or punishment appears to have evolved to be able to detect many different varieties of non-reward, including those that can occur in reciprocal altruism situations. Consistent with this, there are neurons in the orbitofrontal cortex that are specialized for different types of non-reward (Thorpe, Rolls and Maddison 1983).

It is likely that there are interesting individual differences in the neural mechanisms for these social propensities (Fehr 2009), consistent with ideas advanced here that there are neurobiologically selected heuristics for different aspects of social behaviour that show variation across individuals as part of the process of evolution by natural selection.

In small societies, it is likely that individuals in that society will have many of the same genes, and rules such as 'help your neighbour' (but 'make war with "foreigners" ') will probably be to the advantage of one's genes. This is likely to be based in part on kin altruism. It is fascinating that in countries such as England it was until relatively recently in evolutionary terms, 100 years or a few generations ago, the case that the majority of individuals would spend almost all their lives within a 20-mile radius of their village, and would find a partner to love and have children with within that population. Thus kin altruism could have been an important heuristic when people in one's neighbourhood are part of the 'tribe', and when favouring them compared to 'foreigners' might have been genetically successful.

Reciprocal altruism could also thrive well in small societies such as this, where the number of individuals to keep track of for reciprocation might be a few tens or hundreds. However, when the society increases in size beyond a small village (in the order of 1000), then it may no longer be possible to keep track of individuals in order to maintain the stability of 'tit-for-tat' co-operative social strategies between individuals (Dunbar 1996, Ridley 1996)[16]. Nevertheless, because we have the built-in heuristics for kin and reciprocal altruism, altruistic behaviour may still occur today when the situations in which it evolved no longer apply. For example, individuals may feel that they should help individuals in distant countries when a disaster strikes, in part because these kin and reciprocal altruistic heuristics become engaged (even when there may be little likelihood of genetical advantage because the genes of those distant are likely to be dissimilar), and because those distant may not be able to know who gave to them and might not be able to reciprocate.

4.2.3 Stakeholder altruism

Many cases of apparent altruism may also be explained by interdependence: many apparently altruistic acts have as their beneficiary an individual in whose welfare the altruist has some interest or stakeŠ (Roberts 2005). Where this is true, altruists can be expected to benefit as a secondary consequence of their behaviour. This means that what may appear as being altruistic could in fact be in one's own long-term best interests. In this way, cooperation could be favoured without reciprocation.

These benefits from stakeholder altruism could arise as a passive consequence of the recipient's welfare, where welfare is a short-hand for the recipient's survival in such a state as to provide such an effect. For there to be such

---

[16] A limit on the size of the group for reciprocal altruism might be set by the ability both to have direct evidence for and remember person–reinforcer associations for large numbers of different individual people. In this situation, reputation passed on verbally from others who have the direct experience of whether an individual can be trusted to reciprocate might be a factor in the adaptive value of language and gossip (Dunbar 1996, Dunbar 1993).

secondary benefits, individuals must be interdependent. This condition is met in many social groups: it is a widespread phenomenon that increasing group size brings fitness benefits, most notably in terms of reduced predation risk, for example through increased vigilance. Another widespread example where individuals are interdependent is in reproductive partnerships. Where rearing offspring depends on contributions from both sexes, the partners' fitnesses will be interdependent and so they can be said to have a stake in each other. For example, a husband may help his wife because doing so may increase his fitness (i.e. the survival of his genes) as a result of the resources provided for his wife. This is different from reciprocal altruism, because there can be an advantage even if the wife does not reciprocate to the husband.

In more detail, the degree of interdependence can be formalized by defining an individual's 'stake' in another as the dependence of its fitness on that of the other. This provides a means of valuing others. Interdependence means that cooperators can benefit as a secondary consequence of helping their recipients. Altruism can then be favoured when its costs are outweighed by the altruist's stake in the recipient's benefits. Whereas the problem of potential exploitation (e.g. by not reciprocating) makes reciprocal altruism inherently unstable, co-operation through interdependence can be stable because whatever others do, it is best to cooperate.

Although stakeholder altruism has been analyzed mainly at the level of individuals (Roberts 2005), it may be a force that promotes tribalism, by encouraging individuals to contribute 'altruistically' to a society which, when one is a member of it, brings benefits to the individual, even if not by reciprocation between particular individuals.

Kin altruism can be considered as a special case of stakeholder altruism in which the stake is at the genetic level, and reciprocal altruism as a special case in which the stake is in a particular individual (Roberts 2005).

4.3   Competition within society, and between societies, in 'tribalism'

We have seen that kin and reciprocal altruism are forces that promote cooperation between individuals. There are many forces that produce the opposite, competition. One such force is intra-sexual selection, which for example promotes competition between males so that they can become strong, wealthy, powerful, and providers for women who compete with each other to obtain large, but also reliable, resources for their children, that is, for their genes (see Section 7.4). The creation of wealth and resources by individuals competing for limited resources may also operate by adaptive selection, by providing the opportunity for gene-specified pleasures such as good-tasting and nutritious food, warmth, and shelter that may be useful for the individuals in promoting their survival and hence reproductive success.

Together, the neural adaptations to social behaviour promote tribalism, which can be conceived as cooperation within a group promoted by kin and / or reciprocal and / or stakeholder altruism; and competition with other groups. This competition will be stimulated by limited or unequal resources. Women may have played a part in promoting this competition, by selecting men who do well in this competition. (Of course, most of these processes are likely to operate unconsciously.) These forces then lead to tribalism, and on the larger scale to war.

Tribalism may thus be seen as a feature of society that arises for biological reasons with neural systems appropriately adapted by natural selection to produce cooperation within the tribe and competition between tribes. Although reciprocal altruism operates most clearly at the level of individuals who identify the individual with whom to reciprocate, at the group level a symbol such as a facial marking specific to a tribe, or a scarf of an English football club, or a religious symbol, appear to enable tribalism to operate without individual recognition, at a group level.

When resources relevant to a tribe such as its land and homesteads are threatened by another tribe, and with sufficient technology and resources, warfare may result.

The impact for ethics and politics of these neurobiological forces that produce altruism and competition are considered in Chapters 9 and 12.

4.4   Mate selection, attractiveness, and love

Important types of social behaviour are related to reproduction. These types of social behaviour include mate selection, pair bonding including love, and parental attachment to their children. In the next two sections I consider how the brain has been shaped by evolution to promote these types of social behaviour by neural mechanisms that are sensitive to certain types of reward. These brain adaptations are not only important in much social behaviour, but also have fascinating implications for factors that are relevant to understanding aesthetics (Chapter 7) and ethics (Chapter 9).

What factors are decoded by our brains to influence mate attractiveness (reward value) and selection? Many factors are involved in mate selection, and they are not necessarily the same for selection of a short term vs a long-term partner. The selection of a long-term partner in species with long-term relatively monogamous relationships is influenced for example by parental investment, which is a major evolutionary adaptive factor in promoting long-term relationships. Thus in humans, males choose females because human males do make a parental investment; and females compete for males. Indeed, the selection of a long-term partner in humans is mutual, and this tends to reduce sex differences in partner choice. Consistent with this, David Buss has shown that in contrast, human sex differences in mate selection are more evident in short-term mating (Buss 1989, Buss 1994, Buss 2008).

Species with shared parental investment are primarily those where two parents can help the offspring to survive better than one, and this includes many birds (where one bird must sit on the eggs to incubate them, while the other finds food), and humans (where the human infant is born so immature[17] that care of the offspring for a number of years could, when humans were evolving, make it more likely that the offspring, containing the father's genes, would survive, and then reproduce).

Most other mammals are not good models of human mate selection and pair bonding, because there is generally less advantage to joint care of the young, and the female, who has made the major investment of the gestation period for the baby, and breast feeding it post-partum (for the whole of which period she will remain relatively infertile), generally assumes most of the responsibility for bringing up her young. In most mammals, females will maximize their reproductive success, given the cost of gestation and lactation, by focusing on the successful rearing of offspring. In contrast, male mammals do not invest by gestation and lactation in their offspring, and the most effective way for males to influence their reproductive success is to maximize the number of fertilizations they achieve, and this is a major factor in mammalian mate selection.

Another factor is that female mammals can be rather sure that there offspring are theirs, whereas this is less true of male mammals because of paternity un-certainty. (Even externally fertilizing male fish such as stickleback and pipefish can be fairly certain of their paternity, and this may promote greater care for the offspring compared to male mammals. For most male mammals, paternity uncertainty may be a factor that reduces their parental investment, and instead their reproductive success is influenced more by competition.

These tendencies are tempered in humans by the advantage of male invest-ment in the offspring, because ensuring that the immature offspring survive sufficiently long that the chances of their reproductive success are high is an adaptive investment.

4.4.1 Female preferences: factors that make men attractive and beautiful to women

Factors that across a range of species influence female selection of male mates include the following.

---

[17]Humans are born secondarily altricial (where altricial is the opposite of precocial) because of the narrow female pelvis of humans associated with bipedality, which results in gestation being shortened from an estimated 21 to 9 months (Gould 1985).

Athleticism

The ability to compete well in mate selection (including being healthy and strong), as this will be useful for her genes when present in her male offspring. Athleticism may be attractive (rewarding) also as an indicator of protection from male marauding (single females are at risk in some species of abuse, and forced copulation, which circumvents female mate choice), from predators, and as an indicator of hunting competency (meat was important in human evolution, although the hunt may also have been co-opted by sexual selection as a mating ritual giving the males a chance to show off). Consistent with these points, women show a strong preference for tall, strong, athletic men (Buss and Schmitt 1993).

Resources, power, and wealth

In species with shared parental investment (which include many birds and humans), having power and wealth may be attractive to the female, because they are indicators of resources that may be provided for her young. Women should desire a man who shows willingness to invest resources (which should be defensible, accruable and controllable) in his partner. (An expensive diamond engagement ring taken by a woman and kept guarded close to her on a finger meets these criteria. At the same time, the ring is a signal to her partner and to others that she is committed, which itself is attractive to her partner.) Women place a greater premium on income or financial prospects than men (Buss 1989). Further, in a cross-cultural study of 37 cultures with 10,047 participants, it was found that irrespective of cultural/political/social background, women consistently placed more value on financial resources (100% more) than men (Buss 1989, Buss 2008). Women value a man's love as an indicator of resource commitment.

Status

Both now and historically, status hierarchies are found in many cultures (and species, for example monkeys' dominance hierarchies, and chickens' pecking order). Status correlates with the control of resources (e.g. alpha male chimpanzees take precedence in feeding), and therefore acts as a good cue for women. Women should therefore find men of high status attractive (e.g. rock stars, politicians, and tribal rulers), and these men should be able to attract the most attractive partners. Consistent with this, cross-culturally women regard high social status as more valuable than do men; and attractive women marry men of high status (Buss 1989, Buss 2008). Status may be attractive because of direct effects (e.g. as an indicator of resources for children), or because of indirect effects (because high status implies good genes for offspring).

Age

Status and higher income are generally only achieved with age, and therefore women should generally find older men attractive. Cross-culturally women prefer older men (3.42 years older on average; and marriage records from 27 countries show that the average age difference was 2.99 years) (Buss 1989).

Ambition and industriousness

These may be good predictors of future occupational status and income, and are attractive. Valued characteristics include those that show a male will work to improve their lot in terms of resources or in terms of rising up in social status. Cross-culturally, women rated ambition/industriousness as highly desirable (Buss 1989).

Testosterone-dependent features

These may be attractive. These features include a strong (longer and broader) jaw, a broad chin, strong cheekbones, defined eyebrow ridges, a forward central face, and a lengthened lower face (secondary sexual characteristics that are a result of pubertal hormone levels). High testosterone levels are immuno-suppressing, so these features may be indicators of immuno-competence (and thus honest indicators of fitness). The attractiveness of these masculinized features increases with increased risk of conception across the menstrual cycle (Penton-Voak, Perrett, Castles, Kobayashi, Burt, Murray and Minamisawa 1999). The implication is that the neural mechanism controlling perception of attractiveness must be sensitive to oestrogen/progesterone levels in women.

Another feature thought to depend on prenatal testosterone levels is the 2nd/4th digit ratio. A low ratio reflects a testosterone-rich uterine environment. It has been found that low ratios correlate with female ratings of male dominance and masculinity, although the relationship to attractiveness ratings was less clear (Swaddle and Reierson 2002).

Symmetry

Symmetry (in both males and females) may be attractive, in that it may reflect good development in utero, a non-harmful birth, adequate nutrition, and lack of disease and parasitic infections (Thornhill and Gangstad 1999). Fluctuating asymmetry (FA) reflects the degree to which individuals deviate from perfect symmetry on bilateral features (e.g. in humans, both ears, both feet, both hands and arms; in other species, bilateral fins, bilateral tail feathers). Greater asymmetry may reflect deviations in developmental design resulting from the disruptive effects of environmental or genetic abnormalities, and in some species is associated with lower fecundity, slower growth, and poorer survival. A low fluctuating asymmetry may thus be a sign of reproductive fitness (Gangestad and Simpson 2000). In humans, more symmetrical men reported more lifetime partners ($r = 0.38$), and more extra-pair partners; and women's choice of extra-pair partners was predicted by male symmetry (Gangestad and Simpson 2000). Moreover, women rate men as more attractive if they have high symmetry (low FA). Intellectual ability (which may be attractive to women) is also correlated with symmetry (Gangestad and Thornhill 1999).

Dependability and faithfulness

These may be attractive, particularly where there is paternal investment in bringing up the young, as these characteristics may indicate stability of resources (Buss, Abbott and Angleitner 1990). Emotionally unstable men may also inflict costs on women, and thus women rate emotional stability and maturity as important. For example, jealousy might lead to abuse.

Risk-taking

Risk-taking by men may be attractive to women, perhaps because it is a form of competitive advertising: surviving the risk may be an honest indicator of high quality genes (Barrett et al. 2002).

Features selected by inter-sexual sexual selection

Characteristics that may not be adaptive in terms of the survival of the male, but that may be attractive because of inter-sexual sexual selection, are common in birds, perhaps less common in most mammals, though present in some primates (Kappeler and van Schaik 2004), and may be present in humans. An example of a sexually selected characteristic that may not increase the survival of the individual, but that may be attractive to females and thus increase the fitness of the male in terms of whether his genes are passed on to the next generation by reproduction, is the peacock's tail. These characteristics may in some cases be an honest indicator of health, in the sense that having a large gaudy tail may be a handicap.

Odour

The preference by women for the odour of symmetrical men is correlated with the probability of fertility of women as influenced by their cycle (Gangestad and Simpson 2000). Another way in which odour can influence preference is by pheromones that are related to major histocompatibility complex (MHC) genes, which may provide a molecular mechanism for producing genetic diversity by influencing those who are considered attractive as mates, as described in Section 7.4.

It is important to note that physical factors such as high symmetry and that are indicators of genetic fitness may be especially attractive when women choose short-term partners, and that factors such as resources and faithfulness may be especially important when women choose long-term partners, in what may be termed a conditional mating strategy (Buss 2008, Buss 2006). This conditionality means that the particular factors that influence preferences alter dynamically, and preferences will often depend on the prevailing circumstances, including the current opportunities and costs.

4.4.2 Male preferences: what makes women attractive and beautiful to men

Males are not always indiscriminate. When a male chooses to invest (for example to produce offspring), there are preferences for the partner with whom he will make the investment. Accurate evaluation of female quality (reproductive value) is therefore important, and a male will need to look out for cues to this, and find these cues attractive, beautiful, and rewarding. The factors that influence attractiveness include the following (Barrett et al. 2002).

Youth

As fertility and reproductive value in females is linked to age (reproductive value is higher when younger, and actual fertility in humans peaks in the twenties), males (unlike females) place a special premium on youth. It is not youth per se that men find attractive, but indicators of youth, for example neotenous traits such as blonde hair and wide eyes. An example of this preference is that male college students preferred an age difference on average of 2.5 years younger (Buss 1989). Another indicator of youth might be a small body frame, and it is interesting that this might contribute to the small body frame of some women in this example of sexual dimorphism.

Beautiful features

Features that are most commonly described as the most attractive tend to be those that are oestrogen-dependent, e.g. full lips and cheeks, and short lower facial features. (Oestrogen caps the growth of certain facial bones.) Like testosterone, oestrogen also affects the immune system, and its effects might be seen as 'honest indicators' of genetic fitness.

For example, when subjects were able to evolve a computer generated image into their ideal standard of female beauty, the beautiful composite had a relatively short lower face, small mouth, and full lips (Johnston and Franklin 1993). There is some agreement across cultures about what constitute beautiful features. For example, in meta-analyses of 11 studies, it has been demonstrated that (a) raters agree about who is and is not attractive, both within and across cultures; (b) attractive children and adults are judged and treated more positively than unattractive children and adults, even by those who know them; and (c) attractive children and adults exhibit more positive behaviours and traits than unattractive children and adults (Langlois, Kalakanis, Rubenstein, Larson, Hallam and Smoot 2000). In an fMRI study, it was found that attractive faces produce more activation of the human medial orbitofrontal cortex (where many pleasant stimuli are represented (Rolls and Grabenhorst 2008)) than unattractive faces (O'Doherty, Winston, Critchley, Perrett, Burt and Dolan 2003b).

Further, small babies were even shown to gaze for longer at slides of the more attractive woman when shown pairs of pictures of women that differed in attractiveness (Langlois, Roggman, Casey, Ritter, Rieserdanner and Jenkins 1987, Langlois, Ritter, Roggman and Vaughn 1991). In another study, 12-month-olds interacted with a stranger. The infants showed more positive affective tone, less withdrawal, and more play involvement with a stranger who wore a professionally constructed attractive than unattractive mask; and played longer with an attractive than an unattractive doll (Langlois et al. 2000). These results extend and amplify earlier findings showing that young infants exhibit visual preferences for attractive over unattractive faces. Both visual and behavioural preferences for attractiveness are evidently exhibited rather early in life.

Women appear to spend more time on fashion and enhancing beauty than men. Why should this be, when in most mammals it is males who may be gaudy to help in their competition for females, given that females make the larger investment in offspring? In humans, there is of course value to investment by males in their offspring, so women may benefit by attracting a male who will invest time and resources in bringing up children together. But nevertheless, women do seem to invest more in bearing and then raising children, so why is the imbalance so marked, with women apparently competing by paying attention to their own beauty and fashion? Perhaps the answer is that males who are willing to make major investments of time and resources in raising the children of a partner are a somewhat limiting resource (as other factors may make it advantageous genetically for men not to invest all their resources in one partner), and because women are competing to obtain and maintain this scarce resource, being beautiful and fashionable is important to women. Faithful men may be a limited resource because there are alternative strategies that may have a low cost, whereas women are essentially committed to a considerable investment in their offspring. These factors lead to greater variability in men's strategies, and thus contribute to making men who invest in their offspring a more limited resource than women who invest in their offspring.

Given that men are a scarce resource, and that women have such a major investment in their offspring that they must be sure of a man's commitment to invest before they commit in any way, we have a scientific basis for understanding why women are reserved and more cautious and shy in their interactions with men, which has been noticed to be prevalent in visual art, in which men look at women, but less vice versa (Berger 1972).

Body fat

The face is not the only cue to a woman's reproductive capacity, and her attractiveness, and beauty. Although the ideal body weight varies significantly with culture (in cultures with scarcity, obesity is attractive, and relates to status, a trend evident in beautiful painting throughout its history), the ideal distribution of body fat seems to be a universal standard, as measured by the waist-to-hip ratio (which cancels out effects of actual body weight). Consistently, across cultures, men preferred an average ratio of 0.7 (small waist/bigger hips) when rating female figures (line drawings and photographic images) for attractiveness (Singh and Luis 1995). Thornhill and Grammer (1999) also found high correlations between rating of attractiveness of nude females by men of different ethnicity. At a simpler level, a low waist to hip ratio is an indication that a woman is not already pregnant, and thus a contributor to attractiveness and beauty.

Fidelity

The desire by males for fidelity in females is most obviously related to her concealed ovulation (see next paragraph and *Emotion Explained* (Rolls 2005a)), and therefore the degree of paternity uncertainty that males may suffer. Males therefore place a premium on a woman's sexual history. Virginity was a requisite for marriage both historically (before the arrival of contraceptives) and cross-culturally (in non-Westernised societies where virginity is still highly valued) (Buss 1989). Nowadays, female monogamy in previous relationships is a sought after characteristic in future long-term partners (Buss and Schmitt 1993). (Presumably with simple genetic methods now available for identifying the father of a child, the rational thought system (Rolls 2005a) might place less value on fidelity as a characteristic in a partner with respect to paternity issues as paternity can be established genetically without having to rely on assessing characteristics of the partner, yet the implicit emotional system may still place high value on fidelity, as during evolution, fidelity was valued as an indicator of paternity probability.) The modern rational emphasis might be especially placed on valuing fidelity because this may indicate less risk of sexually transmitted disease, and perhaps the emotional value and attractiveness of fidelity will be a help in this respect.

Attractiveness and the time of ovulation

Although ovulation in some primates and in humans is concealed, it would be at a premium for men to pick up other cues to ovulation, and find women highly desirable (and beautiful) at these times. Possible cues include an increased body temperature reflected in the warm glow of vascularized skin (vandenBerghe and Frost 1986), and pheromonal cues. Indeed, male raters judged the odours of T-shirts worn during the follicular phase as more pleasant and sexy than odours from T-shirts worn during the luteal phase (Singh and Bronstad 2001). Women generally do not know when they are ovulating (and in this sense ovulation may be double blind), but there is a possibility that ovulation could unconsciously affect female behaviour. In fact, Event-Related Potentials (ERPs) were found to be greater to sexual stimuli in ovulating women, and these could reflect increased affective processing of the stimuli (Krug, Plihal, Fehm and Born 2000). This in turn might affect outward behaviour of the female, helping her to attract a mate at this time. Another possibly unconscious influence might be on the use of cosmetics and the types of clothes worn, which may be different close to the time of ovulation.

In most species, females invest heavily in the offspring in terms of providing the eggs and providing the care (from gestation until weaning, and far beyond weaning in the case of humans). Females are therefore a 'limited resource' for males allowing the females to be the choosier sex during mate choice. In humans, male investment in caring for the offspring means that male choice has a strong effect on intrasexual selection in women. Female cosmetic use and designer clothing could be seen as weapons in this competition, and perhaps are reflected in extreme female self-grooming behaviour such as cosmetic surgery, or pathological disorders such as anorexia, bulimia and body dysmorphic disorder. The modern media, by bombarding people with images of beautiful women, may heighten intrasexual selection even further, pushing women's competitive mating mechanisms to a major scale.

4.5 Pair-bonding and love

We have now seen some of the brain adaptations that underlie attractiveness in men and women, and that thereby help to promote social and reproductive behaviour. We now consider some interesting processes that help pair-bonding, and that may promote love.

Attachment to a particular partner by pair-bonding in a monogamous relationship, which in humans becomes manifest in love between pair-bonded parents, and which occurs in humans in relation to the advantage to the man of investing in his offspring, may have special mechanisms to facilitate it.

Species in which attachment has been investigated include the prairie vole. In monogamous species of prairie voles, mating can increase pair-bonding (as measured by partner preference). Oxytocin, a hormone released from the posterior pituitary, whose other actions include the milk let-down response, is released during mating (Lee, Macbeth, Pagani and Young 2009). Exogenous administration of oxytocin facilitates pair bonding in both female and male prairie voles (Carter 1998). In female prairie voles, antagonists of oxytocin interfere with partner preference formation. In female prairie voles, the endogenous release of oxytocin is thus important in partner preference and attachment. Thus oxytocin has been thought of as the 'hormone of love'. Oxytocin gene knockout mice fail to recognize familiar conspecifics after repeated social exposures, and injection of oxytocin in the medial amygdala restores social recognition (Winslow and Insel 2004). In males, the effects of oxytocin are facilitated by vasopressin, another posterior pituitary hormone whose other effects include promoting the retention of water by the kidney. In the case of vasopressin, it has been possible to show that the vasopressin V1a receptor (V1aR) is expressed in higher concentration in the ventral forebrain of monogamous prairie voles than in promiscuous (i.e. polygamous) meadow voles, and that viral vector V1aR transfer into the forebrain of the meadow mouse increases its partner preference (i.e. makes it more like a monogamous prairie vole) (Lim, Wang, Olazabal, Ren, Terwilliger and Young 2004, Young 2008). Thus a single gene may be important in influencing monogamy vs promiscuity in voles. Stress, or the administration of the hormone corticosterone which is released during stress, can facilitate the onset of new pair bonds (DeVries, DeVries, Taymans and Carter 1996).

Are similar mechanisms at work in humans to promote pair-bonding and love? There is as yet no definitive evidence, but in humans, oxytocin is released by intercourse, and especially at the time of orgasm, in both women and men (Meston and Frohlich 2000, Kruger, Haake, Chereath, Knapp, Janssen, Exton, Schedlowski and Hartmann 2003). It has also been reported that women desiring to become pregnant are more likely to have an orgasm after their partner ejaculates (Singh, Meyer, Zambarano and Hurlbert 1998).

An implication is that there may be hormonal, and other biological, mechanisms that promote a bifurcation in a state space (Rolls and Deco 2010), and have an effect of cementing attraction and love of a particular person after the process has been started. This may have an effect on (partly) blinding a person to a partner's imperfections, and may thus contribute to each individual's judgements about the beauty of a partner, an aesthetic judgement.

Given this Darwinian approach rooted in selfish genes (Dawkins 1989), should we describe the state of love as selfish? I suggest that the answer is that although individual acts can be truly altruistic (and non-adaptive), even the altruism implied by love must have its origins in selfish genes, which shape human behaviour to in this case produce a state that promotes the production of and survival of offspring. Overall, for a characteristic (such as falling in love, or reciprocal altruism, or kin altruism) that is influenced by genes to remain in a breeding population, the characteristic must be good for the (selfish) gene or it would be selected out (Section 4.2).

Even love guided by rational thought must not overall detract too much over generations from the wish to produce offspring, or it would tend (other things being equal) to be selected out of the gene population.

4.6 Parental attachment: beautiful children

Much social behaviour is related to parent–child social interactions. What neural adaptations are at work here?

Many mammal females make strong attachments to their own offspring, and this is also facilitated in many species by oxytocin. One model is the sheep, in which vaginal-cervical stimulation and suckling, which release both oxytocin and endogenous opioids, facilitate maternal bonding (Keverne, Nevison and Martel 1997). Oxytocin injections can cause ewes to become attached to an unfamiliar lamb presented at the time oxytocin is released or injected, and oxytocin antagonists can block filial bonding in sheep. Perhaps oxytocin had an initial role in evolution in the milk let-down reflex, and then became appropriate as a hormone that might facilitate mother–infant attachment.

In humans the evidence is much more correlative, but oxytocin release during natural childbirth, and rapid placing of the baby to breast feed and release more oxytocin (Uvnas-Moberg 1998), might facilitate maternal attachment to her baby. Prolactin, the female hormone that promotes milk production, may also influence maternal attachment – and how beautiful a mother thinks her child is. It is certainly a major factor in humans that bonding can change quite suddenly at the time that a child is born, with women having a strong tendency to shift their interests markedly towards the baby as soon as it is born (probably in part under hormonal influences), and this can result in relatively less attachment behaviour to the husband. In men, oxytocin may also be involved in paternal behaviour (Wynne-Edwards 2001).

Another aspect of parental care is that there is competition between the mother and child, for example over weaning (Trivers 1974). The mother may wish to devote resources to preparing for her next offspring (by building herself up); and continuing to breast feed delays the onset of fertility and cycling. In contrast, it is to the offspring's genetic advantage to demand milk and attention. The infant's scream can be seen as part of trying to wring resources out of its mother, potentially to an extent that is unfavourable for the mother's genes (Buss 2008).

As described above, females generally have a greater investment in their offspring, and tend to provide more parental care and perhaps become more attached than fathers. This situation is not as extreme in humans as in most other mammals, because human offspring are born relatively immature, and a father who helps to rear the offspring can help to increase the reproductive fitness of his genes.

Lack of parental care in step-fathers is evident in many species, and can be as extreme as the infanticide by a male lion of the cubs of another father, so that his new female may come into heat more quickly to have babies by him (Bertram 1975). Infanticide also occurs in non-human primates (Kappeler and van Schaik 2004). In humans, the statistics indicate that step-fathers are much more likely to harm or kill children in the family than are real fathers (Daly and Wilson 1988).

The tendency to find babies beautiful is not of course restricted to parents of their own children. Part of the reason for this is that in the societies in which our genes evolved with relatively small groups, babies encountered might often be genetically related, and the tendency to find babies beautiful is probably a way to increase the success of selfish genes. One may still make these aesthetic judgements of babies in distant countries with no close genetic relationship, but this does not of course mean that such judgements do not have their evolutionary origin in kin-related advantageous behaviour.

4.7 Sperm competition and its consequences for sexual behaviour: a sociobiological approach

It turns out that sperm competition helps us to understand some aspects of sexual and social behaviour, as I now describe.

Monogamous primates well spread out over territory have small testes, for example gibbons and some tarsiers. Polygamous primates living in groups with several males in the group have large testes and frequent copulation, e.g. chimpanzees and monkeys (see Ridley (1993b), Harcourt, Purvis and Liles (1995) and Barrett et al. (2002)). The reason for this appears to be sperm warfare – in order to pass his genes on to the next population, a male in a polygamous society with competition between males needs to increase the probability that he will fertilize a female, and the best way to do this is to copulate often, and swamp the female with sperm, so that his sperm have a greater probability of getting to the egg to fertilize it. Therefore in polygamous groups with more than one male, males should have large testes, to produce large numbers of sperm and large quantities of seminal fluid[18]. The largest testis size in relation to body weight is found in chimpanzees, who live in multimale groups, are highly promiscuous, and have on average 13 partners per birth (Wrangham 1993). Sperm competition can be seen as a form of non-combative, non-injurious, male–male intrasexual competition, which has evolved by intrasexual sexual selection (see Section 7.4). Not only testis size, but also seminal vesicle size, is large in species where other males mate with the females (Dixson 1998). In monogamous societies, with little competition between sperm, the male should just pick a good partner, produce only enough sperm to fertilize an egg and not enough to compete with others' sperm, stay with her to bring up the children, and guard them because they are his genetic investment (Ridley 1993b).

---

[18]Competition between males is the key factor here, for in gorillas which have one male in a polygamous (or strictly polygynous, meaning multiple females) group the testis size is small. The term that describes a multimale multifemale group is polygynandry, and it is in groups of this type that there are high levels of sperm competition, and the testes are large.

What about humans? Despite being apparently mainly monogamous, they are intermediate in testis size and penis size – bigger than expected for a monogamous species (Harcourt, Harvey, Larson and Short 1981, Harcourt et al. 1995, Parker, Ball, Stockley and Gage 1997, Barrett et al. 2002). Why? Maybe there is some sperm competition? Remember that although humans usually do pair, and are often apparently monogamous, humans do live in groups or colonies. Can we get hints from other animals that are paired, but also live in colonies?

A problem with comparing humans with most other primates in this respect is that in most primates (and indeed in most mammals), the main parental investment is by the female (in producing the egg, in carrying the foetus, and in feeding the baby until it can become independent). The male does not have to invest in his children for them to have a reasonable chance of surviving. For this reason, the typical pattern in mammals is that the female is choosy in order to obtain healthy and fit males, and to complement this the males compete for females. However, in humans, because the children must be reared for a number of years before they become independent, there is an advantage to paternal investment in helping to bring up the children, in that the paternal resources (e.g. food, shelter, and protection) can increase the chances of the male's genes surviving into the next generation to reproduce again. Part of the reason why investment by both parents is needed in humans is that because of the large final human brain size, at birth the brain is not fully developed, and for this reason the infant needs to be looked after, fed, protected, and helped for a considerable period while the infant's brain develops, favouring pair-bonding between the parents.

A more useful comparison can therefore be made with some birds, such as the swallow, which live in colonies but in which the male and the female pair, and both invest in bringing up the offspring, taking it in turns for example to bring food back to the nest. If checks are made in swallows using DNA techniques for determining paternity, it is found that actually approximately one third of a pair's young are not sired by the 'father', the male of the pair (Birkhead and Moller 1992, Ridley 1993b, Birkhead 2000). What happens is that the female mates sometimes with other males – she commits adultery. She probably does not do this just with a random male either – she may choose an 'attractive' male, in which the signals that attract her are signals that indicate health, strength, and fitness. One well-known example of such a signal is the gaudy 'tail' of the male peacock. One argument is that, given that the tail is a real handicap in life, any male that can survive with such a large tail must be very healthy or fit. Another argument is that if his tail is very attractive indeed, then the female should choose him, because her sons with him would probably be attractive too, and also chosen by females[19]. (It is interesting that if a male were popular with females, then even if he had genes that were not better in terms of survival, etc., it would be advantageous for a female to have offspring with him, as her sons would be more likely to be attractive to other females, and thus maximize her inclusive fitness. This is an example of Fisherian selection (Fisher 1958), see Section 7.4.)

In such a social system, such as that of the swallow, the wife needs a reliable husband with whom she mates (so that he thinks the offspring are his, which for the system to be stable they must be sometimes) to help provide resources for 'their' offspring. (Remember that a nest must be built, the eggs must be incubated, and the hungry young must be well fed to help them become fit offspring. Here fit means successfully passing on genes into the next generation – see Dawkins (1986)). But the wife (or at least her genes) also benefits by obtaining as fit genes as possible, by sometimes cheating on her husband. To ensure that her husband does not find out and therefore leave her and stop caring for the young, she deceives the husband by committing her adultery as much as possible secretly, perhaps hiding behind a bush to mate with her lover. So the (swallow) wife maximizes care for her children using her husband, and maximizes her genetic potential by finding a lover with fit genes that are likely to be attractive in her sons to other females (see Ridley (1993b)).

Could anything like the situation just described for birds such as swallows also apply to humans? It appears that it might apply, at least in part, and that similar evolutionary factors might influence human sexual behaviour, and hence make obtaining particular stimuli in the environment rewarding to humans. We need to understand whether this is the case, in order to understand the rewards that drive sexual behaviour in humans. One line of evidence already described is the large testis and penis size of men. In humans, it has been shown that the number of sperm ejaculated is related to testis size (Simmons, Firman, Rhodes and Peters 2004). Figure 4.1 shows this relation, and also indicates the large variation in both testis size and number of sperm contained in an ejaculate in humans, indicating that there is the potential for variation between humans to play a role in sperm competition. This potential may not be fully realised in modern humans perhaps because of contraceptive practices, in that extrapair paternity rates are estimated at around 4% in modern times (see below), though double paternity in some dizygotic twins does show that conditions for sperm competition in humans can occur (Simmons et al. 2004). (We will get on to the reason for the large penis size soon.) A second line is that some studies in humans of paternity using modern DNA tests suggest that in fact the woman's partner (e.g. husband) is not the father of about 14% of their children. Although surprising, this has been claimed in a study in Liverpool, and in another in

---

[19]This is an example of the use of the intentional stance in the description, when no real propositional state is likely to occur at all.

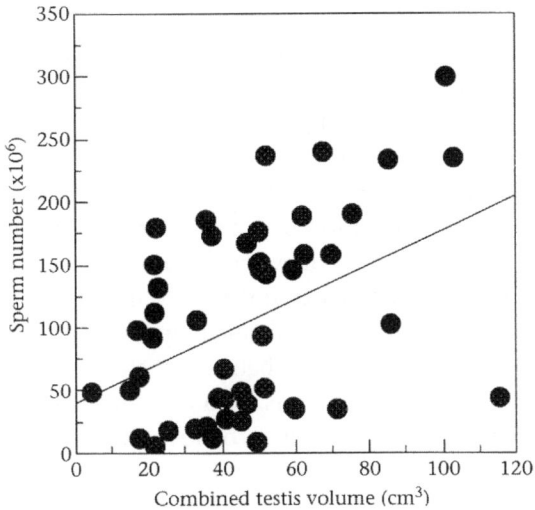

Fig. 4.1 Relation between number of sperm contained in human ejaculate volume and the size of the testes. The relation was significant at P=0.002 in the sample of 50 men. (From Simmons, Firman, Rhodes and Peters 2004.)

the south of England (Baker and Bellis 1995) (see Ridley (1993b)). In other studies, the extrapair paternity rate has been estimated at closer to 2%, although 28% of men and 22% of women did report extrapair copulations (Simmons et al. 2004). Further, an approximate estimate across 53,619 humans in 5 studies yields 24% of men and 20% of women reporting extrapair copulations (estimate made from data in Simmons et al. (2004)). In the data reviewed by Simmons et al. (2004), an approximate estimate of the extra-pair paternity rate is 4%, but there were great variations, with estimates in some traditional cultures of 10–11.8%. The latter estimates are particularly interesting, because estimates of extrapair paternity rates from North American and European cultures, practised in birth control and taught by particular mores, may not reflect the behaviour of our ancestors, selection between whom has shaped the behaviour of modern humans. On balance, these data suggest that while sperm competition may not be a major factor in modern humans, it may be to some extent, and might have been much more important in our ancestors, and have shaped our behaviour to at least some extent. So the possible effects of sperm competition in influencing modern human behaviour are worth exploring further.

So might men produce large amounts of sperm, and have intercourse quite regularly, in order to increase the likelihood that the children produced are theirs, whether by their wife or by their mistress? When women choose men as their lovers, do they choose men who are likely to produce children who are fit, that is children good at passing on their genes, half of which originate from the woman? It appears that women might choose like this, as described below, and that this behaviour may even select genetically for certain characteristics in men, because the woman finds these characteristics rewarding during mate selection. Of course, if such a strategy were employed (presumably mainly unconsciously) all the time in women the system would break down (be unstable), because men would not trust their wives, and the men would not invest in making a home and bringing up their children[20]. So we would not expect this to be the only selective pressure on what women find attractive and rewarding as qualities in men. Pursuing this matter further, we might expect women to find reliability, stability, provision of a home, and help with bringing up her children to be rewarding when selecting a husband; and the likelihood of producing genetically fit children, especially sons who can themselves potentially have many children by a number of women, to be rewarding when selecting a lover (i.e. short-term mate).

What is even more extraordinary is that women may even have evolved ways of influencing whether it is the woman's lover, as opposed to her husband, who fathers her children. Apparently she may be able to do this even while deceiving her husband, even to the extent of having regular intercourse with him. The ways in which this might work have been investigated in research described next (Baker and Bellis 1995, Baker 1996). Although much of the research on the sociobiological background of human sexual behaviour, including sperm warfare in humans, is quite new, and many of the hypotheses remain to be fully established and some can now be rejected (Birkhead 2000, Moore, Martin and Birkhead 1999), this research does have interesting potential implications for understanding the rewards that control human behaviour, and for this reason the research, and its potential implications, are considered here.

In the research by Baker and Bellis (1995) and others (see also Ridley (1993b)), it was claimed that if a woman has no orgasm, or if she has an orgasm more than a minute before the male ejaculates, relatively little sperm is retained in the woman. This is a low-retention orgasm. The claim was that if she has an orgasm less that a minute before him or up to 45 minutes after him, then much more of the sperm stays in the woman, and some of it is essentially sucked up by the cervix during and just following the later stages of her perceived orgasm. This is a high-retention orgasm. After a high-retention orgasm, it was claimed (Baker and Bellis 1995) that the woman is more likely to conceive (by that intercourse) even if she already has sperm in her reproductive tract left from intercourse in the previous 4 or so days. Thus there is a possibility that women can influence to some extent who is the father of their children, not

---

[20]In some tribes, brothers help to bring up their sisters' children, because these children share some of the mother's brother's genes. The brother and sister of course will share some of the same genes, so the behaviour of the brother is appropriate in terms of increasing the fitness of his genes in a promiscuous society.

only by having intercourse with lovers, but also by influencing whether they will become pregnant by their lover. Some clear evidence for such selection is that domestic fowls (hens) appear to select which sperm fertilize their eggs, in that when inseminated with sperm of different cocks, the fertilization was non-random (Birkhead, Chaline, Biggins, Burke and Pizzari 2004).

These data are at present quite controversial, although there is evidence that women have orgasms more frequently after their partner has ejaculated when they desire to become pregnant (Singh et al. 1998), and this might be a mechanism for cryptic choice (Birkhead and Pizzari 2002) when mating with an attractive lover, in that a woman is more likely to have an orgasm with an attractive (symmetric) partner (Thornhill, Gangestad and Comer 1995). If female orgasm is involved in influencing who the father is of a baby, then it might be expected that female orgasm might be somewhat variable in whether it occurs, as part of a putative selection process. Another possible contributory factor in the evolution of female orgasm is that it provides motivation to solicit multiple partners, for example if she does not have an orgasm with one partner, or if she has an orgasm with one partner who then enters a refractory state after ejaculation (cf. Hrdy (1999) and Hrdy (1996)). Indeed, polyandrous mating situations make it adaptive (in order to conceal paternity) for a female to be able to have orgasms without a long refractory period between each, that is to be able to have multiple orgasms. Although a similar argument might be applied to men, a refractory period might nevertheless be adaptive in men in part because of limited sperm resources, and the utility of competing with adequate sperm numbers when insemination does occur. Indeed, dominant males may release limited sperm because of their multiple matings, and this indeed is a factor cited as accounting for females competing for the first mating (Wedell, Gage and Parker 2002).

Another finding by Baker and Bellis (1995) indicates that men have evolved strategies to optimize the chances of their genes in this sperm selection process. One is that men were reported to ejaculate more sperm if they have not been for some time with the woman with whom they are having intercourse. An effect consistent with this is that a man who spends a greater (relative to a man who spends a lesser) proportion of time apart from the partner since the couple's last copulation report (a) that his partner is more attractive, (b) that other men find his partner more attractive, (c) greater interest in copulating with the partner, and (d) that his partner is more sexually interested in him (Shackelford, Le Blanc, Weekes-Shackelford, Bleske-Rechek, Euler and Hoier 2002). (This effect is not just dependent on the time since he has last inseminated his partner, but is related to the time the couple have been apart, so the effects may be interpreted as being related to possible insemination of the partner while away, and not just to sexual frustration (Shackelford et al. 2002).) The (evolutionary, adaptive) function of this may be for the man to increase the chances of his sperm in what could be a sperm war with the sperm of another man. The aim would be to outnumber the other sperm. Moreover, the man should do this as quickly as possible after returning from an absence, as time could be of the essence in determining which sperm get to the egg first if the woman has had intercourse with another man recently. The implication of this for reward mechanisms in men is that after an absence, having intercourse quite soon with the woman from whom the man has been absent should be very rewarding (and this is what is reported (Shackelford et al. 2002)).

There is good evidence that processes of this type do occur in some species. For example, Pizzari, Cornwallis, Lovlie, Jakobsson and Birkhead (2003) found in domestic fowl that males show status-dependent investment in females according to the level of female promiscuity: they progressively reduce sperm investment in a particular female but, on encountering a new female, instantaneously increase their sperm investment; and they preferentially allocate sperm to females with large sexual ornaments signalling superior maternal investment. These results indicate that female promiscuity leads to the evolution of sophisticated male sexual behaviour.

It even appears that there should be some reward value in having intercourse very soon with the woman after an absence, because the action of the glans penis, with its groove behind the head, may be to pull sperm already in the vagina out of it using repeated thrusting and pulling back (Baker and Bellis 1995) (at least in some ancestors, Birkhead (2000)). The potential advantage to this in the sperm warfare may be the biological function that, as a result of evolution, leads to thrusting and withdrawal of the penis during intercourse being rewarding (perhaps to both men and women). Such thrusting and withdrawal of the penis during intercourse should occur especially vigorously (and should therefore have evolved to become especially rewarding) after an absence by the man. The possible advantage in the sperm warfare that shaped our evolution could also result in its being rewarding for a man to have intercourse with a woman if he has just seen her having intercourse with another man. (This could be part of the biological background of why some men find videos showing sex involving women rewarding.) However, large numbers of sperm from a previous man usually remain in the vagina for only up to 45 minutes after intercourse, after which a flowback of sperm and other fluids (the discharge of semen and other secretions) from the vagina usually occurs. Thus the evolutionary shaping of the glans penis, and the rewards produced by thrusting and withdrawing it and the shaft of the penis in the vagina, are likely to have adaptive value more in our ancestors than in us.

Baker and Bellis (1995) proposed that a second human male strategy might be to ejaculate not only sperm that can potentially fertilize an egg, but also killer (or kamikaze) sperm that kill the sperm of another male, and blocker sperm that remain in the mucus at the entrance to the cervix blocking access through the channels in the mucus to other sperm. However, this hypothesis of different sperm types in humans is now strongly criticised, and there is little evidence for different sperm types in humans, and for kamikaze-like effects (Moore et al. 1999, Birkhead 2000, Short 1998). (Nevertheless, there are many interesting sperm adaptations for competition in other species. For example, in the wood mouse the sperm from one individual form 'trains' that increase the motility of the sperm twofold, thus facilitating fertilization by that individual's sperm (Moore, Dvorakova, Jenkins and Breed 2002).) Independently of this argument, given that (the majority of) sperm remain viable once out of the vagina and in the uterus or Fallopian tubes for up to about four days, it becomes important (read adaptive) for a man to have intercourse with his partner at least as often as say twice per week. This would ensure that at least some of the male partner's sperm were present to compete with any other sperm that might arrive as a result of an extra-pair copulation. So because of this function of sperm warfare, the brain should be built to make male intercourse with his partner rewarding at least approximately twice per week.

A third argument of Baker and Bellis (1995) (see also Smith (1984) and Baker and Bellis (1993)) is that it is important (in the fitness sense) that a male should help his lover to have an orgasm, which should occur only just before or for 45 minutes after he has ejaculated for the upsuck effect to operate. (The orgasm was proposed to cause the upsuck effect of sperm into the uterus. This might give the woman some control over whether she will accept new sperm. Whether female orgasm does facilitate sperm retention and the likelihood of fertilization is not yet clear (Levin 2002).) For this reason, men should find intercourse with a lover very exciting and rewarding if this tends to increase the likelihood that their lover will have an orgasm. This biological function of an orgasm in women provides a general reason why men should find it rewarding when a woman has an orgasm (and should therefore try to produce orgasms in women, and be disappointed when they do not occur). This process might be helped if one factor that tended to produce ejaculation in men was knowledge that the woman was having an orgasm. Of course, the reverse may occur – a woman may be triggered to orgasm especially just after she receives an ejaculation, especially if it is from a male with whom (presumably mainly subconsciously) she wishes to conceive. (Indeed, there is evidence that women have orgasms more frequently when they desire to become pregnant (Singh et al. 1998), and with men with an index of healthy genes, low asymmetry (Thornhill et al. 1995, Thornhill and Gangestad 1996).) If she does not have an orgasm, she may fake an orgasm (as this is rewarding to the male), and this may be part of her deception mechanism.

Having now outlined some of the functions at play in sexual behaviour we are in a better position to understand the neural adaptations that may underlie the social interactions involved in sexual behaviour.

4.8 Concealed ovulation and its consequences for sexual behaviour

Women, and a few non-human primate species, have concealed ovulation. It is not clear to males, or to themselves, when they are fertile. Why do women conceal their ovulation?

Diamond (1997) considers evidence that a first process that occurs in evolution is that promiscuity or harems in the mating system give rise to concealed ovulation. This is the 'many fathers' theory. The concealed ovulation (concealed even from the woman, so that she can deceive better – what might be termed 'deceiving conceiving') makes sure that men do not know who the father is (because they do not know when ovulation has occurred), and thus will not attack the young. (It frequently occurs in the animal kingdom that males kill their female's children if they have been born of other males, a process that enables genes to maximize their own reproductive potential. This occurs because the female will stop lactating and will come back into a reproductive state so that the new male can reproduce. Moreover, it will minimize potential use of his resources in helping to bring up children without his genes.)

A second process can then occur in evolution: monogamy evolves. Monogamy, Diamond (1997) argues, has never evolved in species that have bold advertisement of ovulation. It usually evolved in species with (i.e. that already have) concealed ovulation – the 'daddy-at-home' theory. The concealed ovulation means that fathers stay at home all the time (and help), because they want to be assured of their paternity; and because they think that they are the father, because they have been at home (Simmen-Tulberg and Moller 1993). Thus the consequences of concealed ovulation may be that fathers find it rewarding (and have emotions about) staying at home with their partner to guard a primarily monogamous relationship. Indeed, monogamy can be thought of as a form of mate guarding.

Consistent with these hypotheses, it has been found that free-living Hanuman langur females do have long periods of receptivity during which the time of ovulation is variable, that there is the opportunity for paternity confusion in that ovulation is concealed from the males, that there is a dominant male who tries to monopolize the females, and that nevertheless non-dominant males father a substantial proportion of the offspring (Heistermann, Ziegler, van Schaik, Launhardt, Winkler and Hodges 2001). This is direct evidence that extended periods of sexual receptivity in catarrhine primates may have evolved as a female strategy to confuse paternity.

Concealed ovulation could also play a role in combination with female orgasm to enable female cryptic choice, which would it has been suggested (see Section 4.7) occur if a woman has an orgasm with a man who she wants to be the father of her children. The contribution of the concealed ovulation would be to promote male–male competition.

Thus the interests of females and males may not be consistent, and this leads to the development of measures and countermeasures. Concealed ovulation can be seen as a protection against infanticide. Concealed ovulation promotes polyandry, and this results in multiple matings, and sperm competition and sperm allocation as a response to this. Females may then counter with mechanisms for cryptic choice, such as for example selective orgasms.

4.9 Possible implications for human pair-bonding and love

The possible implications of the advances described in this chapter for understanding human pair-bonding have been discussed (Buss 2008, Buss 2003, Fisher 1992, Fisher 2004, Fisher 2009, Wright 1994). The processes may operate at least largely unconsciously. All are just general effects that may be promoted to an extent that depends on the genes in an individual and so may not operate in every individual; that may be modified by upbringing (effects of the environment); and may be to differing extents in different individuals under control of the reasoning system and open to conscious control. Some implications include the following:

Women may wish to pair-bond (and this may of course include feelings of love) with men because of the resources that women may obtain for their children, that is for their genes. The process may operate especially when women are very fertile, that is, relatively young. The desire to remain in a stable relation that is promoted by love may last for the years when the resources are especially needed for the children, that is for the first few years of her youngest child's life. Women who have their own independent resources (whether or not children are present – these are gene-specified feelings that may have been programmed to last for a few years of a relationship because there would normally be children then) might be expected to stay in relationships for less long, and to be willing to try to find another partner (and gain a different set of genes for more children). This factor could be one of those that contribute to the evidence that divorce rates are more than 39% after 5 years in those with no children (this is perhaps an indication that love for and pair-bonding with another might lead to children), and 4% for those with 4 or more children. Women may find liaisons with other partners attractive (though risky) because in the past this might have been a way for females to gain genetic variety in their offspring, so that some of the children might have a good gene combination for circumstances that could arise in the future.

Men will have some advantage in staying to bring up their children (because human children are born so immature that an investment by men can increase the chances that their genes in their children will survive to in turn reproduce). But the fathers gain less than women by investing in the offspring of their pair-bonding, and after a few years the strategy of having liaisons with other women may be attractive to their genes (or to the brains built by their genes!). Indeed, we can distinguish long-term highly selective mating strategies in which individuals invest in their children, and short-term less selective strategies where a short liaison might increase the chances of a man's genes coming into the next generation, even if he makes no continuing investment in the offspring.

Given these factors that favour a joint investment in their own genes, free love, and no particular responsibility for your own children, does not generally work, as perhaps in communes.

A consequence of this asymmetric advantage for men and women in staying pair-bonded may be that men are less faithful as the years progress than women. But fidelity is important to both men and women. For women, faithfulness is highly valued because they are concerned about the continuing availability of resources. For men, faithfulness is highly valued because otherwise there is a risk that the children in whom he is investing will not be his. (There is evidence that up to 1 in 10 children of stable pair-bonded relationships are not the children of the male in the relationship. One in three wives may be unfaithful to their husbands.) This is a major fuel for the emotion of jealousy.

A consequence of these factors may be that men and women may be serially monogamous: there are factors promoting pair-bonding for a period, with joint investment in offspring, but there may also be genetic advantages to changing partners after a number of years, producing serial monogamy. Indeed, in our evolutionary past, there may have been an advantage for pairs to stay together, to be monogamous, for several years, the time it takes to bring up their child until its chances of survival, helped for example by other kin, is reasonable. Is it a coincidence that many couples who divorce do so after around 4 years after marriage? However, the probability of divorce does depend on whether there is a major and continuing investment in joint genes. In one study, within 5 years of marriage, 39% of couples with 0 children divorce, 26% of couples with 1 child, 19% of couples with 2 children, and 4% of couples with 4 children (Buss 2008, Buss 2003).

But another situation that can arise occurs when a man with many resources (and these need not be financial – they might be scientific or artistic creativity which could be indications of genes that will do well in the next generation) has multiple partners (perhaps a wife and mistresses, or in many societies in the world many wives): polygyny. (For a woman, some, including possibly genetic, resources from a powerful man might be a competitive alternative to committed but lower quality resources.) In this situation, the first wife or partner may benefit most, and may even be helped by the other partners.

Polygyny raises the issue that some men may be left without a partner, and without an easy opportunity to reproduce. This may be a desperate plight for genes, and this may be a factor that could underlie the terrible reality that rape is not very uncommon in society (Thornhill and Palmer 2000). (It is estimated that there are more than 600,000 rapes in the USA each year.) Indeed,the concept that rape is a desperate resource for men's genes is that those who are raped are typically younger women: women who are very fertile. Further, the comment has been made that part of the biological underpinning of the high incidence of rape during war is that the biological imperative for men during war may be desperate: they may not survive for long, and that may increase their priority to reproduce, even if they cannot contribute to the survival of their children. Here I issue the reminder that what is natural is not the same as what is right (that would be the naturalistic fallacy): we are trying here to understand what may be biological underpinnings, and the hope is that by understanding better, we may be in a better position to ensure that what societies agree should be right does prevail.

Serial monogamy may be worse than polygyny for the children. The natural father, who has a genetic reason to invest in his children, may no longer be present with serial monogamy. A stepfather (or for that matter a stepmother) has less of a biological investment in the child than a natural parent, and may not provide as many resources as a biological parent (or worse: remember the male lion who kills his new lioness's cubs by a previous mate). (A child is in the first two years of life 60 times more likely to be killed by a stepfather than by a natural father (Daly and Wilson 1988).) Again, there is no necessity in any of this: these are just biological factors that could in our past have been selected for, and that rational thought can ensure do not apply in particular cases.

Finally, once we understand some of this underlying biology that has evolved in the past because of the genetic advantage, we can with our human reasoning minds take decisions about how we behave. The great feelings of love that biology has built into us should be valued and cherished. But it may help us to understand these feelings, both so that individuals may be able to benefit from such good feelings, but also so that individuals can discuss wisely what a rational solution would be when those biological underpinnings promote other types of behaviour than stable pair-bonding. When Margaret Mead was criticized for three failed marriages, her reply was that each was a success. And we should also be aware that our biology might promote different types of pair-bonding at different stages of life: perhaps infatuation, romantic love, and sex early on (in a relationship and in life); stable pair-bonding where we have a mate to bring up children; and perhaps companionship (including sharing of intellectual resources) later on.

5 Neuroreason

In Chapter 3 I described some of the processes by which genes sculpt brains to like some rewards, and avoid some punishers. This is in the self-interest of the genes. In this chapter I show how this system may often work unconsciously, and may enable us to obtain immediate, gene-specified, rewards. But the main aim of this chapter is to show how humans, and perhaps some closely related animals, have a reasoning or rational system that by the use of linguistic processing enables us to plan ahead for the long term, and to defer or redefine our gene-specified emotional goals. It then becomes an interesting and important question about how we choose between our emotional and rational systems. What are the mechanisms of this decision-making? Further, why in our decision-making do we not always choose the same thing? Do our brains operate deterministically, or probabilistically?

5.1 Dual routes to action

According to my formulation (Rolls 2005a, Rolls 2011b), there are two major types of route to action performed in relation to reward or punishment in humans. Examples of such actions include those associated with emotion and motivation.

5.2 The implicit gene-specified goal route to action

The first ('implicit') route is via the brain systems that have been present in non-human primates such as monkeys, and to some extent in other mammals, for millions of years. These systems include the amygdala and, particularly well-developed in primates, the orbitofrontal cortex. These systems control behaviour in relation to previous associations of stimuli with reinforcement. The computation which controls the action thus involves assessment of the reinforcement-related value of a stimulus. This assessment may be based on a number of different factors:

One is the previous reinforcement history, which involves stimulus–reinforcer association learning using the amygdala, and its rapid updating especially in primates (which of course includes humans) using the orbitofrontal cortex. This stimulus–reinforcer association learning may involve quite specific information about a stimulus, for example of the energy associated with each type of food, by the process of conditioned appetite and satiety (Booth 1985).

A second is the current motivational state, for example whether hunger is present, whether other needs are satisfied, etc.

A third factor that affects the computed reward value of the stimulus is whether that reward has been received recently. If it has been received recently but in small quantity, this may increase the reward value of the stimulus. This is known as incentive motivation or the 'salted nut' phenomenon. The adaptive value of such a process is that this positive feedback of reward value in the early stages of working for a particular reward tends to lock the organism on to behaviour being performed for that reward. This means that animals that are for example almost equally hungry and thirsty will show hysteresis in their choice of action, rather than continually switching from eating to drinking and back with each mouthful of water or food. This introduction of hysteresis into the reward evaluation system makes action selection a much more efficient process in a natural environment, for constantly switching between different types of behaviour would be very costly if all the different rewards were not available in the same place at the same time. (For example, walking half a mile between a site where water was available and a site where food was available after every mouthful would be very inefficient.) The amygdala is one structure that may be involved in this increase in the reward value of stimuli early on in a series of presentations, in that lesions of the amygdala (in rats) abolish the expression of this reward-incrementing process which is normally evident in the increasing rate of working for a food reward early on in a meal (Rolls and Rolls 1973, Rolls 2005a).

A fourth factor is the computed absolute value of the reward or punishment expected or being obtained from a stimulus. One example is the sweetness of the stimulus (set by evolution so that sweet stimuli will tend to be rewarding, because they are generally associated with energy sources). Another example is the pleasantness of touch. This is set by evolution to be pleasant according to the extent to which it brings animals of the opposite sex together, and depending on the investment in time that the partner is willing to put into making the touch pleasurable, a sign that indicates the commitment and value for the partner of the relationship, as in social grooming. Touch is also set to be rewarding by evolution because of the value to the mother's genes of holding and protecting an infant in which she has invested so much, and because of the value to the infant's genes of the caring, touching protection being provided by a mother. Painful touch is set by evolution to be aversive to help protect the person from harm.

After the reward value of the stimulus has been assessed in these ways, behaviour is then initiated based on approach towards or withdrawal from the stimulus. A critical aspect of the behaviour produced by this type of 'implicit'

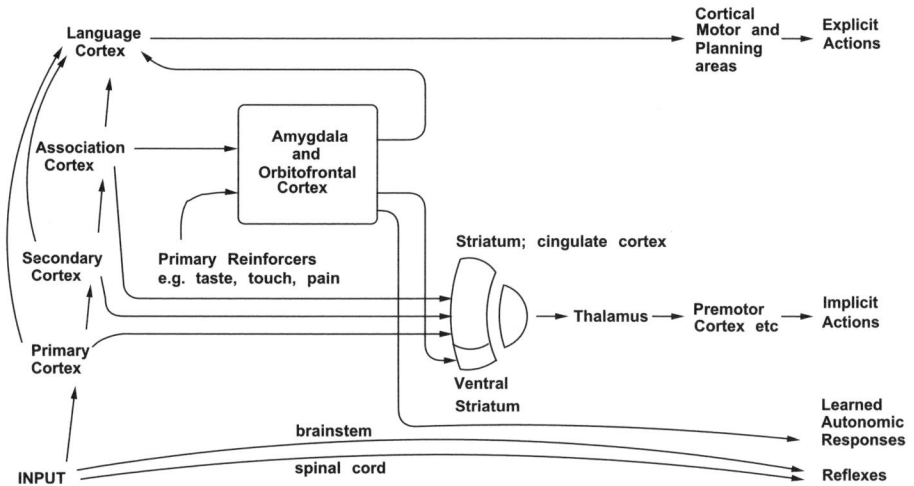

Fig. 5.1 Dual routes to the initiation of actions in response to rewarding and punishing stimuli. The inputs from different sensory systems to brain structures such as the orbitofrontal cortex and amygdala allow these brain structures to evaluate the reward- or punishment-related value of incoming stimuli, or of remembered stimuli. The different sensory inputs enable evaluations within the orbitofrontal cortex and amygdala based mainly on the primary (unlearned) reinforcement value for taste, touch, and olfactory stimuli, and on the secondary (learned) reinforcement value for visual and auditory stimuli. In the case of vision, the 'association cortex' that outputs representations of objects to the amygdala and orbitofrontal cortex is the inferior temporal visual cortex. One route for the outputs from these evaluative brain structures is via projections directly to structures such as the basal ganglia (including the striatum and ventral striatum) and cingulate cortex to enable implicit, direct behavioural responses based on the reward- or punishment-related evaluation of the stimuli to be made. The second route is via the language systems of the brain, which allow explicit (verbalizable) decisions involving multistep syntactic planning to be implemented.

system is that it is aimed directly towards obtaining a sensed or expected reward, by virtue of connections to brain systems such as the basal ganglia which are concerned with the initiation of actions (see Fig. 5.1). The expectation may of course involve behaviour to obtain stimuli associated with reward, which might even be present in a fixed chain or sequence.

Now part of the way in which the behaviour is controlled with this first ('implicit') route is according to the reward value of the outcome. At the same time, the animal may only work for the reward if the cost is not too high. Indeed, in the field of behavioural ecology animals are often thought of as performing optimally on some cost–benefit curve (see, e.g. Krebs and Kacelnik (1991)). This does not at all mean that the animal thinks about the long-term rewards, and performs a cost–benefit analysis using a lot of thoughts about the costs, other rewards (short and long term) available and their costs, etc. (see Section 8.4.2). Instead, it should be taken to mean that in evolution the system has evolved in such a way that the way in which the reward varies with the different energy densities or amounts of food and the delay before it is received can be used as part of the input to a mechanism which has also been built to track the costs of obtaining the food (e.g. energy loss in obtaining it, risk of predation, etc.), and to then select given many such types of reward and the associated cost, the current behaviour that provides the most 'net reward'. Part of the value of having the computation expressed in this reward-minus-cost form is that there is then a suitable 'currency', or net reward value, to enable the animal to select the behaviour with currently the most net reward gain (or minimal aversive outcome) (Rolls 2005a, Grabenhorst and Rolls 2011).

Part of the evidence that this implicit route often controls emotional behaviour in humans is that humans with orbitofrontal cortex damage have impairments in selecting the correct action during visual discrimination reversal, yet can state explicitly what the correct action should be (Rolls, Hornak, Wade and McGrath 1994a, Rolls 1999b). The implication is that the intact orbitofrontal cortex is normally involved in making rapid emotion-related decisions, and that this emotion-related decision system is a separate system from the explicit system, which by serial reasoning can provide an alternative route to action. The explicit system may simply comment on the success or failure of actions that are initiated by the implicit system, and the explicit system may then be able to switch in to control mode to correct failures of the implicit system. Consistent evidence that an implicit system can control human behaviour is that in psychophysical and neurophysiological studies, it has been found that face stimuli presented for 16 ms and followed immediately by a mask are not consciously perceived, yet produce above chance identification (Rolls and Tovee 1994, Rolls, Tovee, Purcell, Stewart and Azzopardi 1994b, Rolls, Tovee and Panzeri 1999, Rolls 2003, Rolls 2006a). In a similar backward masking paradigm, it was found that happy vs angry face expressions could influence how much beverage was wanted and consumed even when the faces were not consciously perceived (Winkielman and Berridge 2005, Winkielman and Berridge 2003). Thus unconscious emotion-related stimuli (in this case face expressions) can influence actions, and there is no need for processing to be conscious for actions to be initiated. Further, in blindsight, humans with damage to the primary visual cortex may not be subjectively aware of stimuli, yet may be able to guess what the stimulus was, or to perform reaching movements towards it (Weiskrantz 1998). Further, humans with striate cortex lesions may be influenced by emotional stimuli which are not perceived consciously (De Gelder, Vroomen, Pourtois and Weiskrantz 1999).

Thus actions and emotions can be initiated without the necessity for the conscious route to be in control, and we should not infer that all actions require conscious processing.

5.3 The reasoning, rational, route to action

The second ('explicit') route in (at least) humans involves a computation with many 'if ... then' statements, to implement a plan to obtain a reward. In this case, the reward may actually be deferred as part of the plan, which might involve working first to obtain one reward, and only then to work for a second more highly valued reward, if this was thought to be overall an optimal strategy in terms of resource usage (e.g. time). In this case, syntax is required, because the many symbols (e.g. names of people) that are part of the plan must be correctly linked or bound. Such linking might be of the form: 'if A does this, then B is likely to do this, and this will cause C to do this ...'. The requirement of syntax for this type of planning implies that an output to language systems in the brain is required for this type of planning (see Fig. 5.1). **Thus the explicit language system in humans may allow working for deferred rewards by enabling use of a one-off, individual, plan appropriate for each situation.** This explicit system may allow immediate rewards to be deferred, as part of a long-term plan. This ability to defer immediate rewards and plan syntactically in this way for the long term may be an important way in which the explicit system extends the capabilities of the implicit emotion systems that respond more directly to rewards and punishers, or to rewards and punishers with fixed expectancies such as can be learned by reinforcement learning (Rolls 2008c).

Consistent with the point being made about evolutionarily old emotion-based decision systems vs a recent rational system present in humans (and perhaps other animals with syntactic processing) is that humans trade off immediate costs/benefits against cost/benefits that are delayed by as much as decades, whereas non-human primates have not been observed to engage in unpreprogrammed delay of gratification involving more than a few minutes (Rachlin 1989, Kagel, Battalio and Green 1995, McClure, Laibson, Loewenstein and Cohen 2004). However this is a potentially interesting area for further investigation, for example in connection with delay of reward (see Section 8.4.2), and with observations that chimpanzees may save stones to throw later at visitors and perhaps show other signs of forethought (Osvath 2009, Osvath and Osvath 2008).

Another building block for such planning operations in the brain may be the type of short-term memory in which the prefrontal cortex is involved. This short-term memory may be, for example in non-human primates, of where in space a response has just been made. A development of this type of short-term response memory system in humans to enable multiple short-term memories to be held in place correctly, preferably with the temporal order of the different items in the short-term memory coded correctly, may be another building block for the multiple step 'if .... then' type of computation in order to form a multiple step plan. Such short-term memories are implemented in the (dorsolateral and inferior convexity) prefrontal cortex of non-human primates and humans

(Goldman-Rakic 1996, Petrides 1996, Rolls and Deco 2002, Deco and Rolls 2003), and may be part of the reason why prefrontal cortex damage impairs planning and executive function (see Shallice and Burgess (1996)).

Of these two routes (see Fig. 5.1), it is the second that I suggest is related to consciousness (Chapter 6). The hypothesis is that consciousness is the state that arises by virtue of having the ability to think about one's own thoughts, which has the adaptive value of enabling one to correct long multistep syntactic plans. This latter system is thus the one in which explicit, declarative, processing occurs. Processing in this system is frequently associated with reason and rationality, in that many of the consequences of possible actions can be taken into account. The actual computation of how rewarding a particular stimulus or situation is, or will be, probably still depends on activity in the orbitofrontal and amygdala, as the reward value of stimuli is computed and represented in these regions. Further, it is found that verbalized expressions of the reward (or punishment) value of stimuli are dampened by damage to these systems. (For example, damage to the orbitofrontal cortex renders painful input still identifiable as pain, but without the strong affective, 'unpleasant', reaction to it.)

This language system that enables long-term planning may be contrasted with the first system in which behaviour is directed at obtaining the stimulus (including the remembered stimulus) which is currently most rewarding, as computed by brain structures that include the orbitofrontal cortex and amygdala. There are outputs from this system, perhaps those directed at the basal ganglia and cingulate cortex, which do not pass through the language system, and behaviour produced in this way is described as implicit, and verbal declarations cannot be made directly about the reasons for the choice made. When verbal declarations are made about decisions made in this first ('implicit') system, those verbal declarations may be confabulations, reasonable explanations or fabrications, of reasons why the choice was made. These reasonable explanations would be generated to be consistent with the sense of continuity and self that is a characteristic of reasoning in the language system.

5.4   The Selfish Gene vs The Selfish Individual

I have provided evidence in the preceding sections that there are two main routes to decision-making and action.

The first route selects actions by gene-defined goals for action, and is closely associated with emotion. The second route involves multistep planning and reasoning which requires syntactic processing to keep the symbols involved at each step separate from the symbols in different steps. (This second route is used by humans and perhaps by closely related animals.) Now the 'interests' of the first and second routes to decision-making and action are different. As argued very convincingly by Richard Dawkins in *The Selfish Gene* (Dawkins 1976), and by others (Hamilton 1964, Williams 1966, Ridley 1993b, Hamilton 1996), many behaviours occur in the interests of the survival of the genes, not of the individual (nor of the group), and much behaviour can be understood in this way.

I have extended this approach by arguing that an important role for some genes in evolution is to define the goals for actions that will lead to better survival of those genes; that emotions are the states associated with these gene-defined goals; and that the defining of goals for actions rather that actions themselves is an efficient way for genes to operate, as it leaves flexibility of choice of action open until the animal is alive (Rolls 2005a). This provides great simplification of the genotype as action details do not need to be specified, just rewarding and punishing stimuli, and also flexibility of action in the face of changing environments faced by the genes. Thus the interests that are implied when the first route to action is chosen are those of the 'selfish genes', not those of the individual.

However, the second route to action allows, by reasoning, decisions to be taken that might not be in the interests of the genes, might be longer term decisions, and might be in the interests of the individual. An example might be a choice not to have children, but instead to devote oneself to science, medicine, music, or literature. Another example might be a person's decision not to sacrifice her resources and wealth by making major provisions at her own expense for her children (as is common in many animals with its adaptive value in enabling the genes in the children and other kin to survive better), but instead using the resources and wealth to continue to provide for herself, for example in old age, or for example in promoting pleasurable experiences.

The reasoning, rational, system presumably evolved because taking longer-term decisions involving planning rather than choosing a gene-defined goal might be advantageous at least sometimes for genes. But an unforeseen consequence of the evolution of the rational system might be that the decisions would, sometimes, not be to the advantage of the genes in the organism. After all, evolution by natural selection operates utilizing genetic variation like a *Blind Watchmaker* (Dawkins 1986). In this sense, the interests when the second route to decision-making is used are at least sometimes those of the 'selfish phenotype', the selfish individual. (Indeed, we might euphonically say that the interests are those of the 'selfish phene' (where the etymology is Gk phaino, 'appear', referring to appearance, hence the thing that one observes, the individual) (Rolls 2011b).) An example is that an individual person, a phenotype, might decide to devote himself or herself to art, scholarship, science, and/or medicine, and might decide for these reasons not to have children. I know a number of academic colleagues who have taken such decisions, highly rationally. Hence the decision-making referred to in the preceding sections is between a first system where the goals are gene-defined, and a second rational system in which the decisions may be made in the interests of the phenotype, of the individual living person with rational foresight, and not in the interests of the genes. Thus we may speak of the choice as sometimes being between the 'Selfish Genes' and the 'Selfish Individual' or 'Selfish Phene' (Rolls 2011b).

To be clear: I am suggesting that in addition to gene-level selection, which is extremely important as a driver of evolution (Hamilton 1964, Williams 1966, Hamilton 1996, Dawkins 1976), there is now a sense in which the reasoning system might make choices that are not in the interests of the genes, but in the interests of the individual, or for that matter not in the interests of the individual or of the genes, in choices that might be described as truly altruistic.

Now what keeps the decision-making between the 'Selfish Genes' and the 'Selfish Individual' more or less under control and in balance? If the second, rational, system chose too often for the interests of the 'Selfish Individual', the genes in that phenotype might not survive over generations, as there might be no descendants. Having these two systems in the same individual will only be stable if their potency is approximately equal, so that sometimes decisions are made with the first route, and sometimes with the second route. If the two types of decision-making, then, compete with approximately equal potency, and sometimes one is chosen, and sometimes the other, then this is exactly the scenario in which stochastic processes in the decision-making mechanism are likely to play an important role in the decision that is taken. The same decision, even with the same evidence, may not be taken each time a decision is made, because of noise in the system.

The system itself may have some properties that help to keep the system operating well. One is that if the second, rational, system tends to dominate the decision-making too much, the first, gene-based emotional system might fight back over generations of selection, and enhance the magnitude of the reward value specified by the genes, so that emotions might actually become stronger as a consequence of them having to compete in the interests of the selfish genes with the rational decision-making process.

However, it is also conceivable that the reasoning system might promote genetic ends. For example, if a person believed that her kin had been persecuted and killed, she might take a reasoned decision to remedy the perceived injustice by having as many children as possible. The net effect of this reason-led individual-level selection on evolution is thus far from clear.

Another property of the system may be that sometimes the rational system cannot gain all the evidence that would be needed to make a rational choice. Under these circumstances the rational system might fail to make a clear decision, and under these circumstances, basing a decision on the gene-specified emotions is an alternative. Indeed, Damasio (1994) argued that under circumstances such as this, emotions might take an important role in decision-making.

In this respect, I agree with him, basing my reasons on the arguments above. He called the emotional feelings gut feelings, and, in contrast to me, hypothesized that actual feedback from the gut was involved. His argument seemed to be that if the decision was too complicated for the rational system, then send outputs to the viscera, and whatever is sensed by what they send back could be used in the decision-making, and would account for the conscious feelings of the emotional states. My reading of the evidence is that the feedback from the periphery is not necessary for the emotional decision-making, or for the feelings, nor would it be computationally efficient to put the viscera in the loop given that the information starts from the brain, but that is a matter considered elsewhere (Rolls 2005a).

Another property of the system is that the interests of the second, rational, system, although involving a different form of computation, should not be too far from those of the gene-defined emotional system, for the arrangement to be stable in evolution by natural selection. One way that this could be facilitated would be if the gene-based goals felt pleasant or unpleasant in the rational system, and in this way contributed to the operation of the second, rational, system. This is something that I propose is the case.

5.5 Decision-making: the roles of the implicit and explicit systems

The question then arises of how decisions are made in animals such as humans that have both the implicit, direct reward-based, and the explicit, rational, planning systems (see Fig. 5.1). One particular situation in which the first, implicit, system may be especially important is when rapid reactions to stimuli with reward or punishment value must be made, for then the direct connections from structures such as the orbitofrontal cortex to the basal ganglia may allow rapid actions. Another is when there may be too many factors to be taken into account easily by the explicit, rational, planning, system, then the implicit system may be used to guide action.

In contrast, when the implicit system continually makes errors, it would then be beneficial for the organism to switch from automatic, direct, action based on obtaining what the orbitofrontal cortex system decodes as being the most positively reinforcing choice currently available, to the explicit conscious control system which can evaluate with its long-term planning algorithms what action should be performed next. Indeed, it would be adaptive for the explicit system to be regularly assessing performance by the more automatic system, and to switch itself in to control behaviour quite frequently, as otherwise the adaptive value of having the explicit system would be less than optimal.

Another factor that may influence the balance between control by the implicit and explicit systems is the presence of pharmacological agents such as alcohol, which may alter the balance towards control by the implicit system, may allow the implicit system to influence more the explanations made by the explicit system, and may within the explicit system alter the relative value it places on caution and restraint vs commitment to a risky action or plan.

There may also be a flow of influence from the explicit, verbal system to the implicit system, in that the explicit system may decide on a plan of action or strategy, and exert an influence on the implicit system that will alter the reinforcement evaluations made by and the signals produced by the implicit system. An example of this might be that if a pregnant woman feels that she would like to escape a cruel mate, but is aware that she may not survive in the jungle, then it would be adaptive if the explicit system could suppress some aspects of her implicit behaviour towards her mate, so that she does not give signals that she is displeased with her situation[21]. Another example might be that the explicit system might, because of its long-term plans, influence the implicit system to increase its response to for example a positive reinforcer. One way in which the explicit system might influence the implicit system is by setting up the conditions in which, for example, when a given stimulus (e.g. person) is present, positive reinforcers are given, to facilitate stimulus–reinforcement association learning by the implicit system of the person receiving the positive reinforcers. Conversely, the implicit system may influence the explicit system, for example by highlighting certain stimuli in the environment that are currently associated with reward, to guide the attention of the explicit system to such stimuli.

However, it may be expected that there is often a conflict between these systems, in that the implicit system is able to guide behaviour particularly to obtain the greatest immediate reinforcement, whereas the explicit system can potentially enable immediate rewards to be deferred, and longer-term, multistep, plans to be formed. This type of conflict will occur in animals with a syntactic planning ability, that is in humans and any other animals that have the ability to process a series of 'if ... then' stages of planning. This is a property of the human language system, and the extent to which it is a property of non-human primates is not yet fully clear. In any case, such conflict may be an important aspect of the operation of at least the human mind, because it is so essential for humans to decide correctly, at every moment, whether to invest

---

[21] In the literature on self-deception, it has been suggested that unconscious desires may not be made explicit in consciousness (or actually repressed), so as not to compromise the explicit system in what it produces; see, e.g. Alexander (1975), Alexander (1979), Trivers (1976), Trivers (1985); and the review by Nesse and Lloyd (1992).

in a relationship or a group that may offer long-term benefits, or whether to pursue immediate benefits directly (Nesse and Lloyd 1992)[22].

Some investigations on deception in non-human primates have been interpreted as showing that animals can plan to deceive others (see, e.g. Griffin (1992), Byrne and Whiten (1988), and Whiten and Byrne (1997)), that is to utilize 'Machiavellian intelligence'. For example, a baboon might 'deliberately' mislead another animal in order to obtain a resource such as food (e.g. by screaming to summon assistance in order to have a competing animal chased from a food patch) or sex (e.g. a female baboon who very gradually moved into a position from which the dominant male could not see her grooming a subadult baboon) (see Dawkins (1993)). The attraction of the Machiavellian argument is that the behaviour for which it accounts seems to imply that there is a concept of another animal's mind, and that one animal is trying occasionally to mislead another, which implies some planning. However, such observations tend by their nature to be field-based, and may have an anecdotal character, in that the previous experience of the animals in this type of behaviour, and the reinforcements obtained, are not known (Dawkins 1993). It is possible, for example, that some behavioural responses that appear to be Machiavellian may have been the result of previous instrumental learning in which reinforcement was obtained for particular types of response, or of observational learning, with again learning from the outcome observed. However, in any case, most examples of Machiavellian intelligence in non-human primates do not involve multiple stages of 'if ... then' planning requiring syntax to keep the symbols apart (but may involve associative learning which might lead to a description of the type 'if the dominant male sees me grooming a subadult male, I will be punished') (see Dawkins (1993)). Nevertheless, the possible advantage of such Machiavellian planning could be one of the adaptive guiding factors in evolution that provided advantage to a multistep, syntactic system that enables long-term planning, the best example of such a system being human language.

Another, not necessarily exclusive, advantage of the evolution of a linguistic multistep planning system could well be not Machiavellian planning, but planning for social co-operation and advantage. Perhaps in general an 'if ... then' multistep syntactic planning ability is useful primarily in evolution in social situations of the type: 'if X does this, then Y does that; then I would/should do that, and the outcome would be ... '. It is not yet at all clear whether such

---

[22] As Nesse and Lloyd (1992) describe, some psychoanalysts ascribe to a somewhat similar position, for they hold that intrapsychic conflicts usually seem to have two sides, with impulses on one side and inhibitions on the other. Analysts describe the source of the impulses as the id, and the modules that inhibit the expression of impulses, because of external and internal constraints, the ego and superego respectively (Leak and Christopher 1982, Trivers 1985, Nesse and Lloyd 1992). The superego can be thought of as the conscience, while the ego is the locus of executive functions that balance satisfaction of impulses with anticipated internal and external costs. A difference of the present position is that it is based on identification of dual routes to action implemented by different systems in the brain, each with its own selective advantage.

planning is required in order to explain the social behaviour of social animals such as hunting dogs, or socializing monkeys (Dawkins 1993).

However, in humans, members of 'primitive' hunting tribes spend hours recounting tales of recent events (perhaps who did what, when; who then did what, etc.), perhaps to help learn from experience about good strategies, necessary for example when physically weak men take on large animals (see Pinker and Bloom (1992)).

Thus, social co-operation may be as powerful a driving force in the evolution of syntactical planning systems as Machiavellian intelligence. What is common to both is that they involve social situations. However, such a syntactic planning system would have advantages not only in social systems, for such planning may be useful in obtaining resources purely in a physical (non-social) world. An example might be planning how to cross terrain given current environmental constraints in order to reach a particular place[23].

This discussion of dual routes to action has been with respect to the behaviour produced. There is of course in addition a third output of brain regions such as the orbitofrontal cortex and amygdala involved in emotion, which is directed to producing autonomic and endocrine responses. Although it has been argued in Chapter 3 that the autonomic system is not normally in a circuit through which behavioural responses are produced (i.e. against the James–Lange and related theories), there may be some influence from effects produced through the endocrine system (and possibly the autonomic system, through which some endocrine responses are controlled) on behaviour, or on the dual systems just discussed that control behaviour. For example, during female orgasm the hormone oxytocin may be released, and this may influence the implicit system to help develop positive reinforcement associations and thus attachment to her lover.

The thrust of the argument thus is that much complex animal, including human, behaviour can take place using the implicit, non-conscious, route to action. We should be very careful not to postulate intentional states (i.e. states with intentions, beliefs, and desires) unless the evidence for them is strong, and it seems to me that a flexible, one-off, linguistic processing system that can handle propositions is needed for intentional states. What the explicit, linguistic,

[23]Tests of whether such multistep planning might be possible in even non-human primates are quite difficult to devise. One example might be to design a multistep maze. On a first part of the trial, the animal might be allowed to choose for itself, given constraints set on that trial to ensure trial unique performance, a set of choices through a maze. On the second part of that trial, the animal would be required to run through the maze again, remembering and repeating every choice just made in the first part of that trial. This part of the design is intended to allow recall of a multistep plan. To test on probe occasions whether the plan is being recalled, and whether the plan can be corrected by a higher-order thought process, the animal might be shown after the first part of a trial, that one of its previous free choices was not now available. The test would be to determine whether the animal can make a set of choices that indicate corrections to the multistep plan, in which the trajectory has to be altered before the now unavailable choice point is reached.

system does allow is exactly this flexible, one-off, multistep planning-ahead type of computation, which allows us to defer immediate rewards based on such a plan.

Emotions as actions, and emotions as affects, are sometimes contrasted. My view on this is that sometimes emotions can lead to actions implicitly, without the need for conscious processing. However, when emotions involve longer term planning, then representation and processing in the explicit system is required, and affective feelings will then be inextricably linked to the processing.

5.5.1 Selection between conscious vs unconscious decision-making, and free will

An application of the type of model of noisy decision-making described in Section 2.12 is to taking decisions between the implicit (unconscious, which might be a gene-specified reward based emotional system) and the explicit (conscious, rational) systems (Rolls 2005a, Rolls 2008c), where the two different systems could provide the biasing inputs λ_1 and λ_2 to the model. An implication is that noise will influence with probabilistic outcomes which system, the implicit or the conscious reasoning system, takes a decision.

When decisions are taken, sometimes confabulation may occur, in that a verbal account of why the action was performed may be given, and this may not be related at all to the environmental event that actually triggered the action (Gazzaniga and LeDoux 1978, Gazzaniga 1988, Gazzaniga 1995, Rolls 2005a, LeDoux 2008). It is accordingly possible that sometimes in normal humans when actions are initiated as a result of processing in a specialized brain region such as those involved in some types of rewarded behaviour, the language system may subsequently elaborate a coherent account of why that action was performed (i.e. confabulate). This would be consistent with a general view of brain evolution in which, as areas of the cortex evolve, they are laid on top of existing circuitry connecting inputs to outputs, and in which each level in this hierarchy of separate input–output pathways may control behaviour according to the specialized function it can perform.

This raises the issue of free will in decision-making (developed further in Section 6.5).

First, we can note that in so far as the brain operates with some degree of randomness due to the statistical fluctuations produced by the random spiking times of neurons, brain function is to some extent non-deterministic, as defined in terms of these statistical fluctuations. That is, the behaviour of the system, and of the individual, can vary from trial to trial based on these statistical fluctuations, in ways that are described in this book. Indeed, given that each neuron has this randomness, and that there are sufficiently small numbers of synapses on the neurons in each network (between a few thousand and 20,000) that these statistical fluctuations are not smoothed out, and that there are a number of different networks involved in typical thoughts and actions each one of which may behave probabilistically, and with 10^{11} neurons in the brain each with this number of synapses, the system has so many degrees of freedom that it operates effectively as a non-deterministic system. (Philosophers may wish to argue about different senses of the term deterministic, but it is being used here in a precise, scientific, and quantitative way, which has been clearly defined.)

Second, do we have free will when both the implicit and the explicit systems have made the choice? Free will would in Rolls' view (Rolls 2005a, Rolls 2008a, Rolls 2008c, Rolls 2010d, Rolls 2011b) involve the use of language to check many moves ahead on a number of possible series of actions and their outcomes, and then with this information to make a choice from the likely outcomes of different possible series of actions. (If, in contrast, choices were made only on the basis of the reinforcement value of immediately available stimuli, without the arbitrary syntactic symbol manipulation made possible by language, then the choice strategy would be much more limited, and we might not want to use the term free will, as all the consequences of those actions would not have been computed.) Rolls' view is that *when this type of reflective, reasoning, conscious, information processing is occurring and leading to action, the system performing this processing and producing the action would have to believe that it could cause the action*, for otherwise inconsistencies would arise, and the system might no longer try to initiate action. *This belief held by the system may partly underlie the feeling of free will.*

At other times, when other brain modules are initiating actions (in the implicit systems), the conscious processor (the explicit system) may confabulate and believe that it caused the action, or at least give an account (possibly wrong) of why the action was initiated. The fact that the conscious processor may have the belief even in these circumstances that it initiated the action may arise as a property of it being inconsistent for a system that can take overall control using conscious verbal processing to believe that it was overridden by another system. This may be the underlying computational reason why confabulation occurs.

The interesting view we are led to is thus that when probabilistic choices influenced by stochastic dynamics are made between the implicit and explicit systems, we may not be aware of which system made the choice. Further, when the stochastic noise has made us choose with the implicit system, we may confabulate and say that we made the choice of our own free will, and provide a guess at why the decision was taken. In this scenario, the stochastic dynamics of the brain plays a role even in how we understand free will (Rolls 2010d, Rolls 2011b) (see further Section 6.5).

5.6 Conclusions

In this chapter I have contrasted the rational, i.e. reasoning, system with the much evolutionarily older reward and punishment based value systems (Chapter 3). These latter provide heuristics for animals to be guided towards reproductive success for their selfish genes. The rational system on the other hand enables these gene-specified rewards and punishers to be over-ridden, for longer term benefits, and for benefits that may be more aligned to those of the individual (the phenotype) than of the individual's genes (the genotype). In this sense, there may be a conflict of interests between these different brain systems (see further Sections 9.2 and 13.2).

I have also argued that when probabilistic choices are made between the implicit and explicit systems, we may not be aware of which system made the choice. Further, when the stochastic noise has made us choose with the implicit system, we may confabulate and say that we made the choice of our own free will, and provide a guess at why the decision was taken. Taking a mechanistic view of brain function, that is understanding the actual computational mechanisms involved in the operation of the brain, thus gives us deep insight into why we behave the way we do, and how this behaviour may be influenced by the stochastic neurodynamics of the brain (Rolls 2008c, Rolls and Deco 2010).

These conclusions have important implications not only for understanding human behaviour and human society, but also for understanding the differences between individuals in the operation of these computational systems in the brain. There are in turn important ethical and related questions that arise, including for example how we evaluate peoples' stated intentions, and how we care for individuals who for example may have brain damage to parts of these systems.

6 Neurophilosophy

6.1 Introduction

We consider here a neuroscience-based approach to the following issues. What is the relation between the mind and the brain? Do mental, mind, events cause brain events? Do brain events cause mental effects? What can we learn from the relation between software and hardware in a computer about mind–brain interactions and how causality operates? The hard problem of consciousness: why does some mental processing feel like something, and other mental processing does not? What type of processing is occurring when it does feel like something? Is consciousness an epiphenomenon, or is it useful? Are we conscious of the action at the time it starts, or later? How is the world represented in the brain?

6.2 The mind–brain problem

What is the relation between the mind and the brain? This is the mind–brain or mind–body problem. Do mental, mind, events cause brain events? Do brain events cause mental effects? What can we learn from the relation between software and hardware in a computer about mind–brain interactions and how causality operates? Neuroscience shows that there is a close relation between mind and matter (captured by the following inverted saying: 'Never matter, no mind').

My view is that the relationship between mental events and neurophysiological events is similar (apart from the problem of consciousness) to the relationship between the program running in a computer and the hardware of the computer. In a sense, the program (the software loaded onto the computer usually written in a high-level language such as C or Matlab) causes the logic gates (TTL, transistor-transistor logic) of the hardware to move to the next state. This hardware state change causes the program to move to its next step or state. Effectively, we are looking at different levels of what is overall the operation of a system, and causality can usefully be understood as operating both within levels (causing one step of the program to move to the next), as well as between levels (e.g. software to hardware and vice versa). This is the solution I propose to this aspect of the mind–body (or mind–brain) problem.

There are alternative ways of treating the mind–brain issue. Another is to consider the process as a mechanism with different levels of explanation. As described in Section 2.14, we can now understand brain processing from the level of ion channels in neurons, through neuronal biophysics, to neuronal firing, through the computations performed by populations of neurons, to behavioural and cognitive effects, and even perhaps to the phenomenal (feeling) aspects of consciousness as described later in this chapter. The whole processing is now specified from the mechanistic level of neuronal firings, etc. up through the computational level to the cognitive and behavioural level. Sometimes the cognitive effects seem remarkable, for example the recall of a whole memory from a part of it, and we describe this as an 'emergent property', but once understood from the mechanistic level upwards, the functions implemented are elegant and wonderful, but understandable and not magical or poorly understood (Rolls 2008c). Different philosophers may choose or not to say that causality operates between these different levels of explanation, but the point I make is that however they speak about causality in such a mechanistic system with interesting 'emergent' computational properties, the system is now well-defined, is no longer mysterious or magical, and we have now from a combination of neuroscience and analyses of the type used in theoretical physics a clear understanding of the properties of neural systems and how cognition emerges from neural mechanisms. There are of course particular problems that remain to be resolved with this approach, such as that of how language is implemented in the brain, but my point is that this mechanistic approach, supported by parsimony, appears to be capable of leading us to a full understanding of brain function, cognition, and behaviour.

A possible exception where a complete explanation may not emerge from the mechanistic approach is phenomenal consciousness, which is treated next.

6.3 Consciousness

6.3.1 Introduction

It might be possible to build a computer that would perform the functions of emotions described in Chapter 3, and yet we might not want to ascribe emotional feelings to the computer. We might even build the computer with some of the main processing stages present in the brain, and implemented using neural networks that simulate the operation of the real neural networks in the brain (see Rolls and Treves (1998), Rolls and Deco (2002), and Rolls (2008c)), yet we might not still wish to ascribe emotional feelings to this computer. This point often arises in discussions with undergraduates, who may say that they follow the types of point made about emotion in Chapter 3, yet believe that almost the most important aspect of emotions, the feelings, have not been accounted for, nor their neural basis described. In a sense, the functions of reward and punishment in emotional behaviour are described in Chapter 3, but what about the subjective aspects of emotion, what about the pleasure?

A similar point also arises in Chapter 3, where parts of the taste, olfactory, and visual systems in which the reward value of the taste, smell, and sight of food is represented are described. Although the neuronal representation in the orbitofrontal cortex is clearly related to the reward value of food, and in humans the activations found with functional neuroimaging are directly correlated with the reported subjective pleasantness of the stimuli, is this where the pleasantness (the subjective hedonic aspect) of the taste, smell, and sight of food is represented and produced? Again, we could (in principle at least) build a computer with neural networks to simulate each of the processing stages for the taste, smell, and sight of food which are described in Chapter 3, and yet would probably not wish to ascribe feelings of subjective pleasantness to the system we have simulated on the computer.

What is it about neural processing that makes it feel like something when some types of information processing are taking place? It is clearly not a general property of processing in neural networks, for there is much processing, for example that in the autonomic nervous system concerned with the control of our blood pressure and heart rate, of which we are not aware. Is it then that awareness arises when a certain type of information processing is being performed? If so, what type of information processing? And how do emotional feelings, and sensory events, come to feel like anything? These 'feels' are called qualia. These are great mysteries that have puzzled philosophers for centuries. They are at the heart of the problem of consciousness, for why it should feel like something at all is the great mystery, the 'hard' problem.

Other aspects of consciousness may be easier to analyse, such as the fact that often when we 'pay attention' to events in the world, we can process those events in some better way. These are referred to as 'process' or 'access' aspects of consciousness, as opposed to the 'phenomenal' or 'feeling' aspects of consciousness referred to in the preceding paragraph (Block 1995a, Chalmers 1996, Allport 1988, Koch 2004, Block 1995b).

The puzzle of qualia, that is of the phenomenal aspect of consciousness, seems to be rather different from normal investigations in science, in that there is no agreement on criteria by which to assess whether we have made progress. So, although the aim of this section is to address the issue of consciousness, especially of qualia, what is written cannot be regarded as being as firmly scientific as most research relating to brain function (Rolls 2008c, Rolls and Deco 2010). For most of the work in those, there is good evidence for most of the points made, and there would be no hesitation or difficulty in adjusting the view of how things work as new evidence is obtained. However, in the work on qualia, the criteria are much less clear. Nevertheless, the reader may well find these issues interesting, because although not easily solvable, they are very important issues to consider if we wish to really say that we understand some of the very complex and interesting issues about brain function, and ourselves.

With these caveats in mind, I consider in this section (6.3) the general issue of consciousness and its functions, and how feelings, and pleasure, come to occur as a result of the operation of our brains. A view on consciousness, influenced by contemporary cognitive neuroscience, is outlined next. I outline a theory of what the processing is that is involved in consciousness, of its adaptive value in an evolutionary perspective, and of how processing in our visual and other sensory systems can result in subjective or phenomenal states, the 'raw feels' of conscious awareness. However, this view on consciousness that I describe is only preliminary, and theories of consciousness are likely to develop considerably. Partly for these reasons, this theory of consciousness, at least, should not be taken to have practical implications.

6.3.2 A theory of consciousness

6.3.2.1 Conscious and unconscious routes to action

A starting point is that many actions can be performed relatively automatically, without apparent conscious intervention. An example sometimes given is driving a car. Such actions could involve control of behaviour by brain systems that are old in evolutionary terms such as the basal ganglia. It is of interest that the basal ganglia (and cerebellum) do not have backprojection systems to most of the parts of the cerebral cortex from which they receive inputs (see Rolls (2005a)). In contrast, parts of the brain such as the hippocampus and amygdala, involved in functions such as episodic memory and emotion respectively, about which we can make (verbal) declarations (hence declarative memory, Squire (1992)) do have major backprojection systems to the high parts of the cerebral cortex from which they receive forward projections (see Fig. 2.14). It may be that evolutionarily newer parts of the brain, such as the language areas and parts of the prefrontal cortex, are involved in an alternative type of control of behaviour, in which actions can be planned with the use of a (language) system that allows relatively arbitrary (syntactic) manipulation of semantic entities (symbols).

The general view that there are many routes to behavioural output is supported by the evidence that there are many input systems to the basal ganglia (from almost all areas of the cerebral cortex), and that neuronal activity in each part of the striatum reflects the activity in the overlying cortical area (see Rolls (2008c)). The evidence is consistent with the possibility that different cortical areas, each specialized for a different type of computation, have their outputs directed to the basal ganglia, which then select the strongest input, and map this into action (via outputs directed, for example, to the premotor cortex). Within this scheme, the language areas would offer one of many routes to action, but a route particularly suited to planning actions, because of the role of the language

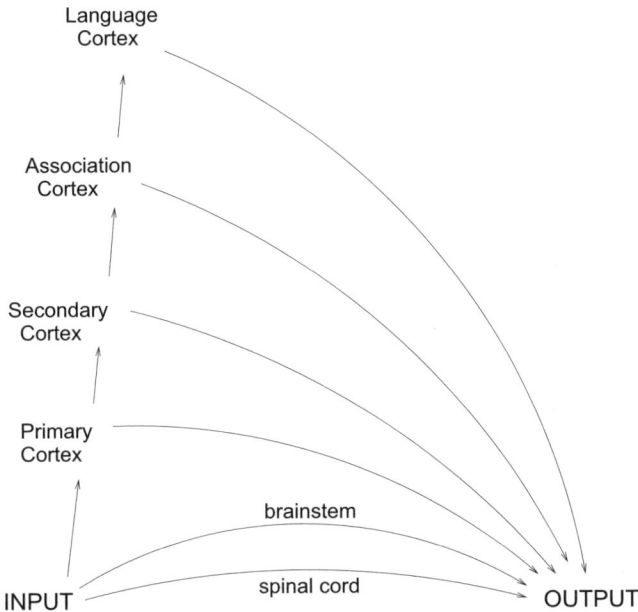

Fig. 6.1 Schematic illustration indicating many possible routes from input systems to action (output) systems. Cortical information-processing systems are organized hierarchically, and there are routes to output systems from most levels of the hierarchy.

areas in the syntactic manipulation of semantic entities that may make long-term planning possible. A schematic diagram of this suggestion is provided in Fig. 6.1.

Consistent with the hypothesis of multiple routes to action, only some of which utilize language, is the evidence that split-brain patients may not be aware of actions being performed by the 'non-dominant' hemisphere (Gazzaniga and LeDoux 1978, Gazzaniga 1988, Gazzaniga 1995). Also consistent with multiple, including non-verbal, routes to action, patients with focal brain damage, for example to the prefrontal cortex, may emit actions, yet comment verbally that they should not be performing those actions (Rolls, Hornak, Wade and McGrath 1994a, Hornak, Bramham, Rolls, Morris, O'Doherty, Bullock and Polkey 2003). In both these types of patient, confabulation may occur, in that a verbal account of why the action was performed may be given, and this may not be related at all to the environmental event that actually triggered the action (Gazzaniga and LeDoux 1978, Gazzaniga 1988, Gazzaniga 1995).

It is accordingly possible that sometimes in normal humans when actions are initiated as a result of processing in a specialized brain region such as those involved in some types of rewarded behaviour, the language system may subsequently elaborate a coherent account of why that action was performed (i.e. confabulate). This would be consistent with a general view of brain evolution in which, as areas of the cortex evolve, they are laid on top of existing circuitry connecting inputs to outputs, and in which each level in this hierarchy of separate input–output pathways may control behaviour according to the specialized function it can perform (see schematic in Fig. 6.1). (It is of interest that mathematicians may get a hunch that something is correct, yet not be able to verbalize why. They may then resort to formal, more serial and language-like, theorems to prove the case, and these seem to require conscious processing. This is a further indication of a close association between linguistic processing, and consciousness. The linguistic processing need not, as in reading, involve an inner articulatory loop.)

6.3.2.2 Higher-order syntactic thoughts and consciousness

We may next examine some of the advantages and behavioural functions that language, present as the most recently added layer to the above system, would confer.

One major advantage would be the ability to plan actions through many potential stages and to evaluate the consequences of those actions without having to perform the actions. For this, the ability to form propositional statements, and to perform syntactic operations on the semantic representations of states in the world, would be important.

Also important in this system would be the ability to have second-order thoughts about the type of thought that I have just described (e.g. I think that she thinks that ..., involving 'theory of mind'), as this would allow much better modelling and prediction of others' behaviour, and therefore of planning, particularly planning when it involves others[24]. This capability for higher-order thoughts would also enable reflection on past events, which would also be useful in planning. In contrast, non-linguistic behaviour would be driven by learned reinforcement associations, learned rules, etc. but not by flexible planning for many steps ahead involving a model of the world including others' behaviour[25]. (For an earlier view that is close to this part of the argument see Humphrey (1980).)

It is important to state that the language ability referred to here is not necessarily human verbal language (though this would be an example). What it is suggested is important to planning is the syntactic manipulation of symbols, and it is this syntactic manipulation of symbols that is the sense in which

---

[24] Second-order thoughts are thoughts about thoughts. Higher-order thoughts refer to second-order, third-order, etc. thoughts about thoughts...

[25] The examples of behaviour from non-humans that may reflect planning may reflect much more limited and inflexible planning. For example, the dance of the honey-bee to signal to other bees the location of food may be said to reflect planning, but the symbol manipulation is not arbitrary. There are likely to be interesting examples of non-human primate behaviour that reflect the evolution of an arbitrary symbol-manipulation system that could be useful for flexible planning, cf. Cheney and Seyfarth (1990), Byrne and Whiten (1988), and Whiten and Byrne (1997).

language is defined and used here. The type of syntactic processing need not be at the natural language level (which implies a universal grammar), but could be at the level of mentalese (Rolls 2005a, Rolls 2004b, Fodor 1994, Rolls 2011b).

It is next suggested that this arbitrary symbol-manipulation using important aspects of language processing and used for planning but not in initiating all types of behaviour is close to what consciousness is about. In particular, consciousness may be the state that arises in a system that can think about (or reflect on) her own (or other peoples') thoughts, that is in a system capable of second- or higher-order thoughts (Rosenthal 1986, Rosenthal 1990, Rosenthal 1993, Dennett 1991). On this account, a mental state is non-introspectively (i.e. non-reflectively) conscious if one has a roughly simultaneous thought that one is in that mental state. Following from this, introspective consciousness (or reflexive consciousness, or self-consciousness) is the attentive, deliberately focused consciousness of one's mental states. It is noted that not all of the higher-order thoughts need themselves be conscious (many mental states are not). However, according to the analysis, having a higher-order thought about a lower-order thought is necessary for the lower-order thought to be conscious.

A slightly weaker position than Rosenthal's on this is that a conscious state corresponds to a first-order thought that has the capacity to cause a second-order thought or judgement about it (Carruthers 1996). Another position that is close in some respects to that of Carruthers and the present position is that of Chalmers (1996), that awareness is something that has direct availability for behavioural control. This amounts effectively for him in humans to saying that consciousness is what we can report about verbally[26]. This analysis is consistent with the points made above that the brain systems that are required for consciousness and language are similar. In particular, a system that can have second- or higher-order thoughts about its own operation, including its

---

[26]Chalmers (1996) is not entirely consistent about this. Later in the same book he advocates a view that experiences are associated with information-processing systems, e.g. experiences are associated with a thermostat (p. 297). He does not believe that the brain has experiences, but that he has experiences. This leads him to suggest that experiences are associated with information-processing systems such as the thermostat in the same way as they are associated with him. "If there is experience associated with thermostats, there is probably experience everywhere: wherever there is a causal interaction, there is information, and wherever there is information, there is experience." (p. 297). He goes on to exclude rocks from having experiences, in that "a rock, unlike a thermostat, is not picked out as an information-processing system". My response to this is that of course there is mutual information between the physical world (e.g. the world of tastants, the chemical stimuli that can produce tastes) and the conscious world (e.g. of taste) – if there were not, the information represented in the conscious processing system would not be useful for any thoughts or operations on or about the world. And according to the view I present here, the conscious processing system is good at some specialized types of processing (e.g. planning ahead using syntactic processing with semantics grounded in the real world, and reflecting on and correcting such plans), for which it would need reliable information about the world. Clearly Chalmers' view on consciousness is very much weaker than mine, in that he allows thermostats to be associated with consciousness, and in contrast to the theory presented here, does not suggest any special criteria for the types of information processing to be performed in order for the system to be aware of its thoughts, and of what it is doing.

planning and linguistic operation, must itself be a language processor, in that it must be able to bind correctly to the symbols and syntax in the first-order system. According to this explanation, the feeling of anything is the state that is present when linguistic processing that involves second- or higher-order thoughts is being performed.

It might be objected that this hypothesis captures some of the process aspects of consciousness, that is, what is useful in an information-processing system, but does not capture the phenomenal aspect of consciousness. I agree that there is an element of 'mystery' that is invoked at this step of the argument, when I say that it feels like something for a machine with higher-order thoughts to be thinking about her own first- or lower-order thoughts. But the return point is the following: if a human with second-order thoughts is thinking about its own first-order thoughts, surely it is very difficult for us to conceive that this would not feel like something? (Perhaps the higher-order thoughts in thinking about the first-order thoughts would need to have in doing this some sense of continuity or self, so that the first-order thoughts would be related to the same system that had thought of something else a few minutes ago. But even this continuity aspect may not be a requirement for consciousness. Humans with anterograde amnesia cannot remember what they felt a few minutes ago, yet their current state does feel like something.)

It is suggested that part of the evolutionary adaptive significance of this type of higher-order thought is that it enables correction of errors made in first-order linguistic or in non-linguistic processing. Indeed, the ability to reflect on previous events is extremely important for learning from them, including setting up new long-term semantic structures. It was shown above that the hippocampus may be a system for such 'declarative' recall of recent memories (see also Squire, Stark and Clark (2004)). Its close relation to 'conscious' processing in humans (Squire has classified it as a declarative memory system) may be simply that it enables the recall of recent memories, which can then be reflected upon in conscious, higher-order, processing. Another part of the adaptive value of a higher-order thought system may be that by thinking about its own thoughts in a given situation, it may be able to understand better the thoughts of another individual in a similar situation, and therefore predict that individual's behaviour better (Humphrey (1980), Humphrey (1986); cf. Barlow (1997)).

As a point of clarification, I note that according to this theory, a language processing system is not sufficient for consciousness. What defines a conscious system according to this analysis is the ability to have higher-order thoughts, and a first-order language processor (which might be perfectly competent at language) would not be conscious, in that it could not think about its own or others' thoughts. One can perfectly well conceive of a system that obeyed the rules of language (which is the aim of much connectionist modelling), and implemented a first-order linguistic system, that would not be conscious. [Possible examples of language processing that might be performed non-consciously include computer programs implementing aspects of language, or ritualized human conversations, e.g. about the weather. These might require syntax and correctly grounded semantics, and yet be performed non-consciously. A more complex example, illustrating that syntax could be used, might be 'If A does X, then B will probably do Y, and then C would be able to do Z.' A first-order language system could process this statement. Moreover, the first-order language system could apply the rule usefully in the world, provided that the symbols in the language system (A, B, X, Y, etc.) are grounded (have meaning) in the world.]

In line with the argument on the adaptive value of higher-order thoughts and thus consciousness given above, that they are useful for correcting lower-order thoughts, I now suggest that correction using higher-order thoughts of lower-order thoughts would have adaptive value primarily if the lower-order thoughts are sufficiently complex to benefit from correction in this way. The nature of the complexity is specific – that it should involve syntactic manipulation of symbols, probably with several steps in the chain, and that the chain of steps should be a one-off (or in American usage, 'one-time', meaning used once) set of steps, as in a sentence or in a particular plan used just once, rather than a set of well learned rules. The first- or lower-order thoughts might involve a linked chain of 'if ... then' statements that would be involved in planning, an example of which has been given above, and this type of cognitive processing is thought to be a primary basis for human skilled performance (Anderson 1996). It is partly because complex lower-order thoughts such as these that involve syntax and language would benefit from correction by higher-order thoughts that I suggest that there is a close link between this reflective consciousness and language.

The hypothesis is that by thinking about lower-order thoughts, the higher-order thoughts can discover what may be weak steps or links in the chain of reasoning at the lower-order level, and having detected the weak link or step, might alter the plan, to see if this gives better success. In our example above, if it transpired that C could not do Z, how might the plan have failed? Instead of having to go through endless random changes to the plan to see if by trial and error some combination does happen to produce results, what I am suggesting is that by thinking about the previous plan, one might, for example, using knowledge of the situation and the probabilities that operate in it, guess that the step where the plan failed was that B did not in fact do Y. So by thinking about the plan (the first- or lower-order thought), one might correct the original plan in such a way that the weak link in that chain, that 'B will probably do Y', is circumvented.

To draw a parallel with neural networks: there is a **'credit assignment'** problem in such multistep syntactic plans, in that if the whole plan fails, how does the system assign credit or blame to particular steps of the plan? [In multilayer neural networks, the credit assignment problem is that if errors are being specified at the output layer, the problem arises about how to propagate back the error to earlier, hidden, layers of the network to assign credit or blame to individual synaptic connection; see Rumelhart, Hinton and Williams (1986) and Rolls (2008c).] **My suggestion is that this solution to the credit assignment problem for a one-off syntactic plan is the function of higher-order thoughts, and is why systems with higher-order thoughts evolved. The suggestion I then make is that if a system were doing this type of processing (thinking about its own thoughts), it would then be very plausible that it should feel like something to be doing this.** I even suggest to the reader that it is not plausible to suggest that it would not feel like anything to a system if it were doing this.

Two other points in the argument should be emphasized for clarity. One is that the system that is having syntactic thoughts about its own syntactic thoughts would have to have its symbols grounded in the real world for it to feel like something to be having higher-order thoughts. The intention of this clarification is to exclude systems such as a computer running a program when there is in addition some sort of control or even overseeing program checking the operation of the first program. We would want to say that in such a situation it would feel like something to be running the higher-level control program only if the first-order program was symbolically performing operations on the world and receiving input about the results of those operations, and if the higher-order system understood what the first-order system was trying to do in the world. The issue of symbol grounding is considered further in Section 6.3.3.

The second clarification is that the plan would have to be a unique string of steps, in much the same way as a sentence can be a unique and one-off string of words. The point here is that it is helpful to be able to think about particular one-off plans, and to correct them; and that this type of operation is very different from the slow learning of fixed rules by trial and error, or the application of fixed rules by a supervisory part of a computer program.

6.3.2.3 Qualia

This analysis does not yet give an account for sensory qualia ('raw sensory feels', for example why 'red' feels red), for emotional qualia (e.g. why a rewarding touch produces an emotional feeling of pleasure), or for motivational qualia (e.g. why food deprivation makes us feel hungry). The view I suggest on such **qualia** is as follows. Information processing in and from our sensory systems (e.g. the sight of the colour red) may be relevant to planning actions using language and the conscious processing thereby implied. Given that these inputs must be represented in the system that plans, we may ask whether it is more likely that we would be conscious of them or that we would not. I suggest that it would be a very special-purpose system that would allow such sensory inputs, and emotional and motivational states, to be part of (linguistically based) planning, and yet remain unconscious (given that the processing being performed by this system is inherently conscious, as suggested above). It seems to be much more parsimonious (Section 7.11) to hold that we would be conscious of such sensory, emotional, and motivational qualia because they would be being used (or are available to be used) in this type of (linguistically based) higher-order thought processing system, and this is what I propose.

The explanation of emotional and motivational subjective feelings or qualia that this discussion has led towards is thus that they should be felt as conscious because they enter into a specialized linguistic symbol-manipulation system, which is part of a higher-order thought system that is capable of reflecting on and correcting its lower-order thoughts involved for example in the flexible planning of actions. It would require a very special machine to enable this higher-order linguistically-based thought processing, which is conscious by its nature, to occur without the sensory, emotional and motivational states (which must be taken into account by the higher-order thought system) becoming felt qualia. The sensory, emotional, and motivational qualia are thus accounted for by the evolution of a linguistic (i.e. syntactic) system that can reflect on and correct its own lower-order processes, and thus has adaptive value.

This account implies that it may be especially animals with a higher-order belief and thought system and with linguistic (i.e. syntactic, not necessarily verbal) symbol manipulation that have qualia. It may be that much non-human animal behaviour, provided that it does not require flexible linguistic planning and correction by reflection, could take place according to reinforcement-guidance. (This reinforcement-guided learning could be implemented using for example stimulus–reinforcer association learning in the amygdala and orbitofrontal cortex followed by action-outcome learning in the cingulate cortex as described in Section 3.11 and elsewhere (Rolls 2005a, Rolls 2009a, Grabenhorst and Rolls 2011); or rule-following using habit or stimulus–response learning in the basal ganglia (Rolls 2005a)). Such behaviours might appear very similar to human behaviour performed in similar circumstances, but need not imply qualia. It would be primarily by virtue of a system for reflecting on flexible, linguistic, planning behaviour that humans (and animals close to humans, with demonstrable syntactic manipulation of symbols, and the ability to think about these linguistic processes) would be different from other animals, and would have evolved qualia.

In order for processing in a part of our brain to be able to reach consciousness, appropriate pathways must be present. Certain constraints arise here. For example, in the sensory pathways, the nature of the representation may change as it passes through a hierarchy of processing levels, and in order to be conscious of the information in the form in which it is represented in early processing stages, the early processing stages must have access to the part of the brain necessary for consciousness. An example is provided by processing in the taste system. In the primate primary taste cortex, neurons respond to taste independently of hunger, yet in the secondary taste cortex, food-related taste neurons (e.g. responding to sweet taste) only respond to food if hunger is present, and gradually stop responding to that taste during feeding to satiety (see Section 3.13 and Rolls (1989c), Rolls (1997b), Rolls (2005a), and Rolls (2012a)). Now the quality of the tastant (sweet, salt, etc.) and its intensity are not affected by hunger, but the pleasantness of its taste is reduced to zero (neutral) (or even becomes unpleasant) after we have eaten it to satiety. The implication of this is that for quality and intensity information about taste, we must be conscious of what is represented in the primary taste cortex (or perhaps in another area connected to it that bypasses the secondary taste cortex), and not of what is represented in the secondary taste cortex. In contrast, for the pleasantness of a taste, consciousness of this could not reflect what is represented in the primary taste cortex, but instead what is represented in the secondary taste cortex (or in an area beyond it) (Rolls 2008c, Grabenhorst and Rolls 2011).

The same argument applies for reward in general, and therefore for emotion, which in primates is not represented early on in processing in the sensory pathways (nor in or before the inferior temporal cortex for vision), but in the areas to which these object analysis systems project, such as the orbitofrontal cortex, where the reward value of visual stimuli is reflected in the responses of neurons to visual stimuli (see Section 3.11 and Rolls (2005a)).

It is also of interest that reward signals (e.g. the taste of food when we are hungry) are associated with subjective feelings of pleasure (see Section 3.13). I suggest that this correspondence arises because pleasure is the subjective state that represents in the conscious system a signal that is positively reinforcing (rewarding), and that inconsistent behaviour would result if the representations did not correspond to a signal for positive reinforcement in both the conscious and the non-conscious processing systems.

Do these arguments mean that the conscious sensation of, e.g. taste quality (i.e. identity and intensity) is represented or occurs in the primary taste cortex, and of the pleasantness of taste in the secondary taste cortex, and that activity in these areas is sufficient for conscious sensations (qualia) to occur? I do not suggest this at all. Instead the arguments I have put forward above suggest that we are only conscious of representations when we have high-order thoughts about them. The implication then is that pathways must connect from each of the brain areas in which information is represented about which we can be conscious (Rolls 2008c, Grabenhorst and Rolls 2011), to the system that has the higher-order thoughts, which as I have argued above, requires language (understood as syntactic manipulation of symbols). Thus,

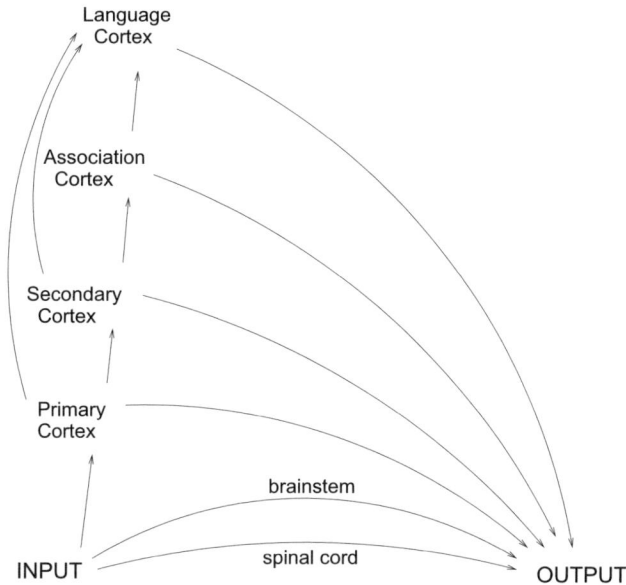

Fig. 6.2 Schematic illustration indicating that early cortical stages in information processing may need access to language areas that bypass subsequent levels in the hierarchy, so that consciousness of what is represented in early cortical stages, and which may not be represented in later cortical stages, can occur. Higher-order linguistic thoughts (HOLTs) could be implemented in the language cortex itself, and would not need a separate cortical area. Backprojections, a notable feature of cortical connectivity, with many probable functions including recall (Rolls and Treves, 1998; Rolls, 2008; Treves and Rolls, 1994), probably reciprocate all the connections shown.

in the example given, there must be connections to the language areas from the primary taste cortex, which need not be direct, but which must bypass the secondary taste cortex, in which the information is represented differently (Rolls 1989c, Rolls 2005a, Rolls 2008c, Rolls 2012a). There must also be pathways from the secondary taste cortex, not necessarily direct, to the language areas so that we can have higher-order thoughts about the pleasantness of the representation in the secondary taste cortex. There would also need to be pathways from the hippocampus, implicated in the recall of declarative memories, back to the language areas of the cerebral cortex (at least via the cortical areas that receive backprojections from the hippocampus, see Fig. 2.14, which would in turn need connections to the language areas). A schematic diagram incorporating this anatomical prediction about human cortical neural connectivity in relation to consciousness is shown in Fig. 6.2.

6.3.2.4 Consciousness and causality

One question that has been discussed is whether there is a causal role for consciousness (e.g. Armstrong and Malcolm (1984)). The position to which the above arguments lead is that indeed conscious processing does have a causal role in the elicitation of behaviour, but only under the set of circumstances when higher-order thoughts play a role in correcting or influencing lower-order thoughts. The sense in which the consciousness is causal is then, it is suggested, that the higher-order thought is causally involved in correcting the lower-order thought; and that it is a property of the higher-order thought system that it feels like something when it is operating. As we have seen, some behavioural responses can be elicited when there is not this type of reflective control of lower-order processing, nor indeed any contribution of language. There are many brain-processing routes to output regions, and only one of these involves conscious, verbally represented processing that can later be recalled (see Fig. 6.1 and Section 5.1).

It is of interest to comment on how the evolution of a system for flexible planning might affect emotions. Consider grief which may occur when a reward is terminated and no immediate action is possible (Rolls 1990c, Rolls 1995b, Rolls 2005a). It may be adaptive by leading to a cessation of the formerly rewarded behaviour, and thus facilitating the possible identification of other positive reinforcers in the environment. In humans, grief may be particularly potent because it becomes represented in a system that can plan ahead, and understand the enduring implications of the loss. (Thinking about or verbally discussing emotional states may also in these circumstances help, because this can lead towards the identification of new or alternative reinforcers, and of the realization that, for example, negative consequences may not be as bad as feared.)

6.3.2.5 Consciousness and free will

This account of consciousness also leads to a suggestion about the processing that underlies the feeling of **free will** (see also Sections 5.5.1 and 6.5). Free will would in this scheme involve the use of language to check many moves ahead on a number of possible series of actions and their outcomes, and then with this information to make a choice from the likely outcomes of different possible series of actions.

In the operation of such a free-will system, the uncertainties introduced by the limited information possible about the likely outcomes of series of actions, and the inability to use optimal algorithms when combining conditional probabilities, would be much more important factors than whether the brain operates deterministically or not. (The operation of brain machinery must be relatively deterministic, for it has evolved to provide reliable outputs for given inputs.) The issue of whether the brain operates deterministically (Section 6.4) is not therefore I suggest the central or most interesting question about free will. Instead, analysis of which brain processing systems are engaged when we are taking decisions (Section 2.12), and which processing systems are inextricably linked to feelings as suggested above, may be more revealing about free will (Section 6.5).

6.3.2.6 Consciousness and self-identity

Before leaving these thoughts, it may be worth commenting on the feeling of continuing self-identity that is characteristic of humans. Why might this arise? One suggestion is that if one is an organism that can think about its own long-term multistep plans, then for those plans to be consistently and thus adaptively executed, the goals of the plans would need to remain stable, as would memories of how far one had proceeded along the execution path of each plan. If one felt each time one came to execute, perhaps on another day, the next step of a plan, that the goals were different, or if one did not remember which steps had already been taken in a multistep plan, the plan would never be usefully executed. So, given that it does feel like something to be doing this type of planning using higher-order thoughts, it would have to feel as if one were the same agent, acting towards the same goals, from day to day, for which autobiographical memory would be important.

Thus it is suggested that the feeling of continuing self-identity falls out of a situation in which there is an actor with consistent long-term goals, and long-term recall. If it feels like anything to be the actor, according to the suggestions of the higher-order thought theory, then it should feel like the same thing from occasion to occasion to be the actor, and no special further construct is needed to account for self-identity. Humans without such a feeling of being the same person from day to day might be expected to have, for example, inconsistent goals from day to day, or a poor recall memory. It may be noted that the ability to recall previous steps in a plan, and bring them into the conscious, higher-order thought system, is an important prerequisite for long-term planning which involves checking each step in a multistep process.

Conscious feelings of self will be likely to be of value to the individual. Indeed, it would be maladaptive if feelings of self-identity, and continuation of the self, were not wanted by the individual, for that would lead to the brain's capacity for feelings about self-identity to leave the gene pool, due for example to suicide. This wish for feelings and thoughts about the self to continue may lead to the wish and hope that this will occur after death, and this may be important as a foundation for religions, as described in Section 11.2.

These are my initial thoughts on why we have consciousness, and are conscious of sensory, emotional, and motivational qualia, as well as qualia associated with first-order linguistic thoughts. However, as stated above, one does not feel that there are straightforward criteria in this philosophical field of enquiry for knowing whether the suggested theory is correct; so it is likely that theories of consciousness will continue to undergo rapid development; and current theories should not be taken to have practical implications.

6.3.3 Content and meaning in representations: How are representations grounded in the world?

In Section 6.3.2 I suggested that representations need to be grounded in the world for a system with higher-order thoughts to be conscious. I therefore now develop somewhat what I understand by representations being grounded in the world.

It is possible to analyse how the firing of populations of neurons encodes information about stimuli in the world (Section 2.3) (Rolls 2008c, Rolls and Treves 2011). For example, from the firing rates of small numbers of neurons in the primate inferior temporal visual cortex, it is possible to know which of 20 faces has been shown to the monkey (Abbott, Rolls and Tovee 1996, Rolls, Treves and Tovee 1997b). Similarly, a population of neurons in the anterior part of the macaque temporal lobe visual cortex has been discovered that has a view-invariant representation of objects (Booth and Rolls 1998). From the firing of a small ensemble of neurons in the olfactory part of the orbitofrontal cortex, it is possible to know which of eight odours was presented (Rolls, Critchley and Treves 1996a). From the firing of small ensembles of neurons in the hippocampus, it is possible to know where in allocentric space a monkey is looking (Rolls, Treves, Robertson, Georges-François and Panzeri 1998). In each of these cases, the number of stimuli that is encoded increases exponentially with the number of neurons in the ensemble, so this is a very powerful representation (Abbott, Rolls and Tovee 1996, Rolls, Treves and Tovee 1997b, Rolls and Treves 1998, Rolls, Aggelopoulos, Franco and Treves 2004, Franco, Rolls, Aggelopoulos and Treves 2004, Aggelopoulos, Franco and Rolls 2005, Rolls 2008c, Rolls and Treves 2011). What is being measured in each example is the mutual information between the firing of an ensemble of neurons and which stimuli are present in the world. In this sense, one can read off the code that is being used at the end of each of these sensory systems.

However, what sense does the representation make to the animal? What does the firing of each ensemble of neurons 'mean'? What is the content of the representation? In the visual system, for example, it is suggested that the representation is built by a series of appropriately connected competitive networks, operating with a modified Hebb-learning rule (Rolls 1992, Rolls 1994, Wallis and Rolls 1997, Rolls 2000a, Rolls and Milward 2000, Stringer and Rolls 2000, Rolls and Stringer 2001a, Rolls and Deco 2002, Elliffe, Rolls and Stringer 2002, Stringer and Rolls 2002, Deco and Rolls 2004, Rolls 2008c, Rolls 2012b). Now competitive networks categorize their inputs without the use of a teacher (Kohonen 1989, Hertz et al. 1991, Rolls 2008c). So which particular neurons fire as a result of the self-organization to represent a particular object or stimulus is arbitrary. What meaning, therefore, does the particular ensemble that fires to an object have? How is the representation grounded in the real world? The fact that there is mutual information between the firing of the ensemble of cells in the brain and a stimulus or event in the world (Rolls 2008c, Rolls and Treves 2011) does not fully answer this question.

One answer to this question is that there may be meaning in the case of objects and faces that it is an object or face, and not just a particular view. This is the case in that the representation may be activated by any view of the object or face. This is a step, suggested to be made possible by a short-term memory in the learning rule that enables different views of objects to be associated together (Wallis and Rolls 1997, Rolls and Milward 2000, Rolls and Stringer 2001a, Rolls 2008c, Rolls 2012b). But it still does not provide the representation with any meaning in terms of the real world. What actions might one make, or what emotions might one feel, if that arbitrary set of temporal cortex visual cells was activated?

This leads to one of the answers I propose. I suggest that one type of meaning of representations in the brain is provided by their reward (or punishment) value: activation of these representations is the goal for actions. In the case of primary reinforcers such as the taste of food or pain, the activation of these representations would have meaning in the sense that the animal would work to obtain the activation of the taste of food neurons when hungry, and to escape from stimuli that cause the neurons representing pain to be activated. Evolution has built the brain so that genes specify these primary reinforcing stimuli, and so that their representations in the brain should be the targets for actions (see Section 3.10). In the case of other ensembles of neurons in, for example, the visual cortex that respond to objects with the colour and shape of a banana, and which 'represent' the sight of a banana in that their activation is always and uniquely produced by the sight of a banana, such representations come to have meaning only by association with a primary reinforcer, involving the process of stimulus–reinforcer association learning.

The second sense in which a representation may be said to have meaning is by virtue of sensory–motor correspondences in the world. For example, the touch of a solid object such as a table might become associated with evidence from the motor system that attempts to walk through the table result in cessation of movement. The representation of the table in the inferior temporal visual cortex might have 'meaning' only in the sense that there is mutual information between the representation and the sight of the table until the table is seen just before and while it is touched, when sensory–sensory association between inputs from different sensory modalities will be set up that will enable the visual representation to become associated with its correspondences in the touch and movement worlds. In this second sense, meaning will be conferred on the visual sensory representation because of its associations in the sensory–motor world.

Thus it is suggested that there are two ways by which sensory representations can be said to be grounded, that is to have meaning, in the real world.

It is suggested that the symbols used in language become grounded in the real world by the same two processes.

In the first, a symbol such as the word 'banana' has meaning because it is associated with primary reinforcers such as the flavour of the banana, and with secondary reinforcers such as the sight of the banana. These reinforcers have 'meaning' to the animal in that evolution has built animals as machines designed to do everything that they can to obtain these reinforcers, so that they can eventually reproduce successfully and pass their genes onto the next generation[27]. In this sense, obtaining reinforcers may have life-threatening 'meaning' for animals, though of course the use of the word 'meaning' here does not imply any subjective state, just that the animal is built as a survival for reproduction machine. This is a novel, Darwinian, approach to the issue of symbol grounding.

In the second process, the word 'table' may have meaning because it is associated with sensory stimuli produced by tables such as their touch, shape, and sight, as well as other functional properties, such as, for example, being load-bearing, and obstructing movement if they are in the way (see Section 6.3.2).

This section (6.3.3) thus adds to Section 6.3.2 on a higher-order syntactic thought (HOST) theory of consciousness, by addressing the sense in which the thoughts may need to be grounded in the world. The HOST theory holds that the thoughts 'mean' something to the individual, in the sense that they may be about the survival of the individual (the phenotype, Section 5.4) in the world, which the rational, thought, system aims to maximize (Chapter 5 and Section 11.2).

6.3.4 Other related approaches to consciousness

Some ways in which the current theory may be different from other related theories (Rosenthal 2004, Gennaro 2004, Carruthers 2000) follow.

The current theory holds that it is higher-order syntactic thoughts, HOSTs, (Rolls 1997a, Rolls 2004b, Rolls 2006a, Rolls 2007a, Rolls 2007b, Rolls 2008a, Rolls 2010d, Rolls 2011b) that are closely associated with consciousness, and this might differ from Rosenthal's higher-order thoughts (HOTs) theory (Rosenthal 1986, Rosenthal 1990, Rosenthal 1993, Rosenthal 2004, Rosenthal 2005) in the emphasis in the current theory on language. Language in the

---

[27]The fact that some stimuli are reinforcers but may not be adaptive as goals for action is no objection. Genes are limited in number, and can not allow for every eventuality, such as the availability to humans of (non-nutritive) saccharin as a sweetener. The genes can just build reinforcement systems the activation of which is generally likely to increase the fitness of the genes specifying the reinforcer (or may have increased their fitness in the recent past).

current theory is defined by syntactic manipulation of symbols, and does not necessarily imply verbal (or natural) language. The reason that strong emphasis is placed on language is that it is as a result of having a multistep, flexible, 'one-off', reasoning procedure that errors can be corrected by using 'thoughts about thoughts'. This enables correction of errors that cannot be easily corrected by reward or punishment received at the end of the reasoning, due to the credit assignment problem. That is, there is a need for some type of supervisory and monitoring process, to detect where errors in the reasoning have occurred. It is having such a HOST brain system, and it becoming engaged (even if only a little), that according to the HOST theory is associated with phenomenal consciousness.

This suggestion on the adaptive value in evolution of such a higher-order linguistic thought process for multistep planning ahead, and correcting such plans, may also be different from earlier work. Put another way, this point is that *credit assignment* when reward or punishment is received is straightforward in a one-layer network (in which the reinforcement can be used directly to correct nodes in error, or responses), but is very difficult in a multistep linguistic process executed once. Very complex mappings in a multilayer network can be learned if hundreds of learning trials are provided. But once these complex mappings are learned, their success or failure in a new situation on a given trial cannot be evaluated and corrected by the network. Indeed, the complex mappings achieved by such networks (e.g. networks trained by backpropagation of errors or by reinforcement learning) mean that after training they operate according to fixed rules, and are often quite impenetrable and inflexible (Rumelhart, Hinton and Williams 1986, Rolls 2008c). In contrast, to correct a multistep, single occasion, linguistically based plan or procedure, recall of the steps just made in the reasoning or planning, and perhaps related episodic material, needs to occur, so that the link in the chain that is most likely to be in error can be identified. This may be part of the reason why there is a close relationship between declarative memory systems, which can explicitly recall memories, and consciousness.

Some computer programs may have supervisory processes. Should these count as higher-order linguistic thought processes? My current response to this is that they should not, to the extent that they operate with fixed rules to correct the operation of a system that does not itself involve linguistic thoughts about symbols grounded semantically in the external world. If on the other hand it were possible to implement on a computer such a high-order linguistic thought–supervisory correction process to correct first-order one-off linguistic thoughts with symbols grounded in the real world (as described at the end of Section 6.3.3), then prima facie this process would be conscious. If it were possible in a thought experiment to reproduce the neural connectivity and operation of a human brain on a computer, then prima facie it would also have the attributes of consciousness[28]. It might continue to have those attributes for as long as power was applied to the system.

Another possible difference from earlier theories is that raw sensory feels are suggested to arise as a consequence of having a system that can think about its own thoughts. Raw sensory feels, and subjective states associated with emotional and motivational states, may not necessarily arise first in evolution.

A property often attributed to consciousness is that it is *unitary*. The current theory would account for this by the limited syntactic capability of neuronal networks in the brain, which render it difficult to implement more than a few syntactic bindings of symbols simultaneously (Rolls and Treves 1998, McLeod, Plunkett and Rolls 1998, Rolls 2008c). This limitation makes it difficult to run several 'streams of consciousness' simultaneously. In addition, given that a linguistic system can control behavioural output, several parallel streams might produce maladaptive behaviour (apparent as, e.g. indecision), and might be selected against. The close relationship between, and the limited capacity of, both the stream of consciousness, and auditory–verbal short-term working memory, may be that both implement the capacity for syntax in neural networks.

The suggestion that syntax in real neuronal networks is implemented by temporal binding (Malsburg 1990, Singer 1999) seems unlikely (Rolls 2008c, Deco and Rolls 2011, Rolls and Treves 2011)[29].

However, the hypothesis that syntactic binding is necessary for consciousness is one of the postulates of the theory I am describing (for the system I describe must be capable of correcting its own syntactic thoughts). The fact that the binding must be implemented in neuronal networks may well place limitations on consciousness that lead to some of its properties, such as its unitary nature. The postulate of Crick and Koch (1990) that oscillations and synchronization are necessary bases of consciousness could thus be related to the present theory if it turns out that oscillations or neuronal synchronization are the way the brain implements syntactic binding. However, the fact that os-

---

[28]This is a functionalist position. Apparently Damasio (2003) does not subscribe to this view, for he suggests that there is something in the 'stuff' (the 'natural medium') that the brain is made of that is also important. It is difficult for a person with this view to make telling points about consciousness from neuroscience, for it may always be the 'stuff' that is actually important.

[29]For example, the code about which visual stimulus has been shown can be read off from the end of the visual system without taking the temporal aspects of the neuronal firing into account; much of the information about which stimulus is shown is available in short times of 30–50 ms, and cortical neurons need fire for only this long during the identification of objects (Tovee, Rolls, Treves and Bellis 1993, Rolls and Tovee 1994, Tovee and Rolls 1995, Rolls and Treves 1998, Rolls and Deco 2002, Rolls 2003, Rolls 2006a) (these are rather short time-windows for the expression of multiple separate populations of synchronized neurons); and stimulus-dependent synchronization of firing between neurons is not a quantitatively important way of encoding information in the primate temporal cortical visual areas involved in the representation of objects and faces (Tovee and Rolls 1992, Rolls and Treves 1998, Rolls and Deco 2002, Rolls, Franco, Aggelopoulos and Reece 2003b, Rolls, Aggelopoulos, Franco and Treves 2004, Franco, Rolls, Aggelopoulos and Treves 2004, Aggelopoulos, Franco and Rolls 2005, Rolls 2008c, Deco and Rolls 2011, Rolls and Treves 2011) – see Section 2.3.

cillations and neuronal synchronization are especially evident in anaesthetized cats does not impress as strong evidence that oscillations and synchronization are critical features of consciousness, for most people would hold that anaesthetized cats are not conscious. The fact that oscillations and stimulus-dependent neuronal synchronization are much more difficult to demonstrate in the temporal cortical visual areas of awake behaving monkeys (Tovee and Rolls 1992, Franco, Rolls, Aggelopoulos and Treves 2004, Aggelopoulos, Franco and Rolls 2005, Rolls 2008c, Rolls and Treves 2011) might just mean that during the evolution of primates the cortex has become better able to avoid parasitic oscillations, as a result of developing better feedforward and feedback inhibitory circuits (Rolls 2008c).

The theory (Rolls 1997a, Rolls 2004b, Rolls 2006a, Rolls 2007a, Rolls 2007b, Rolls 2008a, Rolls 2010d, Rolls 2011b) holds that consciousness arises by virtue of a system that can think linguistically about its own linguistic thoughts. The advantages for a system of being able to do this have been described, and this has been suggested as the reason why consciousness evolved. The evidence that consciousness arises by virtue of having a system that can perform higher-order linguistic processing is however, and I think might remain, circumstantial. [Why must it feel like something when we are performing a certain type of information processing? The evidence described here suggests that it does feel like something when we are performing a certain type of information processing, but does not produce a strong reason for why it has to feel like something. It just does, when we are using this linguistic processing system capable of higher-order thoughts.] The evidence, summarized above, includes the points that we think of ourselves as conscious when, for example, we recall earlier events, compare them with current events, and plan many steps ahead. Evidence also comes from neurological cases, from, for example, split-brain patients (who may confabulate conscious stories about what is happening in their other, non-language, hemisphere); and from cases such as frontal lobe patients who can tell one consciously what they should be doing, but nevertheless may be doing the opposite. (The force of this type of case is that much of our behaviour may normally be produced by routes about which we cannot verbalize, and are not conscious about.)

This raises discussion of the *causal role of consciousness* (Section 6.3.2.4). Does consciousness cause our behaviour? The view that I currently hold is that the information processing that is related to consciousness (activity in a linguistic system capable of higher-order thoughts, and used for planning and correcting the operation of lower-order linguistic systems) can play a causal role in producing our behaviour (see Fig. 5.1). It is, I postulate, a property of processing in this system (capable of higher-order thoughts) that it feels like something to be performing that type of processing. It is in this sense that I suggest that consciousness can act causally to influence our behaviour –

consciousness is the property that occurs when a linguistic system is thinking about its lower-order thoughts, which may be useful in correcting plans.

The hypothesis that it does feel like something when this processing is taking place is at least to some extent testable: humans performing this type of higher-order linguistic processing, for example recalling episodic memories and comparing them with current circumstances, who denied being conscious, would prima facie constitute evidence against the theory. Most humans would find it very implausible though to posit that they could be thinking about their own thoughts, and reflecting on their own thoughts, without being conscious. This type of processing does appear, for most humans, to be necessarily conscious.

Finally, I provide a short specification of what might have to be implemented in a neuronal network to implement conscious processing. First, a linguistic system, not necessarily verbal, but implementing syntax between symbols implemented in the environment would be needed. This system would be necessary for a multi-step one-off planning system. Then a higher-order thought system also implementing syntax and able to think about the representations in the first-order linguistic system, and able to correct the reasoning in the first-order linguistic system in a flexible manner, would be needed. The system would also need to have its representations grounded in the world, as discussed in Section 6.3.3. So my view is that consciousness can be implemented in neuronal networks (and that this is a topic worth discussing), but that the neuronal networks would have to implement the type of higher-order linguistic processing described in this chapter, and also would need to be grounded in the world.

Further, less related, approaches to consciousness are discussed in Section 6.3.6.

6.3.5 Monitoring and consciousness

An attractor network in the brain with positive feedback implemented by excitatory recurrent collateral connections between the neurons can implement decision-making (Wang 2002, Deco and Rolls 2006, Wang 2008, Rolls and Deco 2010) (see Section 2.12). As explained in detail elsewhere (Rolls and Deco 2010), if the external evidence for the decision is consistent with the decision taken (which has been influenced by the noisy neuronal firing times), then the firing rates in the winning attractor are supported by the external evidence, and become especially high. If the external evidence is contrary to the noise-influenced decision, then the firing rates of the neurons in the winning attractor are not supported by the external evidence, and are lower than expected (Fig. 2.27). In this way the confidence in a decision is reflected in, and encoded by, the firing rates of the neurons in the winning attractor population of neurons (Section 2.12.2) (Rolls and Deco 2010).

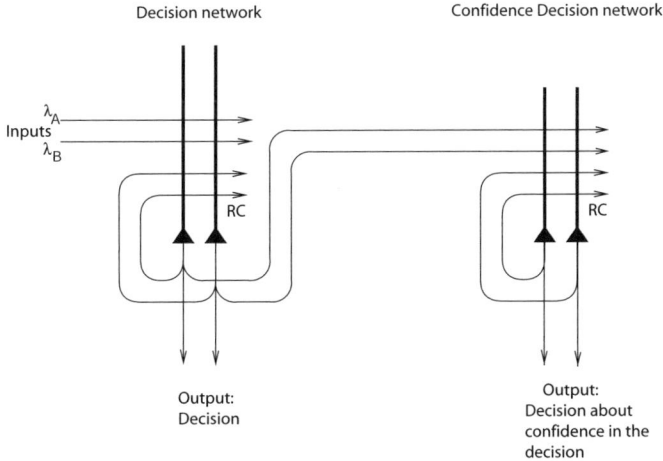

Fig. 6.3 Network architecture for decisions about confidence estimates. The first network is a decision-making network, and its outputs are sent to a second network that makes decisions based on the firing rates from the first network, which reflect the decision confidence. In the first network, high firing of neuronal population (or pool) DA represents decision A, and high firing of population DB represents decision B. Pools DA and DB receive a stimulus-related input (respectively λ_A and λ_B), the evidence for each of the decisions, and these bias the attractor networks, which have internal positive feedback produced by the recurrent excitatory connections (RC). Pools DA and DB compete through inhibitory interneurons. The neurons are integrate-and-fire spiking neurons with random spiking times (for a given mean firing rate) which introduce noise into the network and influence the decision-making, making it probabilistic. The second network is a confidence decision attractor network, and receives inputs from the first network. The confidence decision network has two selective pools of neurons, one of which (C) responds to represent confidence in the decision, and the other of which responds when there is little or a lack of confidence in the decision (LC). The C neurons receive the outputs from the selective pools of the (first) decision-making network, and the LC neurons receive $\lambda_{Reference}$ which is from the same source but saturates at 40 spikes/s, a rate that is close to the rates averaged across correct and error trials of the sum of the firing in the selective pools in the (first) decision-making network. (After Insabato, Pannunzi, Rolls and Deco, 2010.)

If we now add a second attractor network to read the firing rates from the first decision-making network, the second attractor network can take a decision based on the confidence expressed in the firing rates in the first network (Insabato, Pannunzi, Rolls and Deco 2010). The second attractor network allows decisions to be made about whether to change the decision made by the first network, and for example abort the trial or strategy (see Fig. 6.3). The second network, the confidence decision network, is in effect monitoring the decisions taken by the first network, and can cause a change of strategy or behaviour if the assessment of the decision taken by the first network does not seem a confident decision. This is described in detail elsewhere (Insabato, Pannunzi, Rolls and Deco 2010, Rolls and Deco 2010), but Fig. 6.3 shows the simple system of two attractor networks that enables confidence-based (second-level) decisions to be made, by monitoring the output of the first, decision-making, network.

Now this is the type of description, and language used, to describe 'monitoring' functions, taken to be a high-level cognitive process, possibly related to consciousness (Block 1995a, Lycan 1997). For example, in an experiment performed by Hampton (2001) (experiment 3), a monkey had to remember a picture over a delay. He was then given a choice of a 'test flag', in which case he would be allowed to choose from one of four pictures the one seen before the delay, and if correct earn a large reward (a peanut). If he was not sure that he remembered the first picture, he could choose an 'escape flag', to start another trial. With longer delays, when memory strength might be lower partly due to noise in the system, and confidence therefore in the memory on some trials might be lower, the monkey was more likely to choose the escape flag. The experiment is described as showing that the monkey is thinking about his own memory, that is, is a case of meta-memory, which may be related to consciousness (Heyes 2008). However, the decision about whether to escape from a trial can be taken just by adding a second decision network to the first decision network. Thus we can account for what seem like complex cognitive phenomena with a simple system of two attractor decision-making networks (Fig. 6.3) (Rolls and Deco 2010).

The implication is that some types of 'self-monitoring' can be accounted for by simple, two-attractor network, computational processes. But what of more complex 'self-monitoring', such as is described as occurring in a commentary that might be based on reflection on previous events, and appears to be closely related to consciousness (Weiskrantz 1997). This approach has been developed into my higher-order syntactic theory (HOST) of consciousness (Section 6.3.2 (Rolls 1997a, Rolls 2004b, Rolls 2005a, Rolls 2007a, Rolls 2008a, Rolls 2007b, Rolls 2010d, Rolls 2011b)), in which there is a credit assignment problem if a multi-step reasoned plan fails, and it may be unclear which step failed. Such plans are described as syntactic as there are symbols at each stage that must be linked together with the syntactic relationships between the symbols specified, but kept separate across stages of the plan. It is suggested that in this situation being able to have higher-order syntactic thoughts will enable one to think and reason about the first-order plan, and detect which steps are likely to be at fault.

Now this type of 'self-monitoring' is much more complex, as it requires syntax. The thrust of the argument is that some types of 'self-monitoring' are computationally simple, for example in decisions made based on confidence in a first decision (Rolls and Deco 2010), and may have little to do with consciousness; whereas higher-order thought processes are very different in terms of the type of syntactic computation required, and may be more closely related to consciousness (Rolls 1997a, Rolls 2003, Rolls 2004b, Rolls 2005a, Rolls 2007a, Rolls 2008a, Rolls 2007b, Rolls 2010d, Rolls 2011b).

6.3.6 Conclusions on consciousness, and comparisons

It is suggested that it feels like something to be an organism or machine that can think about its own (syntactic and semantically grounded) thoughts.

It is suggested that qualia, raw sensory, and emotional, 'feels', arise secondarily to having evolved such a higher-order thought system, and that sensory and emotional processing feels like something because once this emotional processing has entered the planning, higher-order thought, system, it would be unparsimonious for it not to feel like something, given that all the other processing in this system I suggest does feel like something.

The adaptive value of having sensory and emotional feelings, or qualia, is thus suggested to be that such inputs are important to the long-term planning, explicit, processing system. Raw sensory feels, and subjective states associated with emotional and motivational states, may not necessarily arise first in evolution.

Reasons why the ventral visual system is more closely related to explicit than implicit processing include the fact that representations of objects and individuals need to enter the planning, hence conscious, system, and are considered in more detail by Rolls (2003) and by Rolls (2008c).

Evidence that explicit, conscious, processing may have a higher threshold in sensory processing than implicit processing is considered by Rolls (2003) and Rolls (2006a), based on neurophysiological and psychophysical investigations of backward masking (Rolls and Tovee 1994, Rolls, Tovee, Purcell, Stewart and Azzopardi 1994b, Rolls, Tovee and Panzeri 1999, Rolls 2003, Rolls 2006a). It is suggested there that part of the adaptive value of this is that if linguistic processing is inherently serial and slow, it may be maladaptive to interrupt it unless there is a high probability that the interrupting signal does not arise from noise in the system. In the psychophysical and neurophysiological studies, it was found that face stimuli presented for 16 ms and followed immediately by a masking stimulus were not consciously perceived by humans, yet produced above chance identification, and firing of inferior temporal cortex neurons in macaques for approximately 30 ms. If the mask was delayed for 20 ms, the neurons fired for approximately 50 ms, and the test face stimuli were more likely to be perceived consciously. In a similar backward masking paradigm, it was found that happy vs angry face expressions could influence how much beverage was wanted and consumed even when the faces were not consciously perceived (Winkielman and Berridge 2005, Winkielman and Berridge 2003). This is further evidence that unconscious emotional stimuli can influence behaviour.

The theory is different from some other higher-order theories of consciousness (Rosenthal 1990, Rosenthal 1993, Rosenthal 2004, Carruthers 2000, Gennaro 2004) in that it provides an account of the evolutionary, adaptive, value of a higher-order thought system in helping to solve a credit assignment problem that arises in a multistep syntactic plan, links this type of processing to consciousness, and therefore emphasizes a role for syntactic processing in consciousness.

The theory described here is also different from other theories of consciousness and affect. James and Lange (James 1884, Lange 1885) held that emotional feelings arise when feedback from the periphery (about for example heart rate) reach the brain, but had no theory of why some stimuli and not others produced the peripheral changes, and thus of why some but not other events produce emotional feelings.

Moreover, the evidence that feedback from peripheral autonomic and proprioceptive systems is essential for emotions is very weak, in that for example blocking peripheral feedback does not eliminate emotions, and producing peripheral, e.g. autonomic, changes does not elicit emotion (Reisenzein 1983, Schachter and Singer 1962, Rolls 2005a) (see Section 3.5.1).

Damasio's theory of emotion (Damasio 1994, Damasio 2003) is a similar theory to the James–Lange theory (and is therefore subject to some of the same objections), but holds that the peripheral feedback is used in decision-making rather than in consciousness. He does not formally define emotions, but holds that body maps and representations are the basis of emotions. When considering consciousness, he assumes that all consciousness is self-consciousness (Damasio 2003) (p. 184), and that the foundational images in the stream of the mind are images of some kind of body event, whether the event happens in the depth of the body or in some specialized sensory device near its periphery (Damasio 2003) (p. 197). His theory does not appear to be a fully testable theory, in that he suspects that "the ultimate quality of feelings, a part of why feelings feel the way they feel, is conferred by the neural medium" (Damasio 2003) (p. 131). Thus presumably if processes he discusses (Damasio 1994, Damasio 2003) were implemented in a computer, then the computer would not have all the same properties with respect to consciousness as the real brain. In this sense he appears to be arguing for a non-functionalist position, and something crucial about consciousness being related to the particular biological machinery from which the system is made. In this respect the theory seems somewhat intangible.

LeDoux's approach to emotion (LeDoux 1992, LeDoux 1995, LeDoux 1996) is largely (to quote him) one of automaticity, with emphasis on brain mechanisms involved in the rapid, subcortical, mechanisms involved in fear. LeDoux, in line with Johnson-Laird (1988) and Baars (1988), emphasizes the role of working memory in consciousness, where he views working memory as a limited-capacity serial processor that creates and manipulates symbolic representations (p 280). He thus holds that much emotional processing is unconscious, and that when it becomes conscious it is because emotional information is entered into a working memory system. However, LeDoux (1996) concedes that consciousness, especially its phenomenal or subjective nature, is not completely explained by the computational processes that underlie working memory (p 281).

Panksepp's (1998) approach to emotion has its origins in neuroethological investigations of brainstem systems that when activated lead to behaviours like fixed action patterns, including escape, flight and fear behaviour. His views about consciousness include the postulate that "feelings may emerge when endogenous sensory and emotional systems within the brain that receive direct inputs from the outside world as well as the neurodynamics of the SELF (a Simple Ego-type Life Form) begin to reverberate with each other's changing neuronal firing rhythms" (Panksepp 1998) (p 309).

Thus the theory of consciousness described in this chapter is different from some other theories of consciousness.

6.4 Determinism

There are a number of senses in which our behaviour might be deterministic. One sense might be genetic determinism, and we have already seen that there are far too few genes to determine the structure and function of our brains, and thus to determine our behaviour (see Section 1.2). Moreover, development, and the environment with the opportunities it provides for brain self-organization and learning, play a large part in brain structure and function, and thus in our behaviour.

Another sense might be that if there were random factors that influence the operation of the brain, then our behaviour might be thought not to be completely predictable and deterministic. It is this that I consider here, a topic developed in *The Noisy Brain: Stochastic Dynamics as a Principle of Brain Function* (Rolls and Deco 2010), in which we show that there is noise or randomness in the brain, and argue that this can be advantageous[30].

Neurons emit action potentials, voltage spikes, which transmit information along axons to other neurons. These all-or-none spikes are a safe way to transmit information along axons, for they do not lose amplitude and degrade along a long axon. In most brain systems, an increase in the firing rate of the spikes carries the information. For example, taste neurons in the taste cortex fire faster if the particular taste to which they respond is present, and neurons in the inferior temporal visual cortex fire faster if for example one of the faces to which they are tuned is seen (Rolls 2005a, Rolls 2008c). However, for a given mean firing rate (e.g. 50 spikes/s), the exact timing of each spike is quite random, and indeed is close to a Poisson distribution which is what is expected for a random process in which the timing of each spike is independent of the other spikes. Part of the neuronal basis of this randomness of the spike firing

---

[30]This randomness is not a property of chaotic systems, which although complex, are not random in the sense that the same trajectory is followed from a given starting position (Peitgen, Jürgens and Saupe 2004).

times is that each cortical neuron is held close to its threshold for firing and even produces occasional spontaneous firing, so that when an input is received, some at least of the cortical neurons will be so close to threshold that they emit a spike very rapidly, allowing information processing to be rapid (Rolls 2008c, Rolls and Deco 2010).

This randomness in the firing time of individual neurons results in probabilistic behaviour of the brain (Section 2.12). For example, in decision-making, if the population of neurons that represents decision 1 has by chance more randomly occurring spikes in a short time, that population may win the competition (implemented through inhibitory interneurons) with a different population of neurons that represents decision 2. Decision-making is by this mechanism probabilistic. For example, if the odds are equal for decision 1 and decision 2, each decision will be taken probabilistically on 50% of the occasions or trials. This is highly adaptive, and is much better than getting stuck between two equally attractive rewards and unable to make a decision, as in the medieval tale of Duns Scotus about the donkey who starved because it could not choose between two equally attractive foods (Rolls and Deco 2010).

However, given that the brain operates with some degree of randomness due to the statistical fluctuations produced by the random spiking times of neurons, brain function is to some extent non-deterministic, as defined in terms of these statistical fluctuations. That is, the behaviour of the system, and of the individual, can vary from trial to trial based on these statistical fluctuations, in ways that are described by Rolls and Deco (2010) (Section 2.12). Indeed, given that each neuron has this randomness, and that there are sufficiently small numbers of synapses on the neurons in each network (between a few thousand and 20,000) that these statistical fluctuations are not smoothed out, and that there are a number of different networks involved in typical thoughts and actions each one of which may behave probabilistically, and with 10^{11} neurons in the brain each with this number of synapses, the system has so many degrees of freedom that it operates effectively as a non-deterministic system. (Philosophers may wish to argue about different senses of the term deterministic, but it is being used here in a precise, scientific, and quantitative way, which has been clearly defined.)

6.5 Free will

Do we have free will when we make a choice? Given the distinction made between the implicit system that seeks for gene-specified rewards, and the explicit system that can use reasoning to defer an immediate goal and plan many steps ahead for longer-term goals (Chapter 5), do we have free will when both the implicit and the explicit systems have made the choice?

Free will would in Rolls' view (Rolls 2005a, Rolls 2008a, Rolls 2008c, Rolls 2010d, Rolls 2011b) involve the use of language to check many moves ahead on a number of possible series of actions and their outcomes, and then with this information to make a choice from the likely outcomes of different possible series of actions. (If, in contrast, choices were made only on the basis of the reinforcement value of immediately available stimuli, without the arbitrary syntactic symbol manipulation made possible by language, then the choice strategy would be much more limited, and we might not want to use the term free will, as all the consequences of those actions would not have been computed.) It is suggested that when this type of reflective, conscious, information processing is occurring and leading to action, the system performing this processing and producing the action would have to believe that it could cause the action, for otherwise inconsistencies would arise, and the system might no longer try to initiate action. This belief held by the system may partly underlie the feeling of free will. At other times, when other brain modules are initiating actions (in the implicit systems), the conscious processor (the explicit system) may confabulate and believe that it caused the action, or at least give an account (possibly wrong) of why the action was initiated. The fact that the conscious processor may have the belief even in these circumstances that it initiated the action may arise as a property of it being inconsistent for a system that can take overall control using conscious verbal processing to believe that it was overridden by another system. This may be the underlying computational reason why confabulation occurs (Section 5.1).

The interesting view we are led to is thus that when probabilistic choices influenced by stochastic dynamics (Rolls and Deco 2010) are made between the implicit and explicit systems, we may not be aware of which system made the choice. Further, when the stochastic noise has made us choose with the implicit system, we may confabulate and say that we made the choice of our own free will, and provide a guess at why the decision was taken. In this scenario, the stochastic dynamics of the brain plays a role even in how we understand free will (Rolls 2010d).

The implication of this argument is that a good use of the term free will is when the term refers to the operation of the rational, planning, explicit (conscious) system that can think many moves ahead, and choose from a number of such computations the multistep strategy that best optimizes the goals of the explicit system with long-term goals. When on the other hand our implicit system has taken a decision, and we confabulate a spurious account with our explicit system, and pronounce that we took the decision for such and such a (confabulated) reason of our own "free will", then my view is that the feeling of free will was an illusion (Rolls 2005a, Rolls 2010d).

6.6 Conclusions

In this chapter we have shown that computational neuroscience provides an important new approach to traditional problems in philosophy such as the relation between mental states and brain states (the mind-body problem), to determinism and free will, and helps one with the 'hard' problem, the phenomenal aspects of consciousness.

I agree that neuropsychology, and functional neuroimaging ('blobology' referring to activation blobs in the brain), do not solve the deep problems of the relation between mental states and brain states (Legrenzi, Umilta and Anderson 2011), although they do provide very important lines of evidence,.

On the other hand, one of the themes of this book is that by understanding the computations performed by neurons and neuronal networks, and the effects of noise in the brain on these, we will gain a true understanding of the mechanisms that underlie brain function. This understanding extends very deep, to the statistical mechanics of networks operating with noise (Rolls and Deco 2010). In this sense, we are developing a truly wonderful 'mechanics of the mind' which is enabling how our brains work to be understood from the level of molecules through neurons with ion channels, through networks of such neurons to the global properties of a system, and thus to an understanding of how processes such as memory, perception, attention, decision-making, and emotion actually are implemented in the brain (Rolls 2008c, Rolls and Deco 2010).

Of course there are still 'hard' problems, such as the problem of phenomenal consciousness, and while I have provided new suggestions about this, one must remain humble in the face of major problems such as this.

7 Neuroaesthetics

7.1 Introduction

What are the foundations of what we appreciate in art? Is art—visual art, literature, music—related to fundamental adaptive capacities that help survival and thus reproduction, or is art a useless ornament, like a peacock's 'tail', shaped by sexual selection?

A theory of the origins of aesthetics is described. This has its roots in emotion, in which what is pleasant or unpleasant, a reward or punisher, is the result of an evolutionary process in which genes define the (pleasant or unpleasant) goals for action (Rolls 2005a) (see Chapter 3). It is argued that combinations of multiple such factors provide part of the basis for aesthetics. To this is added the operation of the reasoning, syntactic, brain system which evolved to help solve difficult, multistep, problems, and the use of which is encouraged by pleasant feelings when elegant, simple, and hence aesthetic solutions are found that are advantageous because they are parsimonious, and follow Occam's Razor (see Section 7.11). The combination of these two systems, and the interactions between them, provides an approach to understanding aesthetics (Rolls 2011h) that is rooted in evolution and its effects on brain design and function.

I have considered in Chapter 3 how affective value is generated in the brain as a solution to the problem of how genes can specify useful goals for actions. This is more efficient and produces more flexible behavior than by specifying the actions themselves. The approach provides part of a theory of how value is placed on some stimuli. Value will be placed according to whether the stimuli activate our reward or punishment systems, themselves tuned during evolution to produce goals that will increase the fitness of our genes. ('Fitness' refers to the reproductive success of genes.) Moreover, we have seen that these gene-defined goals may include a wide range of reinforcers, including many involved in social behaviour, and define some of the things that make people and objects attractive. We have seen that humans by reasoning can define a wider range of goals, or at least can place different values on goals as a result of reasoning, and use reasoning as a second route to action (Chapter 5). We have also seen that cognition can influence the representation of affective value in the orbitofrontal cortex. The analysis of the evolutionary basis of reward value provides a fundamental and Darwinian way to understand emotion (Rolls 2005a).

7.2 Outline of a neurobiological approach to aesthetics

I now explore whether the same approach can provide a neurobiological basis for understanding aesthetics. Now that we have a fundamental, Darwinian, approach to the value of people, objects, relationships etc., I propose that this provides a fundamental approach to understanding aesthetics and aesthetic value, that is, what we value aesthetically. I propose that while the gene-specified rewards and punishers define many things that have aesthetic value, the value that we place on items is enhanced by the reasoning, rational, system, which enables what produces aesthetic value to become highly intellectualized, as in music. However, even here I argue that there are certain adaptive principles that influence the operation of our rational system that provide a systematic way to understand aesthetics and aesthetic value.

I emphasize at the outset that this does not at all reduce aesthetics to a common denominator. Genetic variation is essential to evolution by natural selection, and this is one reason why we should expect different people to assign aesthetic value differently. But rational thought, which will lead in different directions in different people, partly because of noise caused by random neuronal firing times in the brain (Rolls and Deco 2010) (Section 2.12), and because of what they have learned from the environment, and because different brain areas will be emphasized in different people, will also be different between individuals. Differences in rational thought will thus also contribute to differences between individuals in what is considered aesthetic.

Indeed, although the theory presented here on the origin of aesthetics is a reductive explanation, in that it treats the underlying bases and causes, it should not be seen at all to 'reduce' aesthetics. Far from it. When we understand the underlying origins and bases of aesthetics, we see that the processes involved are elegant and beautiful, as part of a Darwinian theory. But the approach also provides important pointers about how to enhance aesthetics. For example, by understanding that verbal level cognitive factors that can be produced by reasoning have a top-down modulatory influence on the first cortical area where value (reward) is made explicit in the representation, the orbitofrontal cortex (De Araujo, Rolls, Velazco, Margot and Cayeux 2005, McCabe, Rolls, Bilderbeck and McGlone 2008, Grabenhorst, Rolls and Bilderbeck 2008a), we can see ways in which we can enhance our aesthetic feelings. (For example, if love be the thing, then it can be heightened by explicitly choosing the musical treatment of it in *Tristan and Isolde*.)

I should also emphasize that aesthetic value judgements will usually be influenced by a number of different value factors, so that while accounting for an aesthetic judgement by just one of the value factors I describe is and will often seem too simple, it does seem that aesthetic value judgements can be understood by combinations of some of the factors I describe.

I also emphasize that this is a theory of the origin of aesthetics. I provide generic examples, but of course cannot cover all factors that influence value. An indication of the range of factors that can provide a basis for aesthetic judgements is shown in Table 3.1, but this is by no means complete. These examples are gene-defined goals for action, and we are built to want to obtain these goals (the basis for motivation), to treat them operationally as rewards or punishers, and to have pleasant or unpleasant affective feelings when they are delivered (the basis of emotion) (Rolls 2005a). It is argued here that these factors contribute to aesthetic judgements, that any one stimulus will often have multiple such attributes, and that these factors are afforced by operations of the reasoning system.

I emphasize that rewards of which examples are provided in Table 3.1 contribute to what makes stimuli or brain processing positively aesthetic, beautiful; and that the punishers contribute to what makes stimuli or processing in the brain aesthetically negative, lacking beauty, ugly, or distasteful. Both rewards and punishers are needed for the theory of aesthetics. The overall theory of the origin of aesthetics I propose is that natural selection, whether operating by 'survival or adaptation selection', or by sexual selection, operates by specifying goals for action, and these goals are aesthetically and subjectively attractive or beautiful (Rolls 2005a), or the opposite, and provide what I argue here is the origin of many judgements of what is aesthetic. Many examples of these rewards and punishers, many of which operate for 'survival or adaptation selection', and many of which contribute to aesthetic experience, are shown in Table 3.1.

In contrast to my theory, Miller (2000, 2001) emphasizes the role of sexual selection. Understanding the mechanisms that drive evolution to make certain stimuli rewarding or punishing can help us to understand the origin of aesthetics, and I therefore summarize the characteristics of these two evolutionary processes in Sections 7.3 and 7.4.

I note first that the term 'natural selection' encompasses in its broad sense both 'survival or adaptation selection', and sexual selection. Both are processes now understood to be driven by the selection of genes, and it is gene competition and replication into the next generation that is the driving force of biological evolution (Dawkins 1986, Dawkins 1989). The distinction can be made that with 'survival or adaptation selection', the genes being selected for make the individual stronger, healthier, and more likely to survive and reproduce; whereas sexual selection operates by sexual choice selecting for genes that may or may not have survival value to the individual, but enable the individual to be selected as a mate (by intersexual selection), or to compete for a mate in intra-sexual selection, and thus pass on the genes selected by inter-sexual selection or intra-sexual selection to the offspring. More generally, we might have other types of selection as further types of natural selection, including selection for good parental care, and kin selection.

7.3 'Survival' or 'adaptation' selection (natural selection in a narrow sense)

Darwin (1871) distinguished natural selection from sexual selection, and this distinction has been consolidated and developed (Fisher 1930, Hamilton 1964, Zahavi 1975, Dawkins 1986, Grafen 1990a, Grafen 1990b, Dawkins 1995, Hamilton 1996, Miller 2000). Natural selection can be used in a narrow sense to refer to selection processes that lead to the development of characteristics that have a function of providing adaptive or survival value to an individual so that the individual can reproduce, and pass on its genes. In its narrow sense, natural selection can be thought of as 'survival or adaptation selection'.

An example might be a gene or genes that specify that the sensory properties of food should be rewarding (and should taste pleasant) when we are in a physiological need state for food. Many of the reward and punishment systems described here and by Rolls (2005a) deal with this type of reward and punishment decoding that has evolved to enable genes to influence behaviour in directions in a high-dimensional space of rewards and punishments that are adaptive for the survival and health of the individual, and thus promote reproductive success or fitness of the genes that build such adaptive functionality.

We can include kin-related altruistic behaviours because the behaviour is adaptive in promoting the survival of kin, and thus promoting the likelihood that the kin (who contain one's genes, and are likely to share the genes for kin altruism) survive and reproduce. We can also include reciprocal altruism as an example of 'survival or adaptation' selection. Tribalism can be treated similarly, for it probably has its origins in altruism. Resources and wealth are also understood at least in part as being selected by natural selection, in that resources and wealth may enable the individual to survive better. As we will see next, resources and wealth can also be attractive as a result of sexual selection. (I note that natural selection in a broad sense includes 'survival or adaptation' selection, sexual selection, selection for good parental care, etc.)

7.4 Sexual selection

Darwin (1871) also recognized that evolution can occur by sexual selection, when what is being selected for is attractive to potential mates (inter-sexual selection), or helps in competing with others of the same sex (intra-sexual selection, e.g. the deer's large antlers, and a strong male physique).

The most cited example of mate selection (inter-sexual selection) is the peacock's large 'tail', which does not have survival value for the peacock (and indeed it is somewhat of a *handicap* to have a very long tail), but, because it is attractive to the peahen, becomes prevalent in the population. Indeed, part of the reason for the long tail being attractive may be that it is an honest signal of phenotypic fitness (a 'revealing signal' that is a 'fitness indicator'), in that having a very long tail is a handicap to survival (Zahavi 1975), though the signalling system that reveals this only operates correctly if certain conditions apply (Grafen 1990a, Grafen 1990b, Maynard Smith and Harper 2003). The fact that the long tail is actually a handicap for the peacock, and so is a signal of general physical fitness in the male, may be one way in which sexual selection can occur stably (Zahavi 1975, Grafen 1990a, Grafen 1990b).

Another account is that the inherited genes for a long tail may be expressed in the female's sons, and they will accordingly be attractive to females in the next generation (Fisher's 'sexy son' account) (Fisher 1930). Although the female offspring of the mating will not express the male father's attractive long-tail genes, these genes are likely to be expressed in her sons. The female has to evolve to find the characteristic being selected for in males attractive for this situation to lead to selection of the characteristic being selected for by the choosiness of females. Indeed, the fact that the female who chose a long-tailed male has children following her mating with genes for liking long-tailed males, and for generating long tails, is part of what leads to the sexual selection.

The peacock tail example is categorized as sexual selection because the long tail is not adaptive to the individual with the long tail, though of course it is useful to the male's genes to have a long tail if females are choosing it because it indicates general physical fitness. However, sexual selection can also occur when a revealing or index signal or fitness indicator is not associated with a handicap, but is hard to fake, so that it is necessarily an honest fitness indicator (Maynard Smith and Harper 2003). An example occurs in birds that may show bare skin as part of their courtship, providing a sign that they are parasite resistant (Hamilton and Zuk 1982). Revealing bare skin in women can be beautiful and may have its origins partly in this, as well as in perhaps displaying secondary sexual characteristics (such as breasts) that may be attractive to men (with an origin as indicators of sexual maturity and of maternal readiness). (Note that this account is very different to that of Sigmund Freud.)

The mechanisms of mate choice evolution include the following (Andersson and Simmons 2006):

(i) Direct phenotypic effects. Female preference for a male ornament can evolve as a result of direct phenotypic benefits if the ornament reflects the ability of the male to provide material advantages, such as high-quality territory, nutrition, parental care, or protection.

(ii) Sensory bias. Female preference favouring a male ornament can initially evolve under natural selection for other reasons, for instance in the context of foraging or predator avoidance. Males evolving traits that exploit this bias then become favoured by mate choice (Ryan 1998).

(iii) Fisherian sexy sons. If there are genetic components to variance in female preference and male trait, a female choosing a male with a large trait bears daughters and sons that can both carry alleles for a large trait, and for the preference for it. This genetic coupling might lead to self-reinforcing coevolution between trait and preference (Fisher 1930, Mead and Arnold 2004). (Sexual election may be identified when females choose sexy mates so that the female's sons will be sexy and attractive. Survival selection may be identified if the choice helps the female's daughters as well as sons.)

(iv) Fitness indicator mechanisms ('good genes' or 'handicap mechanisms') suggest that attractive male traits reflect broad genetic quality (Zahavi 1975, Grafen 1990a, Grafen 1990b). Female preference for such traits can provide genetic benefits to those of her offspring that inherit favourable alleles from their father.

(v) Genetic compatibility mechanisms. As well as additive genetic benefits reflected by indicator traits, there might be non-additive benefits from choosing a mate with alleles that complement the genome of the chooser. Examples have been found for instance in major histocompatibility complex genes, which may be associated with odour preferences for potential mates (Dulac and Torello 2003). These genes are involved in the process by which a cell infected with an antigen (from a virus or bacterium) displays short peptide sequences of it at the cell surface, and the T lymphocytes of the immune system then recognize the fragment, and build an antibody to it. This MHC gene system must maintain great diversity to help detect uncommon antigens, with an advantage arising from mating with an individual with different MHC genes. At least some of the MHC genes are very closely associated with gene-specified pheromone receptors, with individual pheromone receptor cells often expressing one or a few MHC genes in a complex with specific V2R-specified olfactory receptors (Dulac and Torello 2003). Thus, a mate may be found attractive (and beautiful) based on odour, and a mechanism such as this may operate in humans (see Rolls (2005a)).

The evolution of mate choice is based either on direct selection of a preference that gives a fitness advantage (mechanisms i–ii) (i.e. there is a survival or adaptation advantage); or on indirect selection of a preference as it becomes genetically correlated with directly selected traits (mechanisms iii, iv) (i.e. the trait has no advantage, and might be thought of as a useless ornament) (Andersson 1994, Mead and Arnold 2004). In addition, rather than favouring any particular display trait, mate choice might evolve because it conveys non-additive genetic benefits (mechanism v). These mechanisms are mutually compatible and can occur together, rendering the evolution of mating preferences a multiple-causation problem, and calling for estimation of the relative roles of individual mechanisms (Andersson 1994).

Some characteristics of sexual selection that help to separate it from survival selection are as follows:

First, the sexually selected characteristic is usually sexually dimorphic, with the male typically showing the characteristic. (For example the peacock but not the peahen has the long tail.) This occurs because it is the female who is being choosy, and is selecting males. The female is the choosy one because she has a considerable investment in her offspring, whom she may need to nurture until birth, and then rear until independent, and for this reason has a much more limited reproductive potential than the male, who could in principle father large numbers of offspring to optimize his genetic potential. This is an example of a sexual dimorphism selected by inter-sexual selection. An example of a sexual dimorphism selected by intra-sexual selection is the deer's antlers. Sexual dimorphism usually reflects sexual selection, but may not, with an example being that the female may be cryptic (hidden against the background, camouflaged) when incubating eggs, in order to be a good parent.

Second, sexually selected characteristics such as ornamentation helpful in identification are typically species-specific, whereas naturally selected characteristics may, because they have survival value for individuals, be found in many species within a genus, and even across genera.

Third, and accordingly, the competition is within a species for sexual selection, whereas competition may be across as well as within species for natural (survival) selection.

Fourth, sexual selection operates most efficiently in polygynous species, that is species where some (attractive) males must mate with two or more females, and unattractive males must be more likely to be childless. Polygyny does seem to have been present to at least some extent in our ancestors, as shown for example by body size differences, with males larger than females. This situation is selected because males compete harder with each other in polygynous species compared to monogamous, where there is less competition. In humans, the male is 10% taller, 20% heavier, 50% stronger in the upper body muscles, and 100% stronger in the hand grip strength than the average female (Miller 2000).

Fifth, the sexually selected characteristics are often apparent after but not before puberty. In humans, one possible example is the deep male voice.

Sixth, there may be marked differences between individuals, as it is these differences that are being used for mate choice. Sexual selection thus promotes genetic diversity. In contrast, when natural or survival selection is operating efficiently, there may be little variation between individuals.

Seventh, the fitness indicator may be costly or difficult to produce, as in this way it can reflect real fitness, and be kept honest (mechanism iv above).

However, sexual selection is not as pure as was once thought: females are less choosy, and more promiscuous, than was once thought (Birkhead 2000).

Overall, Darwinian natural or survival selection increases health, strength, and potentially resources, and survival of the individual, and thus the ability to mate and reproduce, and to look handsome or beautiful. *Inter-sexual sexual selection does not make the individual healthier, but does make the individual more attractive as a mate*, as in female choice, an example of inter-sexual selection. *Intra-sexual sexual selection does not necessarily help survival of the individual, but does help in competition for a mate, for example in intimidation of one male by another* (Darwin 1871, Kappeler and van Schaik 2004). The behaviours and characteristics involved in sperm competition (Section 4.7), which itself may influence what is judged to be attractive and beautiful, are produced by intra-sexual sexual selection (Rolls 2005a, Andersson and Simmons 2006).

It turns out that many of the best examples of inter-sexual sexual selection are in birds (for example the peacock's tail, and the male lyre bird's tail). In mammals, including primates, the selection is often by size, strength, physical prowess, and aggressiveness, which provide for direct physical (and other types of) competition, and are examples of intra-sexual selection (in males) (Kappeler and van Schaik 2004).

It has been suggested that sexual selection is important for further types of characteristic in humans. For example, it has been suggested that human mental abilities that may be important in courtship such as kindness, humour, and telling stories, are the type of characteristic that may be sexually selected in humans (Miller 2000). Before assessing this (in Section 7.9), and illuminating thus some of what may be sexually selected rewards and punishers that therefore contribute to human affective states and aesthetics, we should note a twist in how sexual selection may operate in humans.

In humans, because babies are born relatively immature and may take years of demanding care before they can look after themselves, there is some advantage to male genes of providing at least some parental care for the children. That is, the father may invest in his offspring. In this situation, where there is a male investment, the male may optimize the chance of his genes faring well by being choosy about his wife. The implication is that in humans, sexual selection may be of female characteristics (by males), as well as of male characteristics (by females). This may mean that the differences between the sexes may not be as large as can often be the case with inter-sexual sexual selection, where the female is the main chooser.

One example of how sexual selection may affect female characteristics is in the selection for large breasts. These may be selected to be larger in humans than is really necessary for milk production, by the incorporation of additional fat. This characteristic may be attractive to males (and hence produce affective responses in males) because it is a symbol relating to fertility and child rearing potential, and not because large breasts have any particular adaptive value. It has even been suggested that the large breast size makes them useful to males as a sign of reproductive potential, for their pertness is maximal when a (young) woman's fertility and reproductive potential is at its highest. Although large breasts may be less pert with age, and it might thus be thought to be an advantage for women not to have large breasts, it may be possible that this is offset by the advantageous signal of a pert but large breast when fertility and reproductive potential is at its maximal when young, as this may attract high status males (even though there may be disadvantages later) (Miller 2000). Thus it is possible that inter-sexual selection contributes to the large breast size of some women. The fact that the variation is quite large is consistent with this being a sexually selected, not survival-selected, characteristic. Thus sexual selection of characteristics may occur in women as well as in men, and may contribute to aesthetic judgements.

7.5 Beauty in men and women

Given this background in the processes that drive evolution to make certain stimuli and types of brain processing rewarding or punishing, in this section I review how they contribute to what factors make men and women aesthetically beautiful. Many of these factors have been described in Chapter 4, and so they are reviewed briefly here in terms of how these factors contribute to aesthetic judgements.

Many of these factors may operate unconsciously, and we may confabulate a rational verbal account about why we judge that something is beautiful. We may not realize that the following factors can influence our aesthetic judgements.

Female preferences: factors that make men attractive and beautiful to women

Factors that across a range of species influence female selection of male mates include the following (Section 4.4.1).

Athleticism

Resources

Power and wealth

Status

Age Status and higher income are generally only achieved with age, and therefore women generally find older men attractive.

Ambition and industriousness These may be good predictors of future occupational status and income, and are attractive. Valued characteristics include those that show a male will work to improve their lot in terms of resources or in terms of rising up in social status.

Testosterone-dependent features These features include a strong (longer and broader) jaw, a broad chin, strong cheekbones, defined eyebrow ridges, a forward central face, and a lengthened lower face (secondary sexual characteristics that are a result of pubertal hormone levels). High testosterone levels are immuno-suppressing, so these features may be indicators of immuno-competence (and thus honest indicators of fitness). The attractiveness of these masculinized features increases with increased risk of conception across the menstrual cycle (Penton-Voak et al. 1999). The implication is that the neural mechanism controlling perception of attractiveness must be sensitive to oestrogen/progesterone levels in women.

Another feature thought to depend on prenatal testosterone levels is the 2nd/4th digit ratio. A low ratio reflects a testosterone-rich uterine environment. It has been found that low ratios correlate with female ratings of male dominance and masculinity, although the relationship to attractiveness ratings is less clear (Swaddle and Reierson 2002).

Symmetry Symmetry (in both males and females) may be attractive, in that it may reflect good development in utero, a non-harmful birth, adequate nutrition, and lack of disease and parasitic infections (Thornhill and Gangstad 1999).

Dependability and faithfulness These may be attractive, particularly where there is paternal investment in bringing up the young, as these characteristics may indicate stability of resources (Buss et al. 1990).

Risk-taking Risk-taking by men may be attractive to women, perhaps because it is a form of competitive advertising: surviving the risk may be an honest indicator of high quality genes (Barrett et al. 2002).

Features selected by inter-sexual sexual selection Characteristics that may not be adaptive in terms of the survival of the male, but that may be attractive because of inter-sexual sexual selection, include the peacock's tail.

Odour The preference by women for the odour of symmetrical men is correlated with the probability of fertility of women as influenced by their cycle (Gangestad and Simpson 2000). Another way in which odour can influence preference is by pheromones that are related to major histocompatibility complex (MHC) genes, which may provide a molecular mechanism for producing genetic diversity by influencing those who are considered attractive as mates, as described in Section 4.4.1.

It is important to note that physical factors such as high symmetry and that are indicators of genetic fitness may be especially attractive when women choose short-term partners, and that factors such as resources and faithfulness may be especially important when women choose long-term partners, in what may be termed a conditional mating strategy (Buss 2008, Buss 2006). This conditionality means that the particular factors that influence preferences and what may be found to be aesthetic alter dynamically, and preferences will often depend on the prevailing circumstances, including the current opportunities and costs.

Male preferences: what makes women attractive and beautiful to men

When a male chooses to invest (for example to produce offspring), there are preferences for the partner with whom he will make the investment. Accurate evaluation of female quality (reproductive value) is therefore important, and a male will need to look out for cues to this, and find these cues attractive, beautiful, and rewarding. The factors that influence attractiveness include the following (Section 4.4.2) (Barrett et al. 2002).

Youth As fertility and reproductive value in females is linked to age (reproductive value is higher when younger, and actual fertility in humans peaks in the twenties), males (unlike females) place a special premium on indicators of youth, for example neotenous traits such as blonde hair and wide eyes. Another indicator of youth might be a small body frame, and it is interesting that this might contribute to the small body frame of some women in this example of sexual dimorphism.

Beautiful features Features that are most commonly described as the most attractive tend to be those that are oestrogen-dependent, e.g. full lips and cheeks, and short lower facial features.

Women appear to spend more time on fashion and enhancing beauty than men. Why should this be, when in most mammals it is males who may be gaudy to help in their competition for females, given that females make the larger investment in offspring? In humans, there is of course value to investment by males in their offspring, so women may benefit by attracting a male who will invest time and resources in bringing up children together. But nevertheless, women do seem to invest more in bearing and then raising children, so why is the imbalance so marked, with women apparently competing by paying attention to their own beauty and fashion? Perhaps the answer is that males who are willing to make major investments of time and resources in raising the children of a partner are a somewhat limiting resource (as other factors may make it advantageous genetically for men not to invest all their resources in one partner), and because women are competing to obtain and maintain this scarce resource, being beautiful and fashionable is important to women. Faithful men may be a limited resource because there are alternative strategies that may have a low cost, whereas women are essentially committed to a considerable investment in their offspring. These factors lead to greater variability in men's strategies, and thus contribute to making men who invest in their offspring a more limited resource than women who invest in their offspring.

Given that men are a scarce resource, and that women have such a major investment in their offspring that they must be sure of a man's commitment to invest before they commit in any way, we have a scientific basis for understanding why women are reserved and more cautious and shy in their interactions with men, which has been noticed to be prevalent in visual art, in which men look at women, but less vice versa (Berger 1972).

Body fat Although the body weight found most attractive varies significantly with culture (in cultures with scarcity, obesity is attractive, and relates to status, a trend evident in beautiful painting throughout its history), the ideal distribution of body fat seems to be a universal standard, as measured by the waist-to-hip ratio (which cancels out effects of actual body weight). Consistently, across cultures, men preferred an average ratio of 0.7 (small waist/bigger hips) when rating female figures (line drawings and photographic images) for attractiveness (Singh and Luis 1995). Thornhill and Grammer (1999) also found high correlations between rating of attractiveness of nude females by men of different ethnicity. At a simpler level, a low waist to hip ratio is an indication that a woman is not already pregnant, and thus a contributor to attractiveness and beauty.

Fidelity The desire for fidelity in females is most obviously related to her concealed ovulation (see next paragraph and *Emotion Explained* (Rolls 2005a)), and therefore the degree of paternity uncertainty that males may suffer. Males therefore place a premium on a woman's sexual history. Virginity was a requisite for marriage both historically (before the arrival of contraceptives) and cross-culturally (in non-Westernised societies where virginity is still highly valued) (Buss 1989). Nowadays, female monogamy in previous relationships is a sought-after characteristic in future long-term partners (Buss and Schmitt 1993). (Presumably with simple genetic methods now available for identifying the father of a child, the rational thought system (Rolls 2005a) might just rely on those to establish paternity, yet the implicit emotional system may still place high value on personality characteristics indicating fidelity, as during evolution, fidelity was valued as an indicator of paternity probability.) The modern rational emphasis might be especially placed on valuing fidelity because this may indicate less risk of sexually transmitted disease, and perhaps the emotional value and attractiveness of fidelity will be a help in this respect.

Attractiveness and the time of ovulation Although ovulation in some primates and in humans is concealed, it would be at a premium for men to pick up other cues to ovulation, and find women highly desirable (and beautiful) at these times. Possible cues include an increased body temperature reflected in the warm glow of vascularized skin (vandenBerghe and Frost 1986), and pheromonal cues. Indeed, male raters judged the odours of T-shirts worn during the follicular phase as more pleasant and sexy than odours from T-shirts worn during the luteal phase (Singh and Bronstad 2001). Women generally do not know when they are ovulating (and in this sense ovulation may be double blind), but there is a possibility that ovulation could unconsciously affect female behaviour. In fact, Event-Related Potentials (ERPs) were found to be greater to sexual stimuli in ovulating women, and these could reflect increased affective processing of the stimuli (Krug et al. 2000). This in turn might affect the outward behaviour of the female, helping her to attract a mate at this time. Another possibly unconscious influence might be on the use of cosmetics and the types of clothes worn, which may be different close to the time of ovulation.

In most species, females invest heavily in the offspring in terms of providing the eggs and providing the care (from gestation until weaning, and far beyond weaning in the case of humans). Females are therefore a 'limited resource' for males allowing the females to be the choosier sex during mate choice. In humans, male investment in caring for the offspring means that male choice has a strong effect on intra-sexual selection in women. Female cosmetic use and designer clothing could be seen as weapons in this competition, and perhaps are reflected in extreme female self-grooming behaviour such as cosmetic surgery, or pathological disorders such as anorexia, bulimia, and body dysmorphic disorder. The modern media, by bombarding people with images of beautiful women, may heighten intra-sexual selection even further, pushing women's competitive mating mechanisms to a major scale.

7.6 Pair-bonding, love, and beauty

We have seen in Section 4.5 some of the factors that promote pair-bonding and love, including oxytocin. These processes influence what we judge as beautiful, and aesthetic. An implication is that there may be hormonal, and other biological, mechanisms that have an effect of cementing attraction and love of a particular person after the process has been started. This may have an effect on (partly) blinding a person to a partner's imperfections, and may thus contribute to each individual's judgements about the beauty of a partner, an aesthetic judgement.

Given this Darwinian approach rooted in selfish (Dawkins 1989) genes, should we describe the aesthetic state of love as selfish? I suggest that the answer is that although individual acts can be truly altruistic (and non-adaptive), even the altruism implied by love must have its origins in selfish genes, which shape human behaviour to in this case produce a state that promotes the production of and survival of offspring. Overall, for a characteristic (such as falling in love, or reciprocal altruism, or kin altruism) that is influenced by genes to remain in a breeding population, the characteristic must be good for the (selfish) gene or it would be selected out.

Even love guided by rational thought must not overall detract too much over generations from the wish to produce offspring, or it would tend (other things being equal) to be selected out of the gene population.

7.7 Parental attachment: beautiful children

Many mammal females make strong attachments to their own offspring, and this is also facilitated in many species by oxytocin, as described in Section 4.6. In humans oxytocin release during natural childbirth, and rapid placing of the baby to breast feed and release more oxytocin (Uvnas-Moberg 1998), might facilitate maternal attachment to her baby. Prolactin, the female hormone that promotes milk production, may also influence maternal attachment – and how beautiful a mother thinks her child is.

It is certainly a major factor in humans that should be understood by all parents that bonding can change quite suddenly at the time that a child is born, with women having a strong tendency to shift their interests markedly towards the baby as soon as it is born (probably in part under hormonal influences), and this can result in relatively less attachment behaviour to the husband/partner.

Lack of parental care in step-fathers is evident in many species, and can be as extreme as the infanticide by a male lion of the cubs of another father, so that his new female may come into heat more quickly to have babies by him (Bertram 1975). Infanticide also occurs in non-human primates (Kappeler and van Schaik 2004). In humans, the statistics indicate that step-fathers are much more likely to harm or kill children in the family than are real fathers (Daly and Wilson 1988).

The tendency to find babies beautiful is not of course restricted to parents of their own children. Part of the reason for this is that in the societies in which our genes evolved with relatively small groups, babies encountered might often be genetically related, and the tendency to find babies beautiful is probably a way to increase the success of selfish genes. One may still make these aesthetic judgements of babies in distant countries with no close genetic relationship, but this does not of course mean that such judgements do not have their evolutionary origin in kin-related advantageous behaviour.

Fig. 7.1 Two Forms (January) 1967. Barbara Hepworth. (See colour plates Appendix B.)

7.8 Synthesis on beauty in humans

We see that many factors are involved in making humans attractive, and beautiful. All may contribute, to different extents, and differently in different individuals, and moreover we may not be conscious of some of the origins of our aesthetic judgements, but may confabulate reasons for what we judge to be aesthetic.

When there is a biological foundation for art, for example when it is figurative, and especially when it is about human figures, there may be a basis for consensus about what is good art – art that stimulates our rational system, and at the same time speaks to what we find beautiful due to our evolutionary history. However, if art becomes totally abstract, we lack the biological foundation for judging whether it is aesthetically beautiful, and judgements may be much more arbitrary, and driven by short-term fashion. Some abstraction away from the very realistic and figurative in art can of course have advantages for it allows the viewer to create in their own experience of a work of art by adding their own interpretation.

There is an important point here about the separation between art and the world. Objects of art can idealize beauty, and enhance it. An example is the emphasis on thin bodies, long limbs, and athletic poses found in some Art Deco sculpture, for example in the works of Josef Lorenzl. Here what is beautiful can be made super-normal, one might say in the literal sense super-natural. Another example is in the emotion in the music of *Tristan and Isolde*. We see that art can emphasize and thus idealize some of the properties of the real world, and lose other details that do not enhance, or distract.

This abstraction of what we find beautiful due to evolution can be seen in some semi-figurative / semi-abstract art, as in some of the line drawings of humans by Henri Matisse and Pablo Picasso. It is also found in the sculptures of human forms of Constantin Brancusi.

What I argue is that if art goes too abstract, then it loses the aesthetic value that can be contributed by tapping into these evolutionary origins. Interesting cases are found in the sculptures of Barbara Hepworth and Henry Moore. In the case of Barbara Hepworth, I now see that she often retains sufficient figurative contribution to her sculpture to tap into evolutionary origins, and I show Fig. 7.1 as an example of a work that after all seems to have some relation to a male and female. Much of the sculpture of Henry Moore is clearly figurative, and where his sculpture becomes apparently very abstract it may lose what is gained by tapping into evolutionary origins, but may gain by association and interpretation in relation to his more figurative work. Where art becomes very abstract, as in some of the work of Mark Rothko, perhaps those especially interested are those who have expertise themselves in what is being achieved technically, such as the painting of colours by Rothko.

I thus argue that figurative or semi-figurative may tend to provide wide appeal, and to continue to do this, as it often taps into the biological underpinnings of art and beauty. With fully abstract art, it may be much less certain that it will have wide and long-lasting appeal. Interesting light on this was shown by monkeys' preferences for fractal images. An example of a fractal image is shown in Fig. 7.2. The monkeys tended to prefer more complex fractal images, but each monkey's preference rankings were very different (Takebayashi and Funahashi 2009). The preference ranking was reflected in the responses of orbitofrontal cortex neurons, with some having firing rates that represented the preference ranking on a continuous scale (Shintaro Funahashi, Kyoto University, personal communication 2011). The implication is that there may be a biological propensity to have different preferences for different types of abstract image, but that there may not be consistency of preference between individuals. In contrast, face beauty (also reflected in the activations in the human orbitofrontal cortex (O'Doherty et al. 2003b)), with its biological underpinnings is much more universally agreed (Thornhill and Gangstad 1999, Jefferson 2004). Even when there are biological underpinnings, we may also expect individual differences in what is found attractive and aesthetic, for individual differences in the sensitivity to different types of reward are an important source of variation that drives evolution (Chapter 3).

Consistent with this analysis, it has been shown that human medial orbito-frontal cortex activations reflect the rated beauty of paintings and music (Ishizu and Zeki 2011). Each participant viewed 60 paintings and listened to 60 musical excerpts. The visual stimuli included paintings of portraits, landscapes, and still lifes, most of them from Western art but three from Oriental art. The auditory stimuli included classical and modern excerpts of mainly Western music with two Japanese excerpts. The activations in the medial orbitofrontal cortex reflected the beauty of the paintings and the music, which were categorized as beautiful, indifferent, or ugly by each participant. Each individual's ratings were used in the analysis, and, because they were figurative stimuli, there was some agreement between participants. As this was a brain imaging study, it could not show that different neurons were activated by the paintings and the musical stimuli, but this is likely, given our knowledge of the principles of the neuronal representations of different rewards in the orbitofrontal cortex (Rolls 2005a, Rolls and Grabenhorst 2008, Rolls 2008b, Grabenhorst and Rolls 2011). This latter is an important point in Rolls' theory of emotion (Rolls 2005a), for each type of reward must be represented independently of other types of reward, so that the appropriate action can be made to obtain each particular reward (Chapter 3).

7.9 Sexual selection of mental ability, survival or adaptation selection of mental ability, and the origin of aesthetics

Miller (2000, 2001) has developed the hypothesis that courtship provides an opportunity for sexual selection to select non-sexual mental characteristics such as kindness, humour, the ability to tell stories, creativity, art, and even language. He postulates that these are "courtship tools, evolved to attract and entertain sexual partners". One mechanism of sexual selection views organisms as advertisers of their phenotypic fitness, and Miller sees these characteristics as such signals. From this perspective, hunting is seen as a costly and inefficient exercise (in comparison with food gathering) undertaken by men to obtain small gifts of meat for women, but at the same time to show how competitive and fit the successful hunter is in relation to other men. Conspicuous waste, and conspicuous consumption, are often signs in nature that sexual selection is at work, with high costs for behaviours that seem maladaptive in terms of survival and natural selection in the narrow sense. The mental characteristics described above are not only costly in terms of time, but may rely on many genes operating efficiently for these characteristics to be expressed well, and so, Miller suggests, may be 'fitness indicators'. Consistent with sexual selection,

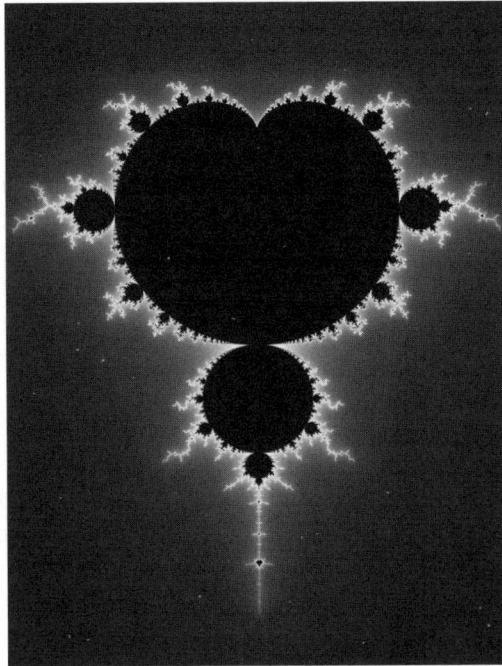

Fig. 7.2 Fractal image. A fractal is a rough or fragmented geometric shape that can be split into parts, each of which is (at least approximately) a reduced-size copy of the whole, a property called self-similarity. This example is of the Mandelbrot set. Fractal images can be produced by mathematical algorithms. (See colour plates Appendix B.)

there is also great individual variability in these characteristics, providing a basis for choice.

One mental characteristic that Miller suggests could have evolved in this way is kindness, which is very highly valued by both sexes (Buss 2008), and is usually judged as aesthetically pleasing. In human evolution, being kind to the mother's children may have been seen as an attractive characteristic in men during courtship, especially when relationships may not have lasted for many years, and the children might not be those of the courting male. Kindness may also be used as an indicator of future cooperation. In a sense kindness thus may indicate potential useful benefits, consistent with the fact that across cultures human females tend to prefer males who have high social status, good income, ambition, intelligence, and energy (Buss 2008). Kindness may also be related to kin altruism (Hamilton 1964) or to reciprocal altruism (Trivers 1971), both of which are genetically adaptive strategies (Section 4.2).

Although the simple interpretation of all these mental characteristics is that they indicate a good provider and potential material and genetic benefits (and thus would be subject to natural or survival selection), Miller (2000) argues that at least kindness is being used in addition as a fitness indicator and is being sexually selected.

Morality can be related in part to kin and reciprocal altruism, which influence survival, and make many of the behaviours described as moral also attractive, because of their evolutionary adaptive value (Ridley 1996, Rolls 2005a) (see Chapter 9). In addition, moral behaviour may bring reproductive benefits and be attractive through the social status that it inspires or by direct mate choice for moralistic displays during courtship (Miller 2000). The suggestion made by Miller (2000) is that the status of moral behaviour helps to attract mates, because it may reflect fitness as the moral behaviour may have costs. In turn, the same effects may influence aesthetic judgements.

Miller (2000, 2001) also suggests that art, language, and creativity can be explained by sexual selection, and that they are difficult to account for by survival selection. He suggests that art develops from courtship ornamentation, and uses bowerbirds as an evolutionary example. Male bowerbirds ornament their often enormous and structurally elaborate nests or bowers with mosses, ferns, shells, berries, and bark to attract female bowerbirds. The nests are used just to attract females, and after insemination the females go off and build their own cup-shaped nests, lay their eggs, and raise their offspring by themselves with no male support. In this sense, the bowers are useless ornamentation, that do not have survival value. Darwin (1871) himself viewed human ornamentation and clothing as outcomes of sexual selection. Sexual selection for artistic ability does not mean of course that the art itself needs to be about sex. This example helps to show that sexual selection can lead to changes in what is valued and found attractive, in areas that might be precursors to art in humans.

In Miller's (2001) view, the fine arts are just the most recent and pretentious manifestations of a universal human instinct for visual self-ornamentation, which in turn is a manifestation of sexual selection's universal tendency to ornament individuals with visual advertisements of their fitness. Thus, the human capacity for visual artistry is viewed as a 'fitness indicator', evolved like the peacock's tail and the bowerbird's bower for a courtship function. So although inherently useless, the bower or work of art is seen as attractive because it is difficult to produce, and might only be made by a brain that is very competent in general, and thus the bower or work of art may act as a fitness indicator.

A useful point (Miller 2001) is that although art-works are now commodified and spread wide so that we may not know the artist producing the ornament, when we seek the evolutionary origins of art, we should remember that any art-work our prehistoric ancestors would have been able to see, would have probably been made by a living individual with whom they could have interacted socially or sexually. The artist was never far from his or her work, or else the work could not have functioned as the artist's extended phenotype.

Miller (2000) also suggests that language evolved as a courtship device in males to attract females. Miller (2000) further suggests that creativity may be related to systems that can explore random new ideas, and also is a courtship device in males to attract females. My view, elaborated here (Chapter 5) and elsewhere (Rolls 2005a, Rolls 2008c, Rolls and Deco 2010, Rolls 2010d, Rolls 2011b), is that language and creativity have functions that have survival value, and thus are not just sexually selected.

Indeed, a criticism of the approach of Miller (2000) is that many of these characteristics (e.g. language, creative solutions, originality, problem solving) may have survival value for both sexes, and are not purely or primarily sexually selected. For example, syntax and language have many uses in problem solving, planning ahead, and correcting multiple-step plans that are likely to be very important to enable immediate rewards to be deferred, and longer term goals to be achieved (Pinker and Bloom 1992, Rolls 2005a, Rolls 2008c). In relation to aesthetics, I argue that when syntax is used successfully to solve a difficult problem, we feel aesthetic pleasure, and I argue that the generation of pleasure generated by the survival value of good ideas contributes to the appeal of those ideas, and that sexual selection of the ideas as mental ornaments is not the only process at work in aesthetics.

Moreover, the notion (Miller 2000, Miller 2001) that art has to do with useless ornaments (useless in the sense that sexual selection is for characteristics that may not have 'survival' value, but may be attractive because they are 'indicators of fitness') does not have much to say about the utilitarian arts such as simplicity of design in architecture. Perhaps the structure of a piece of music can appeal, and be pleasing, because it taps into our reasoning system that finds that elegant and simple solutions to problem-solving produce pleasure. As I argue, interest in social relations and knowledge about them is adaptive as it may help to understand who is doing what to whom, and more generally to understand what can happen to people, and much fictional literature addresses these issues, and is not primarily ornamental and without inherent value. Thus although Miller (2000, 2001) may well be right that there are aspects of art that may be primarily ornamental and useless, and are just indicators of general mental fitness, though attractive to members of the opposite sex in courtship, I suggest that much art has its roots in goals that have been specified as pleasurable or unpleasurable because of their adaptive or survival value, whether as primary reinforcers, other stimuli associated by learning with these, or rewards of a more cognitive origin that accrue when difficult cognitive, syntactic reasoning, problems are solved (see Table 3.1).

Another problem with Miller's approach is that traits that become sexually selected often have survival value in the first place, so it is often not possible to fully dissociate sexual selection from survival or adaptation selection (Andersson 1994, Andersson and Simmons 2006).

Another potential problem with Miller's approach is that some of the processes involved in sexual selection favour fast runaway evolution, because sexual preferences are genetically correlated with the ornaments they favour. Why does mental capacity not develop more rapidly, and with larger sex differences, in humans, if Miller (2000) is right? Why is there not a faster runaway? Miller suggests a number of possible reasons.

1. There is a high genetic correlation between human males and females, with 22/23 chromosomes the same.

2. The female's brain must evolve to be able to appreciate the male's mental adornment – and might even be one step ahead to judge effectively. Further, similar or partly overlapping brain mechanisms may be used to produce (in males) and perceive (in females). In addition, male self-monitoring (and female practice) may help appraisal. Males may even internalize female's appreciation systems, to predict their responses.

3. There is mutual choice in humans: males choose females because human males do make a parental investment; and females compete for males. Indeed, the selection of a long-term partner is mutual, and this tends to reduce sex differences. Consistent with this, Buss (1989, 2008) has shown that, in contrast to the situation with long-term selection of a partner, human sex differences are more evident in short-term mating. It is likely in fact that sexual selection works mainly through long-term relationships, because of concealed ovulation in women. This means that only in a relatively long-term relationship is it likely that a man will become the father of a woman's child, because only if he mates with her regularly is there a reasonable probability that he will hit her fertile time.

Miller might predict that men should be specialized to have artistic creativity, to provide an ornament that women might find attractive because it is a fitness indicator. Evidence on this is difficult to evaluate, because there have been fewer opportunities available for women in the past, as argued for so beautifully by Virginia Woolf in *A Room of One's Own* (Woolf 1928), and I come to no conclusions, but have the following thoughts.

Whereas Virginia Woolf argues about circumstances, one can consider in addition the possibility that women's and men's brains have been subject to different selective pressure in evolution, and that this might contribute to differences in the ways in which they are creative. In terms of artists, composers of music, poets, writers of drama and non-fiction, there appears to be on average a preponderance of men relative to women. This is on average, and there are individual women who given the distribution around the average are undoubtedly highly creative in these areas, and have made enormous contributions. If this is the case (and it might take a long time into the future to know, given the imbalance of opportunity in the past), does this mean that sexual selection is the underlying process?

I suggest that sexual selection would not necessarily be the major, and certainly not the only, driving factor. Such a 'sexual dimorphism' could occur by natural (adaptation) selection, not by sexual selection, in that women might have specialized for an environmental niche to emphasize child rearing, cultivation including food gathering and preparation, fashioning of clothing, and creating peaceful order among siblings and parents. On the other hand, men might have specialized for an environmental niche to emphasize spatial problem solving, useful for producing and using tools, building shelters, creating structures, etc., and navigational problem solving useful for hunting, all of which would be good for survival. Interestingly, the same (narrow) natural selection pressure might have provided a survival advantage for men to have a stronger physique which is likely be advantageous when manufacturing items useful for survival such as shelters. Thus interestingly, one of the predictions of sexual selection, sexual dimorphism, including human mental problem solving as well as physique, could in this case have its origin at least partly in 'survival and adaptation selection' (Section 7.3).

There is however a possible exception to the generalization that at least in the past men have been more likely to be creative in 'art' than women, and this is the area of literary fiction, where there are many women with high reputations as novelists (e.g. Jane Austen, George Eliot, Virginia Woolf). If women take more to this area of creative art, might this be because of the adaptive value of gossip to women, so knowing about who is doing what to whom, and having an interest and expertise in this, could be adaptive, perhaps helping a woman, and her children, to survive better (Dunbar 1996)? If this were the case, there might even be a prediction that women might be relatively more excellent, on average, in areas of fiction, such as novels, where this interest and expertise in mind-reading and gossip, might be especially engaged. (The fact that autism, which is associated with problems with mind reading, is several times more prevalent in men than in women (Baron-Cohen 2008) does fit with this general approach about adaptations suitable for different environmental niches.) More generally, the evolutionary survival value approach might argue that women have adapted to relational, social, caring, problem-solving, and that the novel, particularly the novel of manners, is ideally suited to displaying this specialization. Indeed, the specialization for a caring role is consonant with Carol Gilligan's argument in *In a Different Voice* (1982) that women's sense of morality concerns itself with the activity of "care, ... responsibility, and relationships".

The overall point I make is that natural selection, sometimes operating by 'survival or adaptation selection', and sometimes by sexual selection (and sometimes both, see above), operates by specifying goals for action, and these goals are aesthetically and subjectively attractive or beautiful (Rolls 2005a), or the opposite, and provide what I argue here is the origin of many judgements of what is aesthetic. Many examples of these rewards and punishers, many of which operate for 'survival or adaptation selection', and many of which contribute to aesthetic experience and judgements, are shown in Table 3.1.

7.10 Fashion, and memes

We have seen that sexual selection can provide runaway selective pressure for what is not something that is produced by 'survival or adaptation' selection. In a sense, a fashion or useless ornament (which may indicate fitness) can be selected-for genetically.

However, fashions are strong characteristics of many human aesthetic judgements, and we may ask if there are further reasons for this that are not to do with genetic variation (which necessarily takes place over generations), but that operate over time-scales of months to years. Such fashions (in for example clothing) may occur because they fit adaptations of the human mind, themselves the result of adaptive pressure in evolutionary history. For example, the human mind will be attracted towards new ideas (of clear adaptive value, for it is only by exploring new ideas that advantage may be gained partly as a result of finding a match with one's own genetically influenced capacities) (Rolls 2005a). In this way, there may be runaway changes that do not necessarily make the individual better adapted to the environment, in a way that some consider could be analogous to Fisherian selection (Section 7.4). Of course, many factors, again frequently of evolutionary origin, influence fashion, including its cost (of which the label is an indicator) which helps to make it attractive as it indicates wealth, resources, and status; and the elegance and simplicity of the idea, which as argued below, the human mind finds attractive because simplicity often is a good indicator of a correct and useful solution to a problem. It is argued that memes (Blackmore 1999), ideas that follow some of the rules of fashion, fit these properties of the human mind.

The mechanism of transmission and function of memes are though very different from those of genes. Memes may fit the mind, and be passed from individual to individual, often as useless ornaments. Genetic evolution on the other hand provides an elegant and efficient way to search a landscape where there may be many separate hills of different heights where the height of the hills represents the optimality of a solution (Ackley 1987, Rolls and Stringer 2000). Recombination of genes during sexual reproduction allows new combinations of genes to be brought together to perform local optimization or hillclimbing. Mutation, perhaps 100 times more rare in nature, allows an occasional jump to a new part of the space where there might be a higher hill that can be climbed by the local hillclimbing performed by sexual reproduction. Thus genetic evolution is very different from the transmission of memes. This, of course, is why we have and like sex.

7.11 The elegance and beauty of ideas, and solving problems in the reasoning system

Solving difficult problems feels good, and we often speak about elegant (and beautiful) solutions. What is the origin of the pleasure we obtain from elegant ideas, what makes them aesthetically pleasing?

I suggest that solving problems should feel good to us, to make us keep trying, as being able to solve difficult problems that require syntactic operations may have survival value (Rolls 2005a). But what is it that makes simple ideas and solutions (those with fewest premises, fewest steps to the solution, and fewest exceptions, for a given level of complexity of a problem) particularly aesthetically pleasing, so much so that physicists may use this as a guide to their thinking? It is suggested that the human brain has become adapted to find simple solutions aesthetically pleasing because they are more likely to be correct (Rolls 2005a), and this is exactly the thrust of parsimony and Occam's Razor. (Occam's Razor is the principle or heuristic that entities and hypotheses should not be multiplied needlessly; the simplest of two competing and otherwise equally effective theories is to be preferred. The principle states that the explanation of any phenomenon should make as few assumptions as possible, eliminating those that make no difference in the observable predictions of the explanation or theory. The principle is also captured by the term parsimony.)

This finds expression in art: in for example the structure of a piece of music; in the solution of how to incorporate perspective into painting (which took hundreds of years and was helped by the camera obscura); and in the interest by Vitruvius and Leonardo da Vinci in the proportions of the human body (tapping into our gene-based appreciation of that) to provide rules for proportions in architecture.

Of course, focus on intellectual aspects of art can lead to art that we may find fascinating and revealing, if not conventionally physically beautiful, as in some of the work of Francis Bacon.

Factors such as cultural heritage and familiarity with the rules of a system can also make a style of architecture more appealing than something very unfamiliar. Some of the history of ecclesiastical architecture in England from the eleventh to the fifteenth century (from Norman through Early English and Decorated to Perpendicular) can also be seen as solutions to difficult architectural problems, of how to increase the light and feeling of space in a building, and its impression of grand and daring height.

7.12   Cognition and aesthetics

Not only can operation of our reasoning, syntactic, explicit, system lead to pleasure and aesthetic value, as just described, but also this cognitive system can modulate activity in the emotional, implicit, gene-identified goal system. This cognitive modulation, from the level of word descriptions, can have modulatory effects right down into the first cortical area, the orbitofrontal cortex, where affective value, including aesthetic value, such as the beauty in a face, is first made explicit in the representation (O'Doherty et al. 2003b, De Araujo et al. 2005, McCabe et al. 2008, Rolls 2010a, Grabenhorst et al. 2008a, Rolls and Grabenhorst 2008, Grabenhorst and Rolls 2011). Indeed, cognition and attention can similarly be used to enhance the emotional aspect of aesthetic experience, as described in Section 7.2.

The human mind may create objects such as sculpture and painting in ways that depend to different extents on the explicit reasoning system and the more implicit emotional system. I know at least one sculptor who intentionally reduces cognitive processing by turning off attention to cognitive processing when creating works of art, and then follows this with an explicit, conscious, reasoning stage in which selections and further changes may be made, with the whole creation involving very many such cycles.

Because cognition can by top-down cortico-cortical backprojections influence representations at lower levels, it is possible that training, including cognitive guidance, can help to make more separate the representations of stimuli and their reward value at early levels of cortical processing (Rolls and Treves 1998, Rolls and Deco 2002, Rolls 2008c). This top-down effect may add to the bottom-up effects of self-organization in competitive networks that also through repeated training help representations of stimuli to be separated from and made more different to each other (Rolls and Treves 1998, Rolls and Deco 2002, Rolls 2008c). These effects may be important in many aesthetic judgements that are affected by training, including the appreciation of fine art, architecture, and wine.

7.13   Wealth, power, resources, and reputation

As described above, wealth, power, resources, and status are attractive qualities, aesthetically attractive, because resources are likely to be beneficial to the survival of genes. Reputation is similar, in that guarding one's reputation can be important in reproductive success: trust is important in a mate, or in reciprocal altruism, and hormones such as oxytocin may contribute to trust (Lee et al. 2009).

This provides some insight into the history of Western art, in which individual and family portraits frequently have as one of their aims the portrayal of wealth, power, and resources. The clothes and background are consistent with a contribution of these underlying origins. Commissioned portraits thus frequently emphasize beauty, status, wealth, and resources. Interestingly, because self-portraits are rarely commissioned, they are less likely to emphasize these characteristics (Cumming 2009), and of course can also reflect subjective knowledge of the person portrayed.

An additional property that can add value judged as aesthetic to a portrait is that an image of someone dear is associated with that person, and what that person means to the viewer, and the attraction of photographic images illustrates this. Religion and its accompanying states aiming often at everlasting happiness must also be recognized as drivers of art.

7.14   The beauty of scenery and places

Many topological features of landscapes may be aesthetically attractive because they tap into brain systems that evolved to provide signals of safety, food, etc. (Orians and Heerwagen 1992). Open space may be attractive because potential predators can be seen; cover may be attractive as a place to hide (Appleton 1975); a verdant landscape may be attractive because it indicates abundant food; flowers may be attractive as predictors of fruit later in the season. Many of the properties of the savanna in which we evolved fit this (Orians and Heerwagen 1992) (the 'savanna hypothesis' of why we find certain types of countryside beautiful), and an English parkland may reflect many of these characteristics. The colour blue is preferred by monkeys, and this may be because blue sky, seen from the canopy, is an indicator of a safe place away from predators on the ground (Humphrey 1971). A clear red/orange sunset may be attractive as a predictor of good weather, and of safety overnight without bad weather.

These factors do not operate alone to produce beauty, but may as origins contribute to aesthetic beauty which I argue is multifactorial, influenced by many of the factors described in this theory of the origins of aesthetics (Rolls 2011h).

7.15   The beauty of music

Vocalization is used for emotional communication between humans, with an origin evident in other primates (Rolls et al. 2006a). Examples include warning calls, warlike encouragement to action, and a soothing lullaby or song to an infant. It is suggested that this emotional communication channel is tapped into by music, and indeed consonant vs dissonant sounds differentially activate the orbitofrontal cortex (Blood, Zatorre, Bermudez and Evans 1999, Blood and Zatorre 2001), which is involved in emotion (Rolls 2005a). Of course, the reasoning system then provides its own input to the development, pleasure, and aesthetic value of music, in ways described in Sections 7.11 and 7.12.

What may underlie the greater pleasure and aesthetic value that many people accord to consonant vs dissonant music? I suggest that consonance is generally pleasant because it is associated with natural including vocal sounds with a single source that naturally has harmonics. A good example is a calm female voice. Dissonance may often occur when there are multiple unrelated sources, such as those that might be produced by a catastrophe such as an earthquake, or boulders grinding against each other (or strings on a violin that are not tuned to be harmonics of each other). Further, a human voice when angry, shouting, etc. (and therefore by evolutionary adaptation affectively unpleasant) might have non-linearities, in for example the vocal cords due to over-exertion, and these may be harmonically much less pure than when the voice is calm and softer.

7.16   Beauty, pleasure, and pain

If a mildly unpleasant stimulus is added to a pleasant stimulus, sometimes the overall pleasantness of the stimulus, its attractive value and perhaps its beauty, can be enhanced. A striking example is the sweet, floral scent of jasmine, which as it occurs naturally in *Jasminum grandiflorum* contains typically 2–3% of indole, a pure chemical which on its own at the same concentration is usually rated as unpleasant. The mixture can, at least in some people (and this may depend on their olfactory sensitivity to the different components), increase the pleasantness of the jasmine compared to the same odour without the indole. Why might this occur?

One investigation has shown that parts of the brain such as the medial orbitofrontal cortex that represent the pleasantness of odors (Rolls, Kringelbach and De Araujo 2003c) can respond even more strongly to jasmine when it contains the unpleasant component indole, compared to when it only contains individually pleasant components (Grabenhorst, Rolls, Margot, da Silva and Velazco 2007). Thus one brain mechanism that may underlie the enhancement effect is a principle that brain areas that represent the pleasantness of stimuli can do this in a way that is at least partly independent of unpleasant components, thereby emphasizing the pleasant component of a hedonically complex mixture.

A second factor that may contribute to the enhanced pleasantness of the mixture of jasmine and indole is that the indole may produce a contrast effect in the brain areas that represent the pleasant components of the mixture. An indication of this was found in increased activations in the medial orbitofrontal cortex (which represents the pleasantness of many stimuli) when the jasmine-indole mixture was being applied, compared to just the jasmine alone (Grabenhorst et al. 2007). To the extent that the pleasantness representation may drive hedonic experience separately from unpleasantness representations

(Grabenhorst et al. 2007), then a factor might be the increased activation of pleasantness representations if there is a component to the stimulus that is unpleasant due to a contrast effect heightening the pleasantness. This contrast effect might be facilitated by paying attention selectively to the pleasantness of a stimulus vs its unpleasantness (Rolls, Grabenhorst, Margot, da Silva and Velazco 2008b, Grabenhorst, Rolls and Parris 2008b, Ge, Feng, Grabenhorst and Rolls 2011), Another example of pleasantness enhancement of pleasant by unpleasant stimuli occurs when an odour become more pleasant if it is preceded by an unpleasant (compared to a pleasant) odour, a 'relative pleasantness' effect represented in the human orbitofrontal cortex (Grabenhorst and Rolls 2009).

A third factor is that the interaction between the pleasant (jasmine) and unpleasant (indole) components makes the complex hedonic mixture (jasmine + indole) capture attention (which in turn may enhance and prolong the activation of the brain by the complex hedonic mixture), and evidence for the capture of attentional mechanisms in the brain by the pleasant-unpleasant mixture has been found (Grabenhorst, Rolls and Margot 2011).

These principles may of course operate in most areas where pleasant and unpleasant stimuli combine. Examples might include the pleasure we get from demanding terrain (high cliffs, high mountains, high seas); from spicy food that activates capsaicin (hot somatosensory) as well as gustatory and olfactory receptors (Rolls 2007d); from tragedy in literature, though empathy makes a large contribution here; from difficult feats, such as those performed by Odysseus illustrated on the front cover of Rolls and Deco (2010), etc.

Let us consider the paradox of Tragedy. For Aristotle, tragedy purged one of anxieties (Herwitz 2008). Somehow the depiction of tragedy in drama, which raises unpleasant emotions such as sadness at the tragedy, can also as drama afford pleasure. Hume's explanation was that the beauty of the language and the eloquence of the artist's depictive talents are the source of pleasure (Hume 1757, Yanal 1991). What more can we say about this?

Schadenfreude, gloating, pleasure at the distress of an envied person, is associated with activation of brain areas that respond to pleasant stimuli (Shamay-Tsoory, Tibi-Elhanany and Aharon-Peretz 2007, Takahashi, Kato, Matsuura, Mobbs, Suhara and Okubo 2009), and I suggest is related to the evolutionary origin of competition between individuals, and winning the competition. It is probably not an important factor in the appreciation of tragedy in drama.

What may be more important is first that we (and this is especially strong in women) always want to know what is happening to whom, and gossip has evolutionary value (Dunbar 1996) in that this can provide information about how others are likely to treat you, and more generally, about the things that can happen to people in life, and from which we can potentially learn.

Second, the ability to empathize with another's emotions, and indeed to be good at this and find it rewarding, may also be important in communities, in order to facilitate kin or reciprocal altruism (Ridley 1996).

Third, the ability to have a theory of other people's minds is adaptive in facilitating prediction of their behaviour (Frith and Singer 2008). I propose that fascination with this should again in an evolutionary context be rewarding, and be associated with pleasure, because of its adaptive value. This may contribute to the enjoyment that many women (who overall relative to men may specialize in social relationships) find in fiction.

It is suggested that these three factors are at least important contributors to the pleasure that people find in tragedy in drama. The same factors also I suggest are important contributors to the popularity of novels.

In the cases of both drama and novels, we know that they are fiction, or at least are not happening to the spectator or reader, and this helps to make them particularly rewarding ways to learn about social relations and life events, because there is no risk to the spectator or reader.

Knowing that the work of art (music, literature, painting, sculpture) is a fiction may also account for why the 'aesthetic' emotions are not as long-lasting, and are not as motivating, as the goals in real life.

7.17 Absolute value in aesthetics and art

The approach described here proposes that what we find aesthetic has its roots and origins in two main processes, gene-specified goals, rewards, and punishers; and the value that is felt when our reasoning system produces, and understands, elegant and simple solutions to problems. What implications does this have for absolute aesthetic value? The implication is that while there is no absolute aesthetic value that is independent of these processes, we will nevertheless find considerable agreement between individuals, especially when the aesthetic value being judged has its roots in the two main processes described.

However, as described in this book, there will be variation for good evolutionary reasons between what different individuals find of value, and there will be variation in individuals' thought processes caused by their cultural heritage, and by noise in the brain which is an important component to creativity (Rolls and Deco 2010) (Section 7.2). For these reasons, and because aesthetic value is multifactorial (i.e. is influenced by multiple conscious and unconscious processes), we must expect variation in aesthetic value across people, time, and place, with no absolute aesthetic value.

7.18 Is what is attractive, beautiful and aesthetic?

I wish to counter a possible objection to the theory of the origin of aesthetics described here. The possible objection is that some of the goals specified by our genes, such as the reward value and pleasantness of a high-energy high-fat diet, might seem rather unsavoury, and not quite aesthetic. The point I make is that it is not just the gene-specified rewards and punishers that make stimuli have aesthetic value. My proposal is that the reasoning (rational) system also contributes to aesthetic value, in a number of ways.

The reasoning system makes rather longer-term goals attractive.

It introduces the further goal that innovation is attractive, as this is likely to help solve difficult problems and move the person into a new part of state space where the person may have an advantage.

It introduces the use of syntactic relational structure to provide another way of computation, and problem solving with this reasoning system is encouraged by simple elegant solutions being rewarding and having aesthetic value, as described above.

These factors would help the sophisticated structure in a Bach partita and fugue to contribute to what we judge as aesthetically pleasing, because such music taps not only into our emotional systems, but also into the systems that provide intellectual pleasure because difficult and complex structural problems are posed, and solutions to these difficult structural problems are provided, which as described provides aesthetic pleasure.

In this sense, aesthetic value may have its roots partly in gene-specified rewards (and punishers), but also in the pleasure that the rational system can provide when it is posed, and finds, elegant and simple solutions (which by parsimony are likely to be correct) to complex problems. For this reason, emotions may not be perfectly aligned with aesthetic value. Although both have their origin in gene-specified rewards, emotions may be produced by any one of a large number of reinforcers, whereas aesthetic value usually includes contributions of the reasoning (rational) system, as just described.

Some art that we regard as good aesthetically may not be beautiful. For example, the paintings of Paul Nash representing scenes of the First World War, or for that matter *Guernica* (1937, in response to the Spanish civil war) by Pablo Picasso, are horrifying. But I think the principle is the same: these paintings make impact because they tap by their representations into our emotional brain systems, in this case our horror produced by the ravages of war. Indeed, they arouse in us moral indignation at what happened, and of course moral indignation at injustice and unfairness is the stuff built into the brain by evolution for our fitness, and makes us feel like doing something about the situation (Chapters 4 and 9). In other cases, where the subject matter may not be pleasant, but is nevertheless Art, such as many of Francis Bacon's paintings, perhaps part of what makes it Art is that it makes us look at the world in a new way, and searching for novelty is itself a gene-specified reward (Chapter 3), and at the phenotypic level, search for new knowledge and understanding by the reasoning system should be rewarding, as that knowledge may come in useful and help with survival or reproduction in the future (Section 5.4).

Art as a whole is a larger issue than aesthetics, and beauty. The content of Art might I suggest be seen as the result of multiple separate trajectories through a state space in which each trajectory is guided by the origins of aesthetics (products of adaptations for survival and of sexual selection for useless sometimes handicapping ornament, and rational thought to develop structure in which an elegant and simple solution is pleasing), and depends on each previous trajectory, the history of art in each culture. Each trajectory is not itself deterministic, because it is influenced by noise (Rolls and Deco, 2010) (as is Darwinian evolution). Thus the particular future trajectories cannot be predicted. In each trajectory though a number of factors guide, including new-ness (novelty is biologically attractive as argued above), wildness (as in Beethoven's late string quartets), as well as what we rationally find aesthetic (as described above), and what survival and sexual selection have also provided in us as some of the origins of aesthetics.

7.19 Comparison with other theories of aesthetics

Much research I have performed shows that there is a perceptual representation of objects formed in cortical areas that is kept separate from the representation of the affective value of objects, which happens further on in processing, in brain regions such as the orbitofrontal cortex (and in an area to which it projects, the anterior cingulate cortex) and the amygdala (see Chapter 3 and Fig. 5.1). For example in the inferior temporal visual cortex there is a representation of objects that is independent of whether an object is associated with reward vs punishment, or is made rewarding or not by hunger (Rolls et al. 1977, Rolls et al. 2003a). In the primary taste cortex in the insula and frontal operculum, there is a representation of what taste is present, and of its intensity, that is independent of its reward value as altered by hunger vs satiety, and that is correlated with the subjective intensity but not subjective pleasantness of taste (Rolls, Scott, Sienkiewicz and Yaxley 1988, Yaxley, Rolls and Sienkiewicz 1988, Grabenhorst et al. 2008b). In the primary olfactory (pyriform) cortex, activations are correlated with the subjective intensity but not subjective pleasantness of odor (Rolls et al. 2003c, Rolls et al. 2008b). On the other hand, the affective value of taste, olfactory, visual, thermal, tactile, and auditory stimuli is represented in the orbitofrontal cortex. This is shown by neuronal responses that are modulated by hunger or occur to stimuli when they are associated with a reward, and by correlations of brain activations with subjective ratings of pleasantness but not intensity (Rolls et al. 1989, Critchley and Rolls 1996a, Kringelbach et al. 2003, Rolls 2005a, Rolls 2007d, Rolls and Grabenhorst 2008, Rolls et al. 2010a, Grabenhorst and Rolls 2011).

There are good functional and adaptive reasons for separate representations of objects and of their affective value. We can still see and recognize objects (including tastes, smell, the sight of objects, etc.) even when they are not rewarding to us, for example if they are foods and we are not hungry. (We do not go blind to objects when they are not rewarding or punishing.) Moreover, it is adaptive to be able to learn about where we have seen objects, people, etc. even if they are not currently rewarding, so that we can find them later. Thus there is strong neuroscientific evidence, and sound biological arguments, for separate representations of perceptual objects and of their affective value.

Baumgarten (1750) expressed this thought in his book *Aesthetica* when he suggested that sensation, the use of the five senses, is separate from sensibility, which is something more, a "kind of intuition/cognition/formulation of the thing which judges it beautiful", and in doing so gave rise to the term aesthetics (Herwitz 2008). Before this, abstract questions such as 'What is beauty?', 'What is art?' had not been treated in philosophy, although before this Aristotle had discussed the social role of drama as purging us of ever present anxiety, and Plato had dismissed poetry as obfuscating by sending the mind reeling into hypnotic trances instead of focusing on rational deductions and argument (Herwitz 2008).

David Hume (1777) takes a broad view of taste (which engages beauty), and argues for five standards of ('delicacy of') taste that might be shown by experts: "Strong sense, united to delicate sentiment, improved by practice, perfected by comparison, and cleared of all prejudice, can alone entitle critics to this valuable character; and the joint verdict of such, wherever they are to be found, is the true standard of taste and beauty".

Hume's difficulty is that he believes taste is objective, because delicacy is the probing instrument for truth; but instead, taste is a circular and constructivist enterprise (Herwitz 2008). My approach has in contrast a clear foundation for aesthetics in brain function and its evolutionary design, with clear views about how it includes rational thought which provides its own pleasures, and about how art can idealize beyond the normal world by building on these foundations and origins.

Immanuel Kant (1724–1804) distinguishes between liking something and finding it beautiful. According to Kant when I find a painting beautiful this is not conditioned by any causal relation between its properties and my pleasures. For Kant, a judgement of beauty carries the weight of 'ought', that others should judge it beautiful too, so his theory has moral implications. His judgement of beauty is a 'disinterested' judgement, one that is not peculiar to him. He wants the beauty to be in the person, but not causally dependent on the properties of the object in the world such as the pleasure it produces (Kant 1790). He thus appears to be committed to an objective and universal view of art, with exactly how this view is arrived at not at all clear.

The biological and neuroscientific view that I propose indicates that in contrast art is not universal or objective, but instead can be judged good art if it taps into many of the human rational and gene-based reward systems (see further Section 7.18), with therefore individual differences expected, as described in Section 7.17.

Darwin (1871) recognized that evolution can occur by sexual selection, when what is being selected for has no inherent adaptive or survival value, but is attractive to potential mates (inter-sexual selection), or helps in competing with others of the same sex (intra-sexual selection). His view was that natural beauty arises through competition to attract a sexual partner. His process of sexual selection through mate choice – the struggle to reproduce, not to survive – drove the evolution of visual ornamentation and artistry, from flowers through bird plumage to human self-adornment. Many have developed or ascribed to this idea (including Thorstein Veblen (1899), Ernst Gombrich (1977), Amotz Zahavi (1978) and Denis Dutton (2009), see Miller (2001)), and Miller (2000, 2001) has proposed a sexual selection theory of art. The implication of this theory is that art has to do with what are frequently useless ornaments (useless in the sense that sexual selection is for characteristics that do not have 'survival' value, but are usually just attractive because they are handicaps and are indicators of fitness).

I agree that useless handicapping ornament produced by sexual selection does play a role in aesthetics. However, the sexual selection theory does not therefore have much to say about the utilitarian arts such as simple design in architecture. Perhaps the structure of a piece of music can appeal, and be pleasing, because it taps into our syntactic system that finds that adaptive, survival value-related, elegant, and simple solutions to problem-solving produce pleasure. As I argued above, interest in social relations and knowledge about them is adaptive and has survival value as it may help to understand who is doing what to whom, and more generally to understand what can happen to people, and much fictional literature addresses these issues, and is not purely ornamental and without inherent value. Thus although Miller may well be right that there are aspects of art that may be primarily ornamental and useless, though attractive to members of the opposite sex in courtship, I suggest that much art has its roots in goals that have been specified as pleasurable or unpleasurable because of their 'survival or adaptive' value, whether as primary reinforcers, other stimuli associated by learning with these, or rewards of a more cognitive origin that accrue when difficult cognitive, syntactic, problems are solved. I also emphasize that some of the characteristics emphasized by sexual selection may have some inherent survival value (mechanisms i-ii in Section 7.4).

7.20 Conclusions

To end this chapter, my theory (Rolls' theory) of aesthetics (Rolls 2011h) thus specifies the roles of (both) Darwinian 'survival or adaptation' selection, and sexual selection, in aesthetics. It is thus thoroughly Darwinian.

A key idea is that many of the things that provide pleasure, or its opposite, do so because they are, or are related to, the gene-specified goals for action. Motivational states arise when trying to obtain these goals, and emotional or affective states when these goals are obtained, or are not obtained. These states are associated with affect and value, and with subjective pleasantness or unpleasantness, because it is an efficient way in which genes can influence their own (reproductive) success ('fitness'), and much more efficient and effective as a Darwinian process than prescribing that the animal should make particular responses to particular stimuli (Rolls 2005a) (Chapter 3).

The theory is that aesthetic value has its roots partly in these gene-specified rewards that have survival or adaptive value; but also in the pleasure that the rational system can provide when it is posed, and finds, elegant and simple solutions (which by parsimony are likely to be correct and hence adaptive) to complex problems; and to some extent in sexual selection. What makes good art can be influenced by many factors, as described here, so is complex and multi-faceted, and these factors must include whether the effect of the art is for good or for harm.

It also follows that attempts in aesthetics to produce a systematic account based on consistent explicit beliefs will not succeed, for many factors that are not necessarily consistent with each other are involved in aesthetic values; because there are individual differences in reward systems as part of the variation necessary for evolution; and because some of these factors operate at least partly unconsciously and non-propositionally / non-syntactically, that is, using computational systems in the brain that do not involve reasoning.

8 Neuroeconomics

8.1 Introduction

To what extent does our understanding of how our brains use heuristics to solve difficult problems involving costs and benefits provide a new way of thinking about economic choice made by individuals? Is the classical model of humans as rational decision-makers satisfactory given the implications of modern neuroscience? What promotes trust between individuals? How stable is reciprocal altruism when the players are not perfectly matched? How do tit-for-tat interactions operate in reciprocal altruism, and how is forgiveness an adaptive heuristic in reciprocal altruism? Why are we so sensitive to fairness, cheating, and defection? How do we respond to the outcome that is received in an interaction compared to what we expected, and how to we respond if there is a mismatch, and also correct our expectations and change our future choices? How is value computed by the brain, and does the value of rewards and punishers change as their magnitude increases? These are some of the issues which modern neuroscience is addressing in the field of neuroeconomics (Glimcher et al. 2009, Glimcher 2011), and which we consider in this chapter.

The approach taken again builds on an understanding of reward systems in the brain, and computational neuroscience. Of particular interest is whether humans behave rationally in their own self-interest, or whether other less rational reward-related heuristics built into us during evolution take a part in our economic decision-making, and choices (Chapter 3 and Section 5.4). Also of interest is whether choices are altruistic (Section 4.2). Also of particular interest is whether our choices are not those that might be taken by a deterministic system, but instead are influenced by noise in the brain (Section 2.12).

8.2 Reciprocal altruism, strong reciprocity, generosity, and altruistic punishment

8.2.1 Economic cooperation vs self-interest

A concept used in economics is that humans take rational and self-interested decisions. However, in a number of games (and also in real-life situations), humans do not follow these principles exactly (Fehr and Rockenbach 2004, Montague, King-Casas and Cohen 2006).

One example is the 'prisoner's dilemma game' (PD). In the PD, two players simultaneously choose between cooperation and defection. If both decide to cooperate, they both earn a high outcome (e.g. 10); if both defect, they both receive a low outcome (e.g. 5); and, if one player cooperates and the other defects, the cooperator obtains a very low outcome (e.g. 1), whereas the defector receives a very high outcome (e.g. 15). Hence, it is always better for a player to defect for any given strategy of the opponent. The PD resembles a generic cooperation dilemma in which purely selfish behaviour leads to the defection of both players, even though mutual cooperation would maximize their joint payoff. Cooperation, however, is vulnerable to exploitation. The PD reflects the cooperation dilemma inherent in the provision of a public good, such as cooperative hunting or group defence, with only two individuals involved. More generally, a 'public good game' (PG) consists of an arbitrary number of players who are endowed with a certain number of tokens that they can either contribute to a project that is beneficial for the entire group (the public good) or keep for themselves. The dilemma arises from the fact that all group members profit equally from the public good, no matter whether they contributed or not, and that each player receives a lower individual profit from the tokens contributed to the public good than from the tokens kept. A purely selfish player refuses to contribute anything to the public good and free rides on the contributions of others. Considerable cooperation (contributions between 40 and 60% of the endowment) is typically observed in PGs with one-shot interactions (that is where the game is played once with a particular group of other players). However, cooperation is rarely stable if the game is played repeatedly, and deteriorates to low levels towards the end of the interaction period (Fehr and Rockenbach 2004).

Why is cooperation observed at all and what are the mechanisms that enable and sustain human cooperation in social dilemma situations, even in an environment with (a considerable number of) selfish subjects? *Strong reciprocity* appears to be crucial for the establishment of cooperation in groups with a share of selfish individuals. A person who is willing to reward fair behaviour and to punish unfair behaviour, even though this is often quite costly and provides no material benefit for the person, is called a 'strong reciprocator'. Strong reciprocity can occur in sequential dilemma situations, in which games are repeatedly played with the same set of players. In these situations, individual players use 'altruistic punishment' heavily, that is they punish at their own cost those who do not contribute to the public good. The effect is that a large increase in cooperation within the group is observed (Fehr and Rockenbach 2004). Perceived fairness can influence brain activations, and indeed in an fMRI investigation it was found that the effect of seeing a person in pain produced less activation in males' anterior insular / fronto-insular (i.e.

agranular insular) cortex and anterior cingulate cortex if that person had acted unfairly (Singer, Seymour, O'Doherty, Stephan, Dolan and Frith 2006).

Another example is that when playing a monetary game with another human player, humans may provide information to the other player that will allow the other player to have a better chance of winning, even though to do this will cost the first player (Gneezy 2005, Gintis 2007). Why does this generosity occur? An explanation is that during evolution humans have developed heuristics for behaviour that, while not necessarily advantageous in every short-term situation, are in the long term useful strategies for promoting reciprocal altruism and hence reproductive fitness. In the case being considered, being seen to be willing to give something to another person can be taken as a sign that will enhance the reputation of the player as being honest and being willing to reciprocate, and of course being in a situation of reciprocal altruism can be more advantageous to each player in at least some situations than acting purely in terms of short-term interest. In this sense, the genes have produced a predisposition for a social behaviour which in the long run can increase (reproductive) fitness, but cannot prescribe for every detailed situation in which there might be a short-term cost.

In the case of altruistic punishment, it has been shown that this strategy can survive in evolutionary models of social cooperation (Boyd, Gintis, Bowles and Richerson 2003, Bowles and Gintis 2004), and this suggests that there could be a genetic predisposition to this type of social behaviour too.

Thus there is an account based in the genetically defined heuristics for the strong reciprocity found in some games even when the player knows that the goods will not be repaid, and the player even knows that there will be only a single game with each other player (Fehr and Fischbacher 2003, Fehr and Rockenbach 2004, Camerer and Fehr 2006). Interestingly, people play selfishly in this situation if they know that the opponent is a computer, emphasizing that the generous strategy is adopted in social situations (in which reputation may be important), and is context-dependent.

In a similar way, being generous and 'forgiving' in tit-for-tat games can be helpful if both players have defected, for being generous occasionally may reinstate reciprocal positive play (altruism), which could be to the advantage of both players (Ridley 1996). Being (seen to be) honest can also be viewed as a useful (gene-based) strategy for promoting reciprocal altruism. Because in society it may be important in the long run to maintain one's reputation as a reciprocator, it may be important to be honest, generous, and forgiving, even though in the short term this may be disadvantageous. On the evolutionary time-scale, it may promote (reproductive) fitness (of the relevant selfish genes) to be generous, forgiving, honest, and an altruistic punisher, because of the advantages reciprocal altruism brings. These heuristic ways of behaving may have evolved because of their long-term benefits, and this is an advance on traditional models in economics that have assumed that humans act selfishly and rationally to reach economic decisions. Of course, there are other advantages to being seen to be generous, including the status this provides partly because one can be seen as a potential provider of resources and protection (Ridley 1996, Rolls 2005a).

Some of these issues have been investigated in trust games (Montague et al. 2006, King-Casas, Tomlin, Anen, Camerere, Quartz and Montague 2005). In a trust game there is an exchange between two players and cooperation and defection can be parametrically encoded as the amount of money sent to one's partner. On each exchange, one player (the investor) is endowed with an amount of money. The investor can keep all the money or decide to invest some amount, which is tripled and sent to the other player (the trustee) who then decides what fraction to send back to the investor. The investor routinely does not keep all the money, but makes offers to the trustee that could be considered close to fair splits. This move is typically met with a reasonable (close to fair) return from the trustee. The initial offer from the investor entails some likelihood of a loss, and can be viewed as a cooperator signal. In repeated games between two players, reputations can be established.

In one such study (King-Casas et al. 2005), activations in the ventral part of the head of the caudate nucleus (which receives inputs from the prefrontal cortex) were correlated with deviations from tit-for-tat reciprocity. In a similar brain region in the trustee, activations were also correlated with whether on the next move the trustee was going to increase or decrease the amount of money sent, thus reflecting the trustee's "intention to change the level of trust" (King-Casas et al. 2005). Thus activations in the brain do reflect how trust games are being played. Where in the brain the computations are being performed that result in the players' moves will be an interesting issue to unravel.

Whether strategies are substitutable or complementary is also important in the behaviour that is selected. Self-regarding preferences and rationality may predict aggregate behaviour well when strategic substitutability applies. [Strategies are complements if agents have an incentive to match the strategies of other players. Strategies are substitutes if agents have an incentive to do the opposite of what the other players are doing. For example, if a firm can earn more profit by matching the prices chosen by other firms, then prices are strategic complements. If firms can earn more profit by choosing a low price when other firms choose high prices (and vice versa), then prices are strategic substitutes.] Other-regarding preferences may predict aggregate behaviour well when strategic complementarity applies (Camerer and Fehr 2006).

8.2.2 Framing issues

Framing issues (that is, how the task is described or framed (Kahneman and Tversky 1984)) are probably very important in the behaviours shown by individuals in these economic games. For example, if the instructions were to explicitly maximize the short-term profit, then a rational selfish strategy would be likely to be adopted. Such a strategy might be more likely if a player was told that each opponent would be played only once; that the opponent was a computer; and/or if it was emphasized that the game had been set to evaluate quantitative economic reasoning, and logic. On the other hand, if the task instructions made it clear that there would be repeated games with another player, and that the choices made were being inspected by the experimenter, then this would tend to promote altruistic heuristics. Being informed that the other player was altruistic might also influence a player's strategy. (Reputations associated with church attendance might have a similar effect.)

8.2.3   Trust, and oxytocin

The biological bases of some of the predispositions for prosocial behaviour related to reciprocal altruism are starting to be investigated. It has been shown for example that the hormone oxytocin, implicated in attachment between mother and offspring and between partners (see Carter (1998), Insel and Young (2001), Winslow and Insel (2004) and Rolls (2005a)), also increases trust in an economic game between humans (Kosfeld et al. 2005). In a bargaining game which is a probe for fairness, the ultimatum game (Montague et al. 2006), insula activation was produced by unfair offers (Sanfey, Rilling, Aronson, Nystrom and Cohen 2003), though such activation might just reflect autonomic activity associated with for example disgust (Nagai, Critchley, Featherstone, Trimble and Dolan 2004). [In the one-round ultimatum game, a pair of players is given an endowment, say $100. The first player proposes a split of the money to the second player, who can respond by either accepting the proposal (take it) or rejecting it (leave it). If the proposal is rejected, neither player receives any money. A rational agent model predicts that proposers should offer as little as possible, and responders should accept whatever they are offered because something is better than nothing. However, responders routinely reject offers less than about 20% of the endowment, and, correspondingly, proposers routinely offer significant amounts. Thus humans routinely act with a sense of 'fairness' in the ultimatum game. While this is not in their short-term interest, it is a strategy or heuristic that may have evolved to promote reciprocation and in the long run mutual benefit for reciprocators.] Activation of the dorsolateral prefrontal cortex and anterior cingulate cortex was also produced by unfair offers. In a prisoner's dilemma game, cooperation with a human partner produced greater activation of the orbitofrontal cortex and related brain systems than did cooperation with a computer partner (Rilling, Gutman, Zeh, Pagnoni, Berns and Kilts 2002).

8.2.4 The neuroeconomics of cooperation

We have seen that there are evolutionary advantages to generosity, forgiveness, and altruistic punishment, and that these could be evolutionarily stable and hence lead to genetically inherited heuristics that influence behaviour (Ridley 1996, Ridley 1993b, Buss 2008, Boyd et al. 2003, Bowles and Gintis 2004, Rolls 2005a). We may note that although the heuristics are universal (cross-cultural), the triggers are culturally defined. Indeed, the centrality of culture and complex social organization to the evolutionary success of *Homo sapiens* implies that individual fitness in humans will depend on the structure of cultural life. Since clearly culture is influenced by human genetic propensities, it follows that human cognitive, affective, and moral capacities are the product of a unique dynamic known as *gene–culture co-evolution*. It is in part this co-evolutionary process that has endowed us with preferences as predispositions that go beyond the self-regarding concerns emphasized in traditional economic and biological theory, and embrace such non-self-regarding values as a taste for cooperation, fairness, and retribution, the capacity to empathize, and the ability to value such constitutive behaviours as honesty, hard work, toleration of diversity, and loyalty to one's reference group (Gintis 2007, Bowles and Gintis 2005, Gintis 2003, Boyd et al. 2003).

However, while there probably are these predispositions, to varying extents in different individuals and even populations (because of genetic variation), this does not exclude at all the possibility that there may be rational reasons within a culture for cooperation, with these strategies operating on top of what is inherited. These predispositions to act altruistically may provide part of a foundation to which ethical principles are related (Ridley 1996). Ethical principles might *naturally* emphasize behaviours such as generosity, forgiveness, and even '*Honi soit qui mal y pense*' (Evil be to him who evil thinks). Of course, ethical principles should not be subject to the naturalistic fallacy, that what is natural is right. However, ethical principles might fare best and gain most acceptance if what they promoted was not too inconsistent with inherited predispositions (Rolls 2005a) (Chapter 9).

8.3 Prospect theory

Prospect theory is a behavioural theory of decision-making under risk, that is when the probabilities are made known explicitly to the decision-maker, and describes important properties of decision-making when there are potential gains and losses (Kahneman and Tversky 1979, Tversky and Kahneman 1992) (see Fox and Poldrack (2009)). Prospect theory was the major work for which Daniel Kahneman, a psychologist, was awarded the 2002 Nobel Prize in economics. (In decisions under uncertainty, the probabilities are not made known explicitly to the decision-maker.)

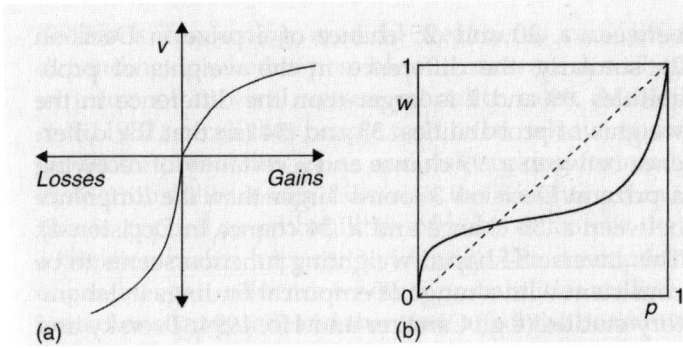

Fig. 8.1 Representative value and weighting functions from prospect theory. (a) A hypothetical value weighting function illustrating concavity for gains, convexity for losses, and a steeper loss than gain limb. v= subjective value. (b) A hypothetical prospect theory weighting function illustrating its characteristic inverse S-shape, the tendency to overweight low probabilities and underweight moderate to high probabilities, and the tendency for weights of complementary probabilities to sum to less than 1. p=probability. w=weight given to a given probability. (After Fox and Poldrack 2009.)

According to prospect theory, the value V of a simple prospect that pays \$$x$ with probability p (and nothing otherwise) is given by:

$$V = w(p)v(x) \tag{8.1}$$

where v measures the subjective value of the consequence x, and w measures the impact of probability p on the attractiveness of the prospect (see Fig. 8.1).

The subjective value v increases from the status quo reference point (i.e. the current value of v is assumed to be 0) with diminishing increases as the gain \$$x$ increases, as shown in Fig. 8.1a. The subjective value v decreases from the status quo reference point as the loss \$$-x$ increases, as shown in Fig. 8.1a, and the rate of decrease of v is high. This captures several points important in neuroeconomics. The first is that as the gains increase, the subjective value tends to saturate. In other words, our subjective value function is very sensitive to small gains, and less sensitive than might be expected for larger gains. This makes biological sense: we are set up so that we are very sensitive to small gains (and losses), which may help us to climb local reward gradient inclines. The concavity for gains contributes to *risk aversion* for gains. The second is that we are more sensitive in terms of subjective value v to losses than to gains. This is referred to as *loss aversion*. This again has biological adaptive value: we should be very sensitive to losses, as even an occasional loss could be life threatening when close to surviving, whereas occasional gains have less impact on our future. In this sense, and for this reason, most people tend to be risk aversive, though there are large differences between individuals, with some gamblers being rather sensitive to gains, and less sensitive to losses. The concavity of the subjective value for increasing gain (Fig. 8.1a) contributes to risk aversion for gains.

What determines the status quo here, where v is assumed to be 0, is important. If we are shifted to a scenario on which there are regularly larger gains, our operating point will increase, and v will be set to be 0 at a higher average gain $x. This adaptation may take some time.

This mirrors the point made about emotion that individuals living in very different circumstances, for example a Western middle class person and a person in an African village, may have not very different levels of average happiness, and this it is suggested reflects a resetting of the reward mechanism so that it approaches the same mean value over time, and is thus very sensitive to any small change of reward value, which is biologically adaptive, as then the sensitivity to a change of reward gradient is likely to be greatest (Rolls 2005a) (Section 3.4).

In the context of decision under risk, loss aversion gives rise to risk aversion for mixed (gain-loss) gambles so that, for example, people typically reject a gamble that offers a 0.5 chance of gaining $100 and a 0.5 chance of losing $100, and require at least twice as much 'upside' as 'downside' to accept such gambles (Fox and Poldrack 2009).

The probability of the outcomes is also not taken into account in the linear way that would be expected of the rational selfish individual of classical economics. Instead, the subjective value is influenced also by a weighting function, as shown in Equation 8.1 and Fig. 8.1b. The weighting function shown in Fig. 8.1b captures diminishing changes in sensitivity to changes in probability. For probability, there are two natural reference points, impossibility ($p=0$), and certainty ($p=1$). Hence, diminishing sensitivity implies an inverse-S shaped weighting function that is concave near $p=0$ and convex near $p=1$, as shown in Fig. 8.1b. The impact of the weighting function is that moderate to high probabilities are underweighted (which reinforces the pattern of risk aversion for gains and risk seeking for losses implied by the value of the value function shown in Fig. 8.1a). Further, low probabilities are overweighted (which reverses the pattern implied by the value function and leads to risk seeking for gains and risk aversion for losses).

Fox and Poldrack (2009) provide an example to illustrate these points in prospect theory and some implications of the functions shown in Fig. 8.1. In the example, the reason that most participants in a study by Tversky and Kahneman (1992) would rather have a 0.95 chance of $100 than $77 for sure is partly because they find receiving $77 nearly as appealing as receiving $100 (i.e. the slope of the value function decreases with dollars gained), and partly because a 0.95 chance 'feels' like a lot less than a certainty (i.e. the slope of the weighting function is high near one). Likewise, most participants would rather face a 0.95 chance of losing $100 than pay $85 for sure is partly because paying $85 is almost as painful as paying $100, and partly because a 0.95 chance feels like it is much less than certain. On the other hand, the reason that most participants would rather have a 0.05 chance of $100 than $13 for sure is that a 0.05 chance 'feels' like much more than no chance at all (i.e. the slope of the weighting function is steep near zero) – in fact it 'feels' like more than its objective probability, and this distortion is more pronounced than the feeling that receiving $13 is more than 13% as attractive as receiving $100. Likewise, the reason most participants would rather lose $7 for sure than face a 0.05 chance of losing $100 is that the 0.05 chance of losing money looms larger than its respective probability, and this effect is more pronounced than the feeling that receiving $7 is more than 7% as attractive as receiving $100.

In sum, prospect theory explains attitudes towards risk via distortions in the shape of the value and weighting functions. The data of Tversky and Kahneman (1992) suggest that the pattern of risk attitudes for simple prospects that offer a gain or a loss with low or high probability is driven primarily by curvature of the weighting function, because the value function is not especially curved for the typical participant in those studies. Pronounced risk aversion for mixed prospects that offer an equal probability of a gain or loss is driven almost entirely by loss aversion, because the curvature of the value function is typically similar for losses versus gains and decision weights are similar for gain versus loss components. Further advances and tests of prospect theory are described by Fox and Poldrack (2009), and its advantages and limitations are discussed by Glimcher (2011).

8.4   Neuroeconomics, reward magnitude, expected value, and expected utility

Reward magnitude and punishment magnitude are represented in the orbitofrontal cortex, as shown by investigations in which reward and punisher magnitude has been parametrically varied (see Section 3.11). One type of evidence has been obtained with reward devaluation produced by sensory-specific satiety, in which activations in the orbitofrontal cortex to the flavour of a food decrease as the food is fed to satiety and becomes less rewarding and less subjectively pleasant (Rolls 2005a, Kringelbach et al. 2003, Rolls and Grabenhorst 2008, Grabenhorst and Rolls 2011). Another has been with trial-by-trial variation of the monetary gain and loss, allowing correlation of activations of different parts of the orbitofrontal cortex with the magnitude of the gain or loss (O'Doherty, Kringelbach, Rolls, Hornak and Andrews 2001, Rolls, McCabe and Redoute 2008e). In the probabilistic monetary reward task of O'Doherty, Kringelbach, Rolls, Hornak and Andrews (2001) in which different amounts of money might be won or lost on each trial, activations in the medial orbitofrontal cortex increased linearly with the amount of money won

Reward correlation : medial orbitofrontal cortex

Punishment correlation : lateral orbitofrontal cortex

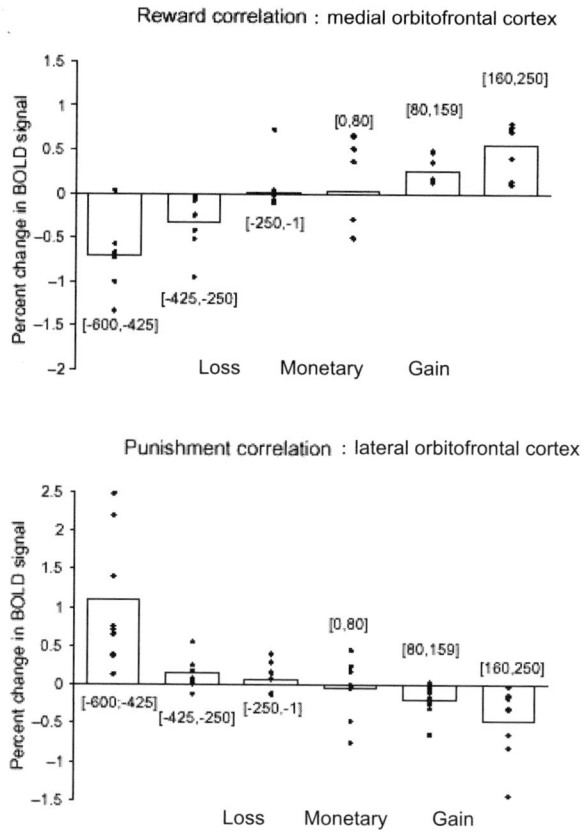

Fig. 8.2 Correlation of brain activations with the amount of money won or lost in a visual discrimination reversal task with probabilistic monetary reward and loss. The mean percent change in the fMRI BOLD signal from the baseline across subjects for 6 different category ranges of monetary gain or loss (plotted along the abscissa). The signal was averaged across a category range within each subject and then the average signal change from each category was averaged across subjects. This is plotted for voxels in the medial orbitofrontal cortex (OFC) that significantly correlated with reward (upper) and for voxels in the lateral OFC that significantly correlated with punishment (lower). The ranges of monetary reward and losses (losses are shown as negative numbers, indicating the amount of money lost on each trial) in each category are shown on the chart and were determined by their relative frequencies, which follow from the experimental design. (After O'Doherty, Kringelbach, Rolls, Hornak and Andrews 2001.)

on each trial, and activations in the lateral orbitofrontal cortex increased linearly with the amount of money lost on each trial, as shown in Fig. 8.2.

The human medial orbitofrontal cortex has neural systems that respond to different types of reward, including not only monetary reward, but also taste, smell, touch, warmth, and visual reward and subjective pleasantness (Rolls 2005a, Rolls and Grabenhorst 2008, Grabenhorst and Rolls 2011). The representations of each type of reward are probably separate, as shown by evidence that different neurons in primates are tuned to respond best to different types of reward. It is very important that there are separate representations of each type of reward, for then the reward value of each can vary independently, so that for example the reward value of food can remain high even after we have drunk water to satiety, and water is no longer rewarding. It has been suggested though that bringing different reward representations close together in the human medial orbitofrontal cortex does allow some mutual inhibition between them, implemented by the short-range cortical inhibitory interneurons. This mutual inhibition provides a form of competition between different rewards, which may be helpful in selecting the reward that at a given time is strongest as the goal for action, and in scaling each type of reward, so that each is chosen sometimes, which is important for the survival value of the genes that specify the different rewards, and thereby for the survival of the individual (Rolls 2005a, Rolls 2008c).

In a similar way, and as illustrated in Fig. 8.2, the human lateral orbitofrontal cortex appears to represent many different types of loss and non-reward (Rolls 2005a, Rolls and Grabenhorst 2008, Grabenhorst and Rolls 2011).

These reward, non-reward, and loss systems in the orbitofrontal cortex thus provide a basis for guiding behaviour to obtain rewards, and to avoid not obtaining expected rewards, and to avoid loss. The benefits and costs of stimuli are what are represented in the orbitofrontal cortex, and we refer to these as the intrinsic costs and benefits (Grabenhorst and Rolls 2011) (Section 3.10.2). A region to which the orbitofrontal cortex connects, the anterior cingulate cortex, may be important in action selection based on the rewards and the costs of the actions performed to obtain each reward. We refer to these as extrinsic costs. We now extend this analysis to situations in which rewards are uncertain, as in many economic choices.

8.4.1 Expected utility ≈ expected value = probability multiplied by reward magnitude

The question arises for example of how a decision is influenced if the reward magnitude is high, but there is a small probability of obtaining the reward. Here we can adopt approaches used in reinforcement learning, and use the terms **reward value** (RV) or **reward magnitude** (RM) for the magnitude of the reward obtained on a trial, that is the **reward outcome**; and **expected value** (EV) as the probability of obtaining the reward multiplied by the reward magnitude (Glimcher 2003, Glimcher 2004, Kahneman and Tversky 1984, Sutton and Barto 1998, Dayan and Abbott 2001, Glimcher 2011) (see Rolls (2008c)). In an approach related to microeconomics, expected utility theory has provided a useful estimate of the desirability of actions, and indicates that, except for very high and very low probabilities and for decisions with framing issues, **expected utility**, indicated by choices, does approximately track expected value (Kahneman and Tversky 1979, Kahneman and Tversky

1984, Tversky and Kahneman 1986, Glimcher and Rustichini 2004, Gintis 2007, Glimcher 2011). (If the probability of obtaining a reward is low, then we are less likely to choose it than when the probability is high.) However, deviations from this linear relation are important, and humans tend to overweigh the value of low probability gambles (and to be risk-seeking in this domain); and to underweigh the value of high probability gambles (and to be risk averse in this domain) (Tversky and Kahneman 1981, Paulus and Frank 2006). Activations in the anterior cingulate cortex appear to reflect individual differences in this weighting, for a lack of appropriate activation in the anterior cingulate cortex was linked to excessive risk-seeking in choices when there was a low probability of success, and excessive risk avoidance in choices when there was a high probability of success (Paulus and Frank 2006).

Dopamine neurons show increasing responses to conditioned stimuli predicting reward with increasing probability (Fiorillo, Tobler and Schultz 2003), and decrease their firing to predicted reward omission (Tobler, Dickinson and Schultz 2003). However, it also appears that at least some dopamine neurons have activity that is high when reward uncertainty is high (which occurs when reward probability is 0.5) (Fiorillo et al. 2003). This dual coding (of reward prediction error, as described below, and of reward uncertainty) raises problems in how receiving neurons might use this multiplexed information.

Parietal cortex neurons with activity that precedes eye movements in for example area LIP (lateral intraparietal cortex) show more activity if the expected value is high (Glimcher 2003, Glimcher 2004, Platt and Glimcher 1999, McCoy and Platt 2005). This modulation by expected value, as influenced by both probability and reward magnitude, of neurons studied in oculomotor tasks has now been found in a number of areas with oculomotor-related activity, including the cingulate cortex and the superior colliculus (McCoy and Platt 2005).

For both the dopamine and the parietal cortex neurons, it seems unlikely that the actual computation of probability multiplied by reward value is performed in those areas, as reward stimuli are not known to be encoded there.

It is therefore of interest to determine where in the brain representations of reward magnitude, known to be present in the orbitofrontal cortex (O'Doherty, Kringelbach, Rolls, Hornak and Andrews 2001), become multiplied by reward probability to yield a signal that encodes expected value and even expected utility. In an fMRI investigation in which the reward value and expected value of monetary reward were altered by altering the probability of obtaining rewards for a particular choice, it was found that activations in the medial orbitofrontal cortex were correlated with Reward Magnitude and with Expected Value (Rolls, McCabe and Redoute 2008e). Moreover, it was found that expected utility, reflected in the choices made by the participants, approximately tracked the expected value, and in this sense expected utility was also reflected in activations in the medial orbitofrontal cortex.

8.4.2 Delay of reward, emotional choice, and rational choice

Another factor that can influence decisions for rewards is the delay before the reward is obtained. If the reward will not be available for a long time, then we discount the reward value. Most models assume an exponential decrease in the reward value as a function of the delay until the reward is obtained, as rational choice entails treating each moment of delay equally (Frederick, Loewenstein and O'Donoghue 2002, McClure, Laibson, Loewenstein and Cohen 2004). Impulsive preference changes may reflect a disproportionate valuation of rewards available in the immediate future (Ainslie 1992, Benabou and Pycia 2002, Rachlin 2000, Montague and Berns 2002, Metcalfe and Mischel 1999). It is possible that there are two systems that influence decisions in these circumstances.

One is a rational, logic-based, system requiring syntactic manipulation of symbols (see Section 5.1, Fig. 5.1, and Rolls (2005a)) that can treat each moment of delay equally, and calculate choice based on an exponential decrease of reward value with increasing delay. This rational decision system might involve language or mathematical systems in the brain, and the ability to hold several items in a working memory while the trade-offs of different long-term courses of action are compared.

A different more emotion-based system that can operate implicitly might operate according to heuristics that have become built into the system during evolution which might value disproportionately immediate rewards compared to delayed rewards. This emotion-based system might involve the orbitofrontal cortex, which as we have seen (Section 3.11) represents different types of reward and punisher (e.g. monetary gain and loss), and lesions of which in humans lead to impairments in changing behaviour when rewards are received less often for particular choices (Hornak, O'Doherty, Bramham, Rolls, Morris, Bullock and Polkey 2004, Berlin, Rolls and Kischka 2004), to impulsive choices (Berlin, Rolls and Iversen 2005), and to impairments in gambling tasks (Bechara et al. 1994, Bechara, Damasio, Tranel and Anderson 1998). My hypothesis is that the impaired choices and impulsive gambling decision-making after lesions of the orbitofrontal cortex are related to damage to the system that represents non-reward and punishment (Thorpe, Rolls and Maddison 1983, Rolls, Hornak, Wade and McGrath 1994a, Rolls 1999b, Kringelbach and Rolls 2003, Hornak, O'Doherty, Bramham, Rolls, Morris, Bullock and Polkey 2004, Berlin, Rolls and Kischka 2004, Rolls and Grabenhorst 2008, Grabenhorst and Rolls 2011). It is also an interesting aspect of the impulsive decision-making of patients with orbitofrontal cortex damage that their perception of time is speeded up (Berlin, Rolls and Kischka 2004). This speeding up may contribute to why they take decisions relatively early, and in this sense act impulsively.

Moreover, individual differences in sensitivity to rewards and punishers could lead to personality differences with respect to impulsive behaviour, and indeed patients with Borderline Personality Disorder behave similarly with respect to their impulsive behaviour to patients with orbitofrontal cortex lesions (Berlin, Rolls and Kischka 2004, Berlin and Rolls 2004, Berlin, Rolls and Iversen 2005).

The suggested dissociation of emotional vs reasoning decision systems is the same concept as that encompassed by the hypothesis of dual routes to action considered in Section 5.1, Fig. 5.1, and elsewhere (Rolls 1999a, Rolls 2003, Rolls 2004b, Rolls 2005a, Rolls 2007a, Rolls 2008a, Rolls 2011b).

Consistent with the point being made about evolutionarily old emotion-based decision systems vs a recent rational system present in humans (see Section 5.1) is that humans trade off immediate costs/benefits against cost/benefits that are delayed by as much as decades, whereas non-human primates have not been observed to engage in unpreprogrammed delay of gratification involving more than a few minutes (Rachlin 1989, Kagel et al. 1995)[31].

Consistent with dual emotional and rational bases for decisions in humans, a 'quasi-hyperbolic' time discounting function that splices together two discounting functions – an emotional one that distinguishes sharply between present and future, and a rational one that discounts exponentially and more shallowly – provides a good fit to experimental data including retirement saving, credit-card borrowing, and procrastination (Laibson 1997, Angeletos, Laibson, Repetto, Tobacman and Weinberg 2001, O'Donoghue and Rabin 1999). This dual mechanism process can be modelled formally by

$$r(t) = \beta \gamma^t r(0) \tag{8.2}$$

where $r(t)$ is the time discounted reward value at time t, and $r(0)$ is the reward value if received immediately at time $t = 0$ (McClure, Laibson, Loewenstein and Cohen 2004). β ($0 < \beta \leq 1$) (or in fact its inverse) represents the uniform down-weighting of future compared to immediate rewards, and is the parameter that encompasses the effects of emotion on decision-making in this formulation. β is 1 at time zero, and is set to a value that scales a reward at any future time relative to the value at time 0. If $\beta = 0.8$, this indicates that relative to a reward of value r at time zero, the reward at any future time would have a value of 0.8. In this sense, it models the role of emotion in decision-making as down-valuing a reward at any future time compared to immediately by a uniform discounting factor β. The γ ($\gamma \leq 1$) parameter is the discount rate in the standard exponential formula that treats a given delay equivalently independently of when it occurs (i.e. in any time interval, the value decreases by a fixed proportion of the value it has already reached), and encompasses the rational route to decision-making. In the model, it produces exponential decay of the value of a reward according

---

[31] Seasonal food storage is not an exception, in that it appears to be stereotyped and instinctive, and hence is unlike the generalizable nature of human planning (McClure, Laibson, Loewenstein and Cohen 2004).

to how long it is delayed. It is used in the model to capture the effects of long-term economic planning for the future.

McClure, Laibson, Loewenstein and Cohen (2004) performed an fMRI investigation in which smaller immediate rewards (today) could be chosen vs larger delayed rewards (given after delays of up to six weeks). (The monetary rewards were in the range \$5–\$40.) Brain areas that showed more activation for immediate vs delayed rewards (and reflected the β emotional parameter) included the medial orbitofrontal cortex, the medial prefrontal cortex/pregenual cingulate cortex, and the ventral striatum. Brain areas where activations reflected the decisions being made and the decision difficulty but which were not preferentially activated in relation to the immediate reward parameter β included the lateral prefrontal cortex (a brain region implicated in higher level cognitive functions including working memory and executive functions (Miller and Cohen 2001, Deco and Rolls 2003)), and a part of the parietal cortex implicated in numerical processing (Dehaene, Dehaene-Lambertz and Cohen 1998). (Activations in these prefrontal and parietal areas reflect the effects of the γ^t variable in Equation 8.2.) Thus emotional decisions that emphasize the importance of immediate rewards may preferentially activate reward-related areas ('β areas') such as the medial orbitofrontal cortex, pregenual cingulate cortex, and the ventral striatum; whereas difficult decisions requiring cost–benefit analysis about the value of long-term rewards preferentially activate a more cognitive system ('γ areas') that may be involved in rational thought and multistep calculation.

8.4.3 Reward prediction error, temporal difference error, and choice

The expected value may alter from time to time, for example during a trial. For example, there is a (negative) reward prediction error when a reward is predicted but not obtained. Similarly, there is a (positive) reward prediction error when a reward is not expected but is obtained. The reward prediction error may be defined as the difference between the reward obtained and the reward predicted (Rolls 2008c). The firing of dopamine neurons may reflect these reward prediction errors (Schultz 1998, Schultz, Dayan and Montague 1997, Waelti, Dickinson and Schultz 2001, Schultz 2004, Schultz 2006) (but see Rolls (2008c) where problems such as the asymmetry of the error signals for positive vs negative error predictions are raised).

O'Doherty, Dayan, Schultz, Deichmann, Friston and Dolan (2004) related reward prediction error correlated activations of the ventral striatum to a 'critic' that learns to predict a future reward because these activations occurred even when no action was required in a Pavlovian conditioning task, and reward prediction error correlated activations of the dorsal striatum to an 'actor' because it showed stronger activation during instrumental learning than Pavlovian association[32].

The hypothesis that dopamine neuron firing provides a reward prediction error signal (Schultz 1998, Schultz et al. 1997, Waelti et al. 2001, Schultz 2004, Schultz 2006) appears to be inconsistent with the evidence that dopamine neuron firing and activations of parts of the striatum are also produced by aversive, novel, or intense/salient stimuli (Zink, Pagnoni, Martin, Dhamala and Berns 2003, Zink, Pagnoni, Martin-Skurski, Chappelow and Berns 2004). Indeed, Zink et al. (2004) argue, from an fMRI investigation in which caudate and nucleus accumbens activations were greater when responses were made to obtain money than when money was given passively, that the activity in these regions is not related to reward value or predictions, but instead to *saliency*, that is to an arousing event to which attentional and/or behavioural resources are redirected (Rolls 2008c).

Reward prediction error encoding is in contrast to that of many neurons in the head of the caudate nucleus, which fire in relation to predicted rewards, that is to expected reward value (Rolls, Thorpe and Maddison 1983). They do this in that they start firing as soon as a cue such as a tone, or a light that precedes the tone, is given to indicate that a trial is starting, and continue to respond if a visual stimulus is shown indicating that a juice reward will be obtained, and stop responding if a different visual stimulus is shown indicating that aversive saline will be obtained (Rolls 2008c). Neurons that reflect negative reward prediction error (i.e. less reward is obtained than was expected) are found in the orbitofrontal cortex (Thorpe, Rolls and Maddison 1983, Rolls and Grabenhorst 2008, Rolls 2011e) (Section 3.11).

The reward prediction error approach to changes in expected value can be developed into a temporal difference learning approach, in which the temporal difference error depends on the difference in the reward value prediction at two successive time steps (Rolls 2008c). This temporal difference error is useful in some temporal difference reinforcement learning algorithms for producing learning that optimizes predictions, and thus how to learn optimal actions as events unfold in time, for example during a trial. Temporal difference models have been applied to model the activity of dopamine neurons (Suri and Schultz 2001), and of fMRI activations related to the anticipation of reward (O'Doherty, Dayan, Friston, Critchley and Dolan 2003a).

Seymour, O'Doherty, Dayan, Koltzenburg, Jones, Dolan, Friston and Frackowiak (2004) took the temporal difference approach in an fMRI analysis of a more complicated, second-order, pain conditioning task, with two successive visual cues to predict either low or high pain. The second cue was fully predictive of the strength of the subsequently experienced pain. The first cue

---

[32]See Sutton and Barto (1998) and Rolls (2008c) for a description of the functions of a 'critic' and an 'actor' in reward prediction learning.

only allowed a probabilistic prediction. Thus in a low proportion (18%) of the trials, the expectation evoked by the first cue was reversed by the second cue. The punisher value (pain) prediction thus alters on some trials after the second cue is delivered, generating a temporal difference prediction error. After many conditioning trials, the punisher prediction value becomes good on the 82% of trials where the first cue does predict the second cue, and during the learning the temporal difference error at the time the second cue is shown becomes low. However, on the 18% of trials where the first cue makes the incorrect prediction of the second cue, temporal difference prediction errors remain when the second cue is presented. The temporal difference error was correlated with activations in the ventral putamen (a part of the ventral striatum), the right insular cortex (probably providing a somatosensory representation of the left hand to which the pain was delivered), the right head of the caudate nucleus, and the substantia nigra (a region where dopamine neurons are located), suggesting that these areas are involved in learning expectations of pain. It should be noted that this was a conditioning procedure, and that although pain expectations were being learned, decisions were not being made by the subjects.

The temporal difference approach was taken in an fMRI study of a decision task by Rolls, McCabe and Redoute (2008e). They showed in a probabilistic decision task in which the expected value was systematically varied that temporal difference (reward prediction) errors were reflected in activity in the ventral striatum, which receives from the orbitofrontal cortex. However, the findings showed that care is needed in interpreting fMRI signals as related to temporal difference (reward prediction) errors, for the correlation with TD error was related to the fact that in the ventral striatum, the activations were related to the reward actually obtained on each trial, and changed at the point in each trial at which this information was made available to the participant. Thus the ventral striatal activation was related to decision-making in so far as its activation reflected the reward actually provided on a given trial (the reward magnitude or outcome).

The temporal difference approach to reinforcement learning has a weakness that although it can be used to predict internal signals during reinforcement learning in some tasks, it does not directly address learning with respect to actions that are not taken. Q-learning is an extension of TD learning which takes signals from actions not taken into account, for example information gained by observation of others (Montague et al. 2006).

8.5 Conclusions

Overall, in this chapter we have seen how modern neuroscience in the field of neuroeconomics is offering an alternative approach to classical economics.

In classical economics choice is analysed as that by a rational self-interested individual. We have seen that choice is not always by a rational system, but may be influenced by non-rational reward and punishment value-based systems that play an important role in humans' choices (Chapter 3). These choices may not be rational in the sense that humans are not good at logic due to the way in which computations are implemented in the brain (Chapter 2), and are instead often guided by reward and punishment systems selected by evolution to perform adaptively in the long term for the survival of the genes. These brain systems have evolved in the context of the limited computational power of the brain, at least in our evolutionary history.

These choices may not be self-interested and rational also in the sense that they may not necessarily be for the individual's direct advantage, but may be heuristics built into the system that may be adaptive for the individual's genes in a population. An example is altruistic punishment. Another example might be choices made at one's expense for one's children.

Economic choices may not be universally rational and self-interested further in the sense that there are large individual differences in for example reward discounting, with some individuals being impulsive and interested in immediate rewards, and others being more willing to consider optimizing economic gain over long periods. Again, these individual differences are related to the adaptive value of variation between individuals in sensitivity to specific rewards and punishers (Chapter 3), and for that matter in the weight placed on the rational vs the emotional systems (Chapter 5).

In addition, our choices are not those that might be taken by a deterministic system, but instead are influenced by noise in the brain (Section 2.12). This may be especially important in influencing whether the choices are made on a particular occasion based more on the affective systems in the brain with their rewards based on the interests of the genes, or more on the rational system which can take into account the interests of the individual and discount those of the genes (Section 5.4). Moreover, it is not just noise that can influence these choices, but also the balance between the systems can be influenced by factors such as top-down cognition, selective attention, drugs such as alcohol that may reduce the effects of punishers and increase impulsivity, etc.

In this situation, it would seem to be important to take into account these neurobiological underpinnings of economic choice and the concomitant differences between individuals so that society does not disadvantage some individuals. At the same time, predictions of economic choice may be considerably improved if they take into account these neurobiological underpinnings and individual differences.

9 Neuroethics

9.1 Biological underpinnings to ethics

In this chapter I focus on the biological underpinnings of ethics. I address particularly the implications of our modern understanding of the computational brain mechanisms for emotion and decision-making, and of the evolution of different brain systems, as underpinnings for our understanding of ethics, free will, determinism, and responsibility. I also suggest reasons why what is encouraged by the heuristics promoted by these biological underpinnings may be incorporated into ethical systems.

Neuroethics is a rapidly developing field that encompasses many ethical questions raised by modern neuroscience including the uses of neurotechnology; aging and dementia; and law and public policy, and these other aspects of neuroethics are treated elsewhere (Farah 2005, Illes 2006, Levy 2007, Illes and Sahakian 2011, Farah 2012).

Rolls (2005a) in *Emotion Explained* has argued that much of the foundation of our emotional behaviour arises from specification by genes of primary reinforcers that provide goals for our actions. We have emotional reactions in certain circumstances, such as when we see that we are about to suffer pain, when we fall in love, or if someone does not return a favour in a reciprocal interaction. What is the relation between our emotions, and what we think is right, that is our ethical principles? If we think something is right, such as returning something that has been on loan, is this a fundamental and absolute ethical principle, or might it have arisen from deep-seated biologically based systems shaped to be adaptive by natural selection operating in evolution to select genes that tend to promote the survival of those genes?

Many principles that we regard as ethical principles *might* arise in this way. For example, as noted in Chapter 3, guilt might arise when there is a conflict between an available reward and a rule or law of society. Jealousy is an emotion that might be aroused in a male if the faithfulness of his partner seems to be threatened by her liaison (e.g. flirting) with another male. In this case the reinforcement contingency that is operating is produced by a punisher, and it may be that males are specified genetically to find this punishing because it indicates a potential threat to their paternity and parental investment, as described in Chapter 4. Similarly, a female may become jealous if her partner has a liaison with another female, because the resources available to the 'wife' useful to bring up her children are threatened. Again, the punisher here may be gene-specified, as described in Chapter 4. Such emotional responses might influence what we build into some of the ethical principles that surround marriage and partnerships for raising children.

Many other similar examples can be surmised from the area of evolutionary psychology (see e.g. Ridley (1993b), Ridley (1996), and Buss (2008)). For example, there may be a set of reinforcers that are genetically specified to help promote social cooperation and even reciprocal altruism (Section 4.2), and that might thus influence what we regard as ethical, or at least what we are willing to accept as ethical principles. Such genes might specify that emotion should be elicited, and behavioural changes should occur, if a cooperating partner defects or 'cheats' (Cosmides and Tooby 1999). Moreover, the genes may build brains with genetically specified rules that are useful heuristics for social cooperation, such as acting with a strategy of 'generous tit-for tat', which can be more adaptive than strict 'tit-for-tat', in that being generous occasionally is a good strategy to help promote further cooperation that has failed when both partners defect in a strict 'tit-for-tat' scenario (Ridley 1996). Genes that specify good heuristics to promote social cooperation may thus underlie such complex emotional states as feeling forgiving.

Another example is cheat detection, identifying a person who is not reciprocating favours, or who may be providing dishonest information and cannot be trusted as a cooperator. Reputation is extremely important in a social group, for if an individual develops a bad reputation in the group, that person may be excluded from the group, from the tribe, and that may have serious consequences for survival and reproductive success. The effect of reputation is enormously amplified in humans, because information about an individual can be spread so easily by language. Rumours about individuals may have devastating effects in human populations. This places a high premium on honesty and social cooperation in humans, and makes them 'virtues', ethically and morally valued, though the underpinning here is genetic (and possibly also phenotypic, see Section 9.2) advantage and disadvantage.

It is suggested that many apparently complex emotional states have their origins in designing animals to perform well in such sociobiological and socioeconomic situations (Ridley 1996, Glimcher 2003, Glimcher 2004). In this way, many principles that humans accept as ethical may be closely related to strategies that are useful heuristics for promoting social cooperation, and emotional feelings associated with ethical behaviour may be at least partly related to the adaptive value of such gene-specified strategies.

The situation is clarified by the ideas I have advanced Chapter 5 about a rational syntactically based reasoning system and how this interacts with an evolutionarily older emotional system with gene-specified rewards. The rational system enables us for example to defer immediate gene-specified rewards, and make longer-term plans for actions that in the long term may have more useful outcomes. This rational system enables us to make reasoned choices, and to reason about what is right. Indeed, it is because of the linguistic system that the naturalistic fallacy becomes an issue. In particular, we should not believe that what is right is what is natural (*the naturalistic fallacy*), because we have a rational system that can go beyond simpler gene-specified rewards and punishers that may influence our actions through brain systems that operate at least partly implicitly, i.e. unconsciously. I now consider further the relation between the biological underpinnings to emotion, and ethics, morals, and morality.

There are many reasons why people have particular moral beliefs, and believe that it is good to act in particular ways. It is possible that biology can help to explain why certain types of behaviour are adopted perhaps implicitly by humans, and become incorporated for consistency into explicit rules for conduct. This approach does not, of course, replace other approaches to what is moral, but it may help in implementing moral beliefs held for other reasons to have some insight into some of the directions that the biological underpinnings of human behaviour might lead. Humans may be better able to decide explicitly what to do when they have knowledge and insight into the biological underpinnings. It is in this framework that the following points are made, with no attempt made to lead towards any suggestions about what is 'right' or 'wrong'. The arguments that follow are based on the hypothesis that there are biological underpinnings based on the types of reward and punishment systems that have been built into our genes during evolution for at least some of the types of behaviour held to be moral.

One type of such biological underpinning is kin selection. This would tend to produce supportive behaviour towards individuals likely to be related, especially towards children, grandchildren, siblings, etc. depending on how closely they are genetically related. This does tend to occur in human societies, and is part of what is regarded as 'right', and indeed it is a valued 'right' to be able to pass on goods, possessions, wealth, etc. to children. The underlying basis here would be genes for kin altruism[33].

Another such underpinning might be the fact that many animals, and especially primates, cooperate with others in order to achieve ends which turn out to be on average to their advantage, including genetic advantage. One example includes the coalitions formed by groups of males in order to obtain a female for one of the groups, followed by reciprocation of the good turn later (see Ridley (1996)). This is an example of altruism, in this case by groups of primates, which is to the advantage of both groups or individuals provided that neither individual or group cheats, in which case the rules for social interaction must

---

[33] Kin selection genes spread because of kin altruism. Such genes direct their bodies to aid relatives because those relatives have a high chance of having the same relative-helping gene. This is a specific mechanism, and it happens to be incorrect to think that genes direct their bodies to aid relatives because those bodies 'share genes' in general (see Hamilton (1964); and the chapter on inclusive fitness in M.S.Dawkins (1995)).

change to keep the strategy stable. Another such underpinning, in this case for property 'rights', might be the territory guarding behaviour that is so common from fish to primates. Another such underpinning might be the jealousy and guarding of a partner shown by males who invest parental care in their partner's offspring. This occurs in many species of birds, and also in humans, with both exemplars showing male parental investment because of the immaturity of the children. This might be a biological underpinning to the 'right' to fidelity in a female partner, or at least to thinking of cheating as unethical. In a similar way, husbands might think of other males who cheat with the husband's wife as being unethical.

The suggestion I make is that in all these cases, and in many others, there are biological underpinnings that determine what we find rewarding or punishing, designed into genes by evolution to lead to appropriate behaviour that helps to increase the fitness of the genes. When these implicit systems for rewards and punishers start to be expressed explicitly (in language) in humans, the explicit rules, rights, and laws that are formalized are those that set out in language what the biological underpinnings 'want' to occur[34].

Clearly in formulating the explicit rights and laws, some compromise is necessary in order to keep the society stable. When the rights and laws are formulated in small societies, it is likely that individuals in that society will have many of the same genes, and rules such as 'help your neighbour' (but 'make war with "foreigners" ') will probably be to the advantage of one's genes. However, when the society increases in size beyond a small village (in the order of 1000), then the explicitly formalized rules, rights, and laws may no longer produce behaviour that turns out to be to the advantage of an individual's genes. In addition, it may no longer be possible to keep track of individuals in order to maintain the stability of 'tit-for-tat' cooperative social strategies (Dunbar 1996, Ridley 1996)[35]. In such cases, other factors doubtless come into play to additionally influence what groups hold to be right. For example, a group of subjects in a society might demand the 'right' to free speech because it is to their economic advantage.

Thus overall it is suggested that many aspects of what a society holds as right and moral, and of what becomes enshrined in explicit 'rights' and laws, are

---

[34] Before the rules are explicitly formalized, conventions may be developed and spread using language, for example in the form of verbal traditions handed down from generation to generation that may provide possible models for behaviour, such as Homer's *Odyssey*.

[35] A limit on the size of the group for reciprocal altruism might be set by the ability both to have direct evidence for and remember person–reinforcer associations for large numbers of different individual people. In this situation, reputation passed on verbally from others who have the direct experience of whether an individual can be trusted to reciprocate might be a factor in the adaptive value of language and gossip (Dunbar 1996, Dunbar 1993). In a very large group, there is also a low probability of encountering the same individual again, and so 'investing' in a potential reciprocal interaction may be economically not useful, though still performed perhaps in relation to the original biological heuristic when, for example, we give a tip in a restaurant that we know we will never visit again.

related to biological underpinnings, which have usually evolved because of the advantage to the individual's genes, but that as societies develop, other factors also start to influence what is believed to be 'right' by groups of individuals, related to socioeconomic factors. In both cases, the laws and rules of the society develop so that these 'rights' are protected, but often involve compromise in such a way that a large proportion of the society will agree to, or can be made subject to, what is held as right.

To conclude this discussion, we note that what is natural does not necessarily imply what is 'right' (the naturalistic fallacy, pointed out by G. E. Moore) (see, e.g. Singer (1981)). However, our notions of what we think of as right may be related to biological underpinnings, and the point of this discussion is that it can only give helpful insight into human behaviour to realize this. Other ways are described next in which a biological approach, based on what our brains have evolved to treat as rewarding or punishing, can illuminate moral issues, and rights.

'Pain is a worse state than no pain'. This is a statement held as true by some moral philosophers, and is said to hold without reference being made to biological underpinnings that may be relevant. It is held to be a self-evident truth, and certain implications for behaviour may follow from the proposition. A biological approach to pain is that the elicitation of pain has to be punishing (in the sense that animals will work to escape or avoid it), as pain is the state elicited by stimuli signalling a dimension of environmental conditions that reduces survival and therefore gene fitness.

'Incest is morally wrong. One should not marry a brother or sister. One should not have intercourse with any close relation.' The biological underpinning is that children of close relations have an increased chance of having double-recessive genes, which are sometimes harmful to the individual and reduce fitness. In addition, breeding out may produce hybrid vigour. It is presumably for this reason that many animals as well as humans have behavioural strategies (influenced by the properties of reward systems) that reduce inbreeding (e.g. philopatry, that is only one sex remaining in the natal unit at the time of puberty; and mate selection influenced by the olfactory receptor/major histocompatibility genes as described in Section 7.4).

At the same time, it may be adaptive (for genes) to pair with another animal that has many of the same genes, for this may help complex gene sequences to be passed intact into the next generation. This may underlie the fact that quails have mechanisms that enable them to recognize their cousins, and make them appear attractive, an example of kin selection (Bateson 1983). In humans, if one were part of a strong society (in which one's genes would have a good chance not to be eliminated by other societies), then it could be advantageous (whether male or female) to invest resources with someone else who would provide maximum genetic and resource potential for one's children, which on average across a society of relatively small size and not too mobile would be a person with relatively similar genes and resources (wealth, status, etc.) to oneself. In an exception to this, in certain societies there has been a tradition of marrying close relations, and part of the reason for this could be maintaining financial and other resources within the (genetic) family (i.e. kin altruism). (An example is the Pharaohs of Egypt, for whom there were disastrous genetic consequences when biological underpinnings against incest were not being followed.)

There may be several reasons why particular behavioural conduct may be selected. A first is that the conduct may be good for the individual and for the genes of the individual, at least on average. An example might be a prohibition on killing others in the same society (while at the same time defending that kin group in times of war). The advantage here could be for one's own genes, which would be less at risk in a society without large numbers of killings. A second reason is that particular codes of conduct might effectively help one's genes by making society stable. An example here might be a prohibition on theft, which would serve to protect property. A third reason is that the code of conduct might actually be to other, powerful, individuals' advantage, and might have been made for that reason into a rule that others in society are persuaded to follow. A general rule in society might be that honesty is a virtue, but the rule might be given a special interpretation or ignored by members of society too powerful to challenge.

As discussed in Chapter 4, different aspects of behaviour could have different importance for males and females. This could lead men and women to put different stress on different rules of society, because they have different importance for men and women. One example might be being unfaithful. Because this could be advantageous to men's genes, this may be treated by men as a less serious error of conduct than by women. However, within men there could be differential condemnation, with men predisposed to being faithful being more concerned about infidelity in other men, because it is a potential threat to them. In the same way, powerful men who can afford to have liaisons with many women may be less concerned about infidelity than less powerful men, whose main genetic investment may be with one woman.

Society may set down certain propositions of what is 'right'. One reason for this is that it may be too difficult on every occasion, and for everyone, to work out explicitly what all the payoffs of each rule of conduct are. A second reason is that what is promulgated as 'right' could actually be to someone else's advantage, and it would not be wise to expose this fully. One way to convince members of society not to do what is apparently in their immediate interest is to promise a reward later. Such deferred rewards are often offered by religions (Chapter 11). The ability to work for a deferred reward using a one-off plan in this way becomes possible, it was suggested in Chapter 5, with the evolution of the explicit, propositional, system.

9.2 Ethical principles arising from advantages for the phenotype versus for the genotype

Many of the examples described above of the biological underpinnings to ethical behaviour, what we have come to regard as right and just, reflect advantages to the genotype. These underpinnings related to advantages to the genotype are present in many non-human animals, though developed to greater or lesser extents. Examples include kin altruism (common in non-human animals), reciprocal altruism including tit-for-tat exchanges and forgiveness (though reciprocal altruism is not easily proved in many cases in non-human animals), and stakeholder altruism (Section 4.2).

However, I wish to introduce here a new concept related to the underlying biology, that some aspects of what we regard as ethical and right are related to advantage to the phenotype. Phenotypic advantage may not necessarily be to the advantage of the genotype, and indeed we can contrast the 'selfish phene' with the 'selfish gene' (see Sections 5.4 and 13.2, and Rolls (2011b)). The concept of phenotypic advantage in relation to ethics is that by reasoning, the multiple step syntactic thought brain system described in Chapter 5 and Section 6.3, may lead people to agree to rules of society that are to the advantage of their bodies, even when there may be no genetic advantage. Consider 'Thou shalt not kill' and 'Thou shalt not steal'. Both rules could be to the advantage of the individual person, who may wish to stay alive and not be robbed of all the things that she or he enjoys, even though that may not necessarily be to the advantage of genes. Indeed, in the animal kingdom, conflict and killings even within species that are driven by genotypic advantage are common, as in intra-sexual competition as part of sexual selection (Chapter 4), and a lion killing the cubs of his new lioness by a previous father. In this context, the rules 'Thou shalt not kill' and 'Thou shalt not steal' may not be to the genetic advantage of an individual, but may well be to the advantage of the individual person who may wish to stay alive to enjoy life and property, even beyond the age of reproduction.

Helping and care for the elderly is another example where individuals may contract into a society where help and care for the elderly is valued, and is even a 'right', because it may be to their own advantage later on in life, and they are willing to pay some cost (e.g. not stealing from others) to be part of the society in which they can enjoy 'rights' such as not having their possessions stolen. This social contract, agreeing to the rules of a society, has similarities with a contract with an insurance company, in which there is a cost, but also a potential benefit to an individual, and where that advantage may not necessarily be genetic, and may never have been selected for genetically.

The concept here is as follows. At least humans (possibly some other animals) have evolved to have a brain that can reason, and that has many advantages in enabling them to pursue long-term goals rather than immediate gene-defined goals, as described in Chapter 5. However, once one has a reasoning system, it can reason that some of the things that one has been built to enjoy (perhaps for genetic reasons originally, such as wealth, power, good food) might be enjoyed even when there is no genetic advantage, for example by prolonging life into old age, which becomes a 'right' where it is possible.

This results in ethical values and 'rights' which have their (biological) basis in the good of the phenotype, and not necessarily the good of the genotype (though they are not mutually exclusive, and may often work together). This phenotypically underpinned system of rights that becomes enshrined in a social contract is somewhat like the 'right' to a pension and to medical care when elderly (and beyond the age at which genetic potential is likely to be enhanced by longer life), which allows the individual person or phenotype to benefit, by having obeyed the rules of the society earlier, contributing to its insurance provisions. Such a social contract has similarities with a contract with an insurance company: it costs a bit, but may protect your body in the long term. A pension is a bit like such a social contract. One agrees rationally to a cost, as it may benefit you (and not necessarily your genes) in the long term.

Another example is the 'right' to the benefits from advances in medicine that prolong life beyond the usual lifespan. Multiple genetic mechanisms program us to die before a very old age, with the genetic advantage that the selfish genes evolve faster if the intergenerational time is not too long (so that the genotype can evolve rapidly), and the resources are made available mainly to those of reproductive age. But it may be to the advantage of the human individual phenotype to continue to enjoy life, friendships, intellectual development and enrichment, and possessions, beyond reproductive age. Individuals might therefore argue that they have a 'right' to benefit from the advantages of modern medicine that prolong life, even though this is for the interests of the phenotype, which are clearly in this case different from those of the genotype that has evolved by natural selection.

These 'rights' driven by advantage to the phenotype and enshrined in a social contract may thus have their origin in advantages to the genotype of being able to reason (see Chapter 5). But that means that these phenotypically driven 'rights' and ethical principles are spandrels without any useful (genetic) advantage of their own. (A spandrel is the part of a cathedral roof between the fan vaulting which arises not for the beauty of the fan vaulting, but just to fill in the spaces between the fan vaulting. The term spandrel was used by S.J. Gould to refer to a characteristic that was present but had not been selected for directly, but arose as a by-product of natural selection for another characteristic (Gould and Lewontin 1979).) Thus my argument is that some ethical principles and rights are for the advantage of the phenotype, the individual person, and not for the genotype, and may in some cases be spandrels, evolutionary changes that have not been directly genetically selected. (Of course, there are advantages to reasoning, which drove its evolution (Chapter 5). But my point here is that the reasoning system may also produce a situation which is not necessarily fully in the interests of the genes.)

I emphasize that the genotypic and phenotypic advantages that may underpin many aspects of what is accepted as ethical and right are not mutually exclusive. At the same time, they may sometimes be in competition. For example, being a dictator, or being ruthless, may be in some senses to the advantage of the individual's genes, but not agreeing to a social contract, to share, to general phenotypic advantage can be dangerous, as Caesar discovered when he became too powerful and was killed by the other senators, including Brutus.

It is interesting to assess where animal welfare fits into this framework. One contributory factor may be as follows. Given that we have a social contract that gives rights to other humans for phenotypic advantage, and contract not to for example harm other people because such a contract is to the phenotypic advantage of individual humans, it may be that anthropomorphic generalization of that principle is extended to other species by processes that include empathy. (Empathy may have evolved to promote kin selection (by being empathetic to those related to you), and theory of mind may have evolved in part to help humans predict other human's behaviour. Anthropomorphism may be an extension to animals of such attributes, normally used by humans in interactions with other humans and which have evolved because they are useful in human–human interactions.) There may of course be other contributory factors that provide bases for animal welfare (Dawkins and Bonney 2008, Dawkins 2012).

9.3 The Social Contract

The view that one is led to is that some of our moral beliefs may be explicit, verbal, formulations of what may reflect factors built genetically by kin selection into behaviour, namely a tendency to favour kin, because they are likely to share some of an individual's genes. In a small society this explicit formulation may be 'appropriate' (from the point of view of the genes), in that many members of that society will be related to that individual. When the society becomes larger, the relatedness may decrease, yet the explicit formulation of the rules or laws of society may not change. In such a situation, it is presumably appropriate for society to make it clear to its members that its rules for what is acceptable and 'right' behaviour are set in place so that individuals can live in safety, and with some expectation of help from society in general, as described in Chapter 12.

It is argued that the second biological underpinning of ethics is the evolution in (at least) humans of a reasoning system, which leads humans to values based on phenotypic advantage, which may not always correspond to genetic advantage, as described in Section 9.2.

The operation of this reasoning system may encourage acceptance of a social contract, in part because it may be judged to be useful for the individual's kin and genetic fitness, and also for the individual's phenotype, which may for example accept a cost of not harming others and benefitting from that, in order not to be harmed.

Indeed, the view to which this approach based on neuroscience and the evolution of the brain leads is that there may be no absolute rights, or god-given rights (Chapter 11), but that instead there may be rules and laws of a society, and indeed perhaps 'universally' across societies, that may be agreed by a social contract, and these rules and laws may imply 'rights' (Chapter 12). Justice may in this approach be used to refer to the implementation of these laws and rules (Hobbes 1651, Locke 1689, Rawls 1971). For example, Thomas Hobbes, beginning from a mechanistic understanding of human beings and the passions, postulates what life would be like without government, a condition which he calls the state of nature. In that state, each person would have a right, or license, to everything in the world. This, Hobbes argues, would lead to a "war of all against all" (*bellum omnium contra omnes*), and thus lives that are "solitary, poor, nasty, brutish, and short". This is close to my argument about what would represent the interests of the genotype (see also Ridley (1996)). In this context, Hobbes then argues that to escape this state of war, men in the state of nature accede to a social contract and establish a civil society, a commonwealth. (Hobbes was writing at the time of the English Civil War, and was a supporter of the revolt led by Cromwell against the sovereign. It is interesting that a major supporter of what is now known as the 'commonwealth' is Queen Elizabeth II.) John Locke (1689) continued the development of Hobbes' argument, though more from a starting point that humans are rational, reasoning people.

Other factors that can influence what is held to be right might reflect socioeconomic advantage to groups or alliances of individuals. It would be then in a sense up to individuals to decide whether they wished to accept the rules, with the costs and benefits provided by the rules of that society, in a form of Social Contract. Individuals who did not agree to the social contract might wish to transfer to another society with a different place on the continuum of costs and potential benefits to the individuals, or to influence the laws and policies of their own society, but acting within its laws. Individuals who attempt to cheat the system or break the laws that operate within a society would be expected to pay a cost in terms of punishment meted out by the society in accordance with its rules and laws, what it considers to be 'right' and 'just', with the underpinnings in genotypic and phenotypic advantage described in this chapter.

9.4 Conclusions

In this chapter, I have emphasized that what is right is not what is natural (or biological). However, I have argued that ethical systems rarely deviate far from what offers biological advantage. When considering biological advantage here, I go beyond primarily advantages to the selfish genes (Ridley 1996), to include what may be of advantage to the individual human, the phenotype (Chapter 5).

Given individual differences in the weightings of these different reward systems and the rational system between different individuals, I have suggested that individual differences in what is considered appropriate and ethical could arise. In this situation, I argue that a social contract may be a useful way to proceed, with individuals agreeing or not with their rational systems to the social contract of a particular society, and being excluded from it or punished if they do not accept its rules. This neurobiological approach moves the discussion away from universal absolute rights given by some external principle or provider, towards what is considered as right within a particular society and as decided by that society, and is implemented in its laws (cf. Griffin (2008)).

Finally, I emphasize that this chapter has focused on one aspect of neuroethics, how our ethical beliefs arise, and what their biological underpinnings are. Other areas of neuroethics, such as ethical questions raised by modern neuroscience including the uses of neurotechnology; aging and dementia; and law and public policy, are covered elsewhere in this rapidly developing field (Farah 2005, Illes 2006, Levy 2007, Illes and Sahakian 2011, Farah 2012).

10 Neuropsychiatry

A 'mechanistic' approach to brain function with some randomness in brain computations leads to the concept of the stability of cognitive processes such as short-term memory, attention, and decision-making (Chapter 2). This concept has fundamental implications for understanding and treating some psychiatric disorders such as schizophrenia and obsessive-compulsive disorder (OCD), and in addition normal aging, as described in this chapter.

The assumption of this rather new approach to understanding some psychiatric disorders as related to the stability of the dynamics of the brain is that attractor dynamics are important in cognitive processes (Sections 2.5.3–2.9 and 2.12). The hypothesis is based on the concept of attractor dynamics in a network of interconnected neurons which in their associatively modified synaptic connections store a set of patterns, which could be memories, perceptual representations, or thoughts (Hopfield 1982, Amit 1989, Rolls and Deco 2002, Rolls 2008c, Rolls and Deco 2010) (Section 2.5.3). The attractor states are important in cognitive processes such as short-term memory, attention, and action selection (Rolls 2008c, Rolls and Deco 2010). The network may be in a state of spontaneous activity, or one set of neurons may have a high firing rate, each set representing a different memory state, normally recalled in response to a retrieval stimulus. Each of the states is an attractor in the sense that retrieval stimuli cause the network to fall into the closest attractor state, and thus to recall a complete memory in response to a partial or partly incorrect cue. Each attractor state can produce stable and continuing or persistent firing of the relevant neurons.

The concept of an energy landscape (Hopfield 1982) (Section 2.5.3) is that each pattern has a basin of attraction, and each is stable if the basins are far apart, and also if each basin is deep, caused for example by high firing rates and strong synaptic connections between the neurons representing each pattern, which together make the attractor state resistant to distraction by a different stimulus. The spontaneous firing state, before a retrieval cue is applied, should also be stable. Noise in the network caused by statistical fluctuations in the stochastic spiking of different neurons can contribute to making the network transition from one state to another. These processes can be investigated by performing integrate-and-fire simulations with spiking activity (Rolls 2008c, Rolls and Deco 2010) (Section 2.5.3).

Some psychiatric states can be considered as the ends of a spectrum of natural variation between individuals in the stability of cortical networks, an important component of the Darwinian processes that guide the evolution of the brain, and this concept is described in this chapter.

10.1 Schizophrenia

Schizophrenia is characterized by three main types of symptom: cognitive dysfunction, negative symptoms, and positive symptoms (Liddle 1987, Baxter and Liddle 1998, Mueser and McGurk 2004). I describe how the basic characteristics of these three categories might be produced by instability in the brain's dynamical systems as follows.

10.1.1 Cognitive symptoms

Dysfunction of working memory, the core of the cognitive symptoms, may be related to instabilities of persistent attractor states (Durstewitz, Seamans and Sejnowski 2000b, Wang 2001) which we have shown can be produced by reduced firing rates in attractor networks, in brain regions such as the prefrontal cortex (Rolls 2008c, Rolls and Deco 2010).

The *cognitive symptoms* of schizophrenia include distractibility, poor attention, and the dysexecutive syndrome in which behaviour is poorly organized in tasks with multiple components such as shopping for different items requiring visits to several shops (Liddle 1987, Green 1996, Mueser and McGurk 2004). The core of the cognitive symptoms is a working memory deficit in which there is a difficulty in maintaining items in short-term memory (Goldman-Rakic 1994, Goldman-Rakic 1999), which could directly or indirectly account for these cognitive symptoms (Rolls 2005a, Rolls 2008c). For example, with poor short-term memory systems in the prefrontal cortex, it becomes difficult to manipulate items in short-term memory which is a major property of what is termed working memory (Baddeley 1986), and which is important in planning (Chapter 5), an important part of executive function. A difficulty with short-term memory would also account for poor attention, in that a short-term memory in the prefrontal cortex holds the subject of attention in mind, e.g. an object or a place, and provides the source for a top-down influence on earlier stages of processing to affect how they operate by biasing the competition between incoming inputs (Section 2.9). A failure to maintain attention properly can be manifest by being very distractible.

We have sought an explanation for these cognitive symptoms by considering what computational processes might account for the dysfunctions of short-term memory and thus attention and executive function (Rolls 2005a, Loh et al. 2007a, Rolls et al. 2008d, Rolls and Deco 2011, Rolls and Deco 2010, Rolls 2011f). I have suggested that if the short-term memory circuitry in the prefrontal cortex was too unstable because the firing in an attractor network became low, this would account for the cognitive symptoms (Rolls 2005a). The concept here is that an attractor network is stable if its firing rates are high, because then the positive feedback caused by the recurrent collateral connections between the neurons in the attractor and the inherent non-linearity of the system makes the system resistant to noise. The noise in the system corresponds to the random firing times of the neurons (for a given mean firing rate), or the firing produced by a distracting stimulus. An example of a trial in which the high firing rate implementing a short-term memory is correctly maintained is shown in Fig. 2.10 (top). Figure 2.10 (bottom) shows another trial in which the firing rate was not correctly maintained because of the noise, and the firing sank down into the spontaneous level of firing.

We tested this hypothesis with integrate-and-fire simulations, and showed that once a short-term memory state was started by a short input, the high firing rate state was less likely to be maintained, if a change that reduced the firing rate was present. The change we investigated was decreasing the efficacy of one class of the excitatory receptors on the neurons, NMDA receptors, which respond to the excitatory transmitter glutamate (Loh et al. 2007a, Rolls et al. 2008d). We chose this change to investigate because there is evidence that these excitatory receptors are less efficacious in patients with schizophrenia (Goff and Coyle 2001, Coyle, Tsai and Goff 2003, Coyle 2006), and other evidence consistent with this hypothesis of the cognitive symptoms of schizophrenia that is described in the next section.

10.1.2 Negative symptoms

The *negative symptoms* refer to the flattening of affect and a reduction in emotion. Behavioural indicators are blunted affect, emotional and passive withdrawal, poor rapport, lack of spontaneity, motor retardation, and disturbance of volition (Liddle 1987, Mueser and McGurk 2004). I propose that these symptoms are related to decreases in firing rates in the orbitofrontal cortex and/or anterior cingulate cortex (Rolls 2005a), where neuronal firing rates and activations in fMRI investigations are correlated with reward value and pleasure, and which is involved in emotion (Rolls 2005a, Rolls 2006b, Rolls 2007d, Rolls and Grabenhorst 2008, Rolls 2008b, Grabenhorst and Rolls 2011) (Chapter 3). Consistent with this, imaging studies have identified a relationship between negative symptoms and prefrontal hypometabolism, i.e. a reduced activation of frontal areas (Wolkin, Sanfilipo, Wolf, Angrist, Brodie and Rotrosen 1992, Aleman and Kahn 2005).

We investigated this computational hypothesis in integrate-and-fire simulations, and showed that the same reduction in NMDA glutamate receptor function that could account for the instability of attractor networks also decreased the firing rates of the neurons in the attractor networks (Loh, Rolls and Deco 2007a, Rolls, Loh, Deco and Winterer 2008d). We thus had a unifying account for the cognitive and negative functions of schizophrenia, in which the common cause was a reduction in NMDA glutamate receptor efficacy. When expressed in the dorsolateral prefrontal cortex, this can decrease the stability of attractor networks implementing short-term memory, by decreasing the firing rates of the neurons. When expressed in the orbitofrontal and cingulate cortex, the same change producing a decrease of firing rates can account for the negative symptoms, in that the amount of emotion, the effects of a reward or punisher, is related to the magnitude of the firing rate responses of orbitofrontal cortex neurons (Rolls 2008c).

In this unifying approach, both the negative and cognitive symptoms thus could be caused by a reduction of the NMDA conductance in attractor networks. The proposed mechanism links the cognitive and negative symptoms of schizophrenia in an attractor framework and is consistent with a close relation between the cognitive and negative symptoms: blockade of NMDA receptors by dissociative anesthetics such as ketamine produces in normal subjects schizophrenic symptoms including both negative and cognitive impairments (Malhotra, Pinals, Weingartner, Sirocco, Missar, Pickar and Breier 1996, Newcomer, Farber, Jevtovic-Todorovic, Selke, Melson, Hershey, Craft and Olney 1999); agents that enhance NMDA receptor function reduce the negative symptoms and improve the cognitive abilities of schizophrenic patients (Goff and Coyle 2001); and the cognitive and negative symptoms occur early in the illness and precede the first episode of positive symptoms (Lieberman, Perkins, Belger, Chakos, Jarskog, Boteva and Gilmore 2001, Hafner, Maurer, Loffler, an der Heiden, Hambrecht and Schultze-Lutter 2003, Mueser and McGurk 2004). Consistent with this hypothesized role of a reduction in NMDA conductances being involved in schizophrenia, postmortem studies of schizophrenia have identified abnormalities in glutamate receptor density in regions such as the prefrontal cortex, thalamus, and the temporal lobe (Goff and Coyle 2001, Coyle et al. 2003), brain areas that are active during the performance of cognitive tasks.

10.1.3 Positive symptoms

The *positive symptoms* of schizophrenia include bizarre (psychotic) trains of thoughts, hallucinations, and (paranoid) delusions (Liddle 1987, Mueser and McGurk 2004). We propose that these symptoms are related to shallow basins of attraction of both the spontaneous and persistent states in the temporal lobe semantic memory networks and to the statistical fluctuations caused by the probabilistic spiking of the neurons. The reduction in the stability of the high firing rate attractor states could result in thoughts moving too freely from one unstable thought to another loosely associated thought. The reduction in the stability of the spontaneous state, which is an attractor state in its own right, would mean that the system would suddenly jump from the spontaneous, quiet, state to a high firing rate attractor state even when no external stimulus is present, and resulting it is suggested in hallucinations, and in a feeling of a lack of control that is frequently present.

In the language of stochastic neurodynamics (Rolls and Deco 2010), these two different forms of instability could result in activations arising sponta-neously, and thoughts moving too freely round the energy landscape, loosely from thought to weakly associated thought, leading to bizarre thoughts and associations, which may eventually over time be associated together in se-mantic memory to lead to false beliefs and delusions. Consistent with this, neuroimaging studies suggest higher activation especially in areas of the tem-poral lobe (Weiss and Heckers 1999, Shergill, Brammer, Williams, Murray and McGuire 2000, Scheuerecker, Ufer, Zipse, Frodl, Koutsouleris, Zetzsche, Wiesmann, Albrecht, Bruckmann, Schmitt, Moller and Meisenzahl 2008).

We suggest that the reduction in the stability of the high firing rate state is produced by a reduction in the NMDA receptor efficacy, which is present in schizophrenia (Goff and Coyle 2001, Coyle et al. 2003). We suggest that the reduction in the stability of the spontaneous (low) firing rate state is produced by a reduction in the inhibitory receptor GABA efficacy, which is present in schizophrenia (Wang, Tegner, Constantinidis and Goldman-Rakic 2004, Lewis, Hashimoto and Volk 2005), and which we suggest is especially related to the positive symptoms (Rolls 2005a, Loh et al. 2007a, Rolls et al. 2008d, Rolls and Deco 2011, Rolls 2011f). Agents that act on cannabinoid receptors (such as cannabis) may decrease the firing of the GABA inhibitory neurons, and by reducing inhibition and therefore the stability of the spontaneous (low) firing state of excitatory neurons in attractor networks, may thereby tend to promote positive-like symptoms (Rolls 2011f).

When both NMDA and GABA are reduced one might think that these two counterbalancing effects (excitatory and inhibitory) would either cancel each other out or yield a tradeoff between the stability of the spontaneous and persist-ent state. However, this is not the case. The stability of both the spontaneous (i.e. low firing rate) state and the persistent (i.e. high firing rate) state is reduced. We relate this pattern to the positive symptoms of schizophrenia, in which both the spontaneous and attractor states are shallow, and the system merely jumps by the influence of statistical fluctuations between the different (spontaneous and high firing rate attractor) states.

To investigate more directly the wandering between spontaneous and several different persistent attractor states, we simulated the condition with decreased NMDA and GABA conductances over a long time period in which no cue stimulus input was given. Figure 10.1 shows the firing rates of the two selective pools S1 and S2. The high activity switches between the two attractors due to the influence of fluctuations, which corresponds to spontaneous wandering in a shallow energy landscape, corresponding for example to sudden jumps between unrelated cognitive processes. These results are consistent with the

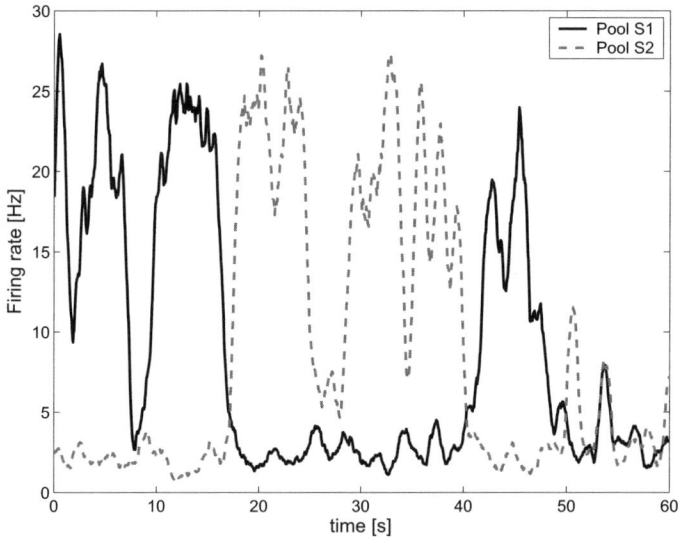

Fig. 10.1 Wandering between attractor states by virtue of statistical fluctuations caused by the randomness of the spiking activity. We simulated a single long trial (60 s) in the spontaneous test condition for the synaptic modification (−NMDA, −GABA). The two curves show the activity of the two selective pools over time smoothed with a 1 s sliding averaging window. The activity moves noisily between the attractor for the spontaneous (low firing) state and the two persistent (high firing) states S1 and S2. (After Loh, Rolls and Deco 2007a.)

mean-field flow analysis and demonstrate that the changes in the attractor landscape influence the behaviour at the stochastic level.

The positive symptoms (Fig. 10.2, right column) of schizophrenia include delusions, hallucinations, thought disorder, and bizarre behaviour. Examples of delusions are beliefs that others are trying to harm the person, impressions that others control the person's thoughts, and delusions of grandeur. Hallucinations are perceptual experiences, which are not shared by others, and are frequently auditory but can affect any sensory modality. These symptoms may be related to activity in the temporal lobes (Liddle 1987, Epstein, Stern and Silbersweig 1999, Mueser and McGurk 2004). The attractor framework approach taken here hypothesizes that the basins of attraction of both spontaneous and persistent states are shallow (Fig. 10.2). Due to the shallowness of the spontaneous state, the system can jump spontaneously up to a high activity state causing hallucinations to arise and leading to bizarre thoughts and associations. This might be the cause for the higher activations in schizophrenics in temporal lobe areas which are identified in imaging experiments (Shergill et al. 2000, Scheuerecker et al. 2008).

Our general hypothesis regarding the attractor landscape is meant to describe the aberrant dynamics in cortical regions which could be caused by several pathways. A strength of this approach is that one can investigate in

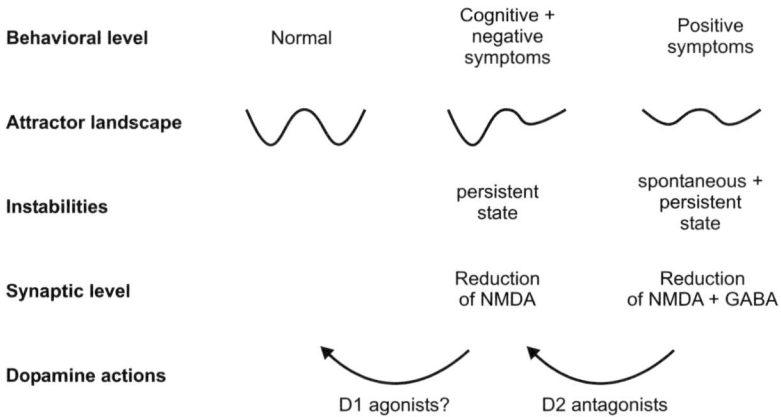

		Cognitive + negative symptoms	Positive symptoms
Behavioral level	Normal		

Fig. 10.2 Summary of the attractor hypothesis of schizophrenic symptoms and simulation results (see text). The first basin (from the left) in each energy landscape is the spontaneous state, and the second basin is the persistent attractor state. The vertical axis of each landscape is the energy potential. (After Loh, Rolls and Deco 2007b.)

the simulations the effects of treatment with particular combinations of drugs designed to facilitate both glutamate transmission (including modulation with glycine or serine of the NMDA receptor) and GABA effects. Drug combinations identified in this way could be useful to explore, as combinations might work where treatment with a single drug on its own may not be effective, and possibilities can be systematically explored with the approach described here.

We thus have a new set of concepts related to the stability of brain states to bring to bear on the symptoms and causes of schizophrenia. We also have new ways of proposing new treatments based on the application of computational hypotheses that link synaptic receptor and neuronal effects of drugs to the effects that could be produced at the global level of the operation of cognitive systems involved in memory, attention, and the stability of memory and thus thought processes (Rolls 2011f).

10.2 Obsessive-compulsive disorder

10.2.1 The symptoms

Obsessive-compulsive disorder (OCD) is a chronically debilitating disorder with a lifetime prevalence of 2–3% (Robins, Helzer, Weissman, Orvaschel, Gruenberg, Burke and Regier 1984, Karno, Golding, Sorenson and Burnam 1988, Weissman, Bland, Canino, Greenwald, Hwu, Lee, Newman, Oakley-Browne, Rubio-Stipec, Wickramaratne et al. 1994). It is characterized by two sets of symptoms, obsessive and compulsive. Obsessions are unwanted, intrusive, recurrent thoughts or impulses that are often concerned with themes of contamination and 'germs', checking household items in case of fire or burglary, order and symmetry of objects, or fears of harming oneself or others. Compulsions are ritualistic, repetitive behaviours or mental acts carried out in relation to these obsessions e.g. washing, household safety checks, counting, rearrangement of objects in symmetrical array or constant checking of oneself and others to ensure no harm has occurred (Menzies, Chamberlain, Laird, Thelen, Sahakian and Bullmore 2008).

Patients with OCD experience the persistent intrusion of thoughts that they generally perceive as foreign and irrational but which cannot be dismissed. The anxiety associated with these unwanted and disturbing thoughts can be extremely intense; it is often described as a feeling that something is incomplete or wrong, or that terrible consequences will ensue if specific actions are not taken. Many patients engage in repetitive, compulsive behaviours that aim to discharge the anxieties associated with these obsessional thoughts. Severely affected patients can spend many hours each day in their obsessional thinking and resultant compulsive behaviours, leading to marked disability (Pittenger, Krystal and Coric 2006). While OCD patients exhibit a wide variety of obsessions and compulsions, the symptoms tend to fall into specific clusters. Common patterns include obsessions of contamination, with accompanying cleaning compulsions; obsessions with symmetry or order, with accompanying ordering behaviours; obsessions of saving, with accompanying hoarding; somatic obsessions; aggressive obsessions with checking compulsions; and sexual and religious obsessions (Pittenger, Krystal and Coric 2006).

10.2.2 A hypothesis about the increased stability of attractor networks and the symptoms of obsessive-compulsive disorder

I now describe an approach to how obsessive-compulsive disorders arise, and of the different symptoms, that has its foundations in a theoretical understanding of the stability of cortical systems in the brain (Rolls and Deco 2010). The aim is to show the way in which complex symptoms in psychiatric states, and their treatment, can be approached by an understanding of how the brain functions, and how neural dysfunctions could lead to psychiatric states. The description thus illustrates the neuro-approach to psychiatry, where the symptoms and their treatment can be approached by a theoretical understanding of cortical function from the level of ion channels through neuronal firing, to neuronal networks with their collective behaviour, to the global operation of the system.

The theory (Rolls, Loh and Deco 2008c, Rolls 2011f) is based on the top-down proposal that there is overstability of attractor neuronal networks in cortical and related areas in obsessive-compulsive disorders. The approach is top-down in that it starts with the set of symptoms and maps them onto the dynamical systems framework, and only after this considers detailed underlying biological mechanisms, of which there could be many, that might produce the effects. (In contrast, a complementary bottom-up approach starts from detailed neurobiological mechanisms, and aims to interpret their implications with a brain-like model for higher level phenomena.) We show by integrate-and-fire neuronal network simulations that the overstability could arise by for example overactivity in glutamatergic excitatory neurotransmitter synapses, which produces an increased depth of the basins of attraction, in the presence of which neuronal spiking-related and potentially other noise is insufficient to help the system move out of an attractor basin. I relate this top-down proposal, related to the stochastic dynamics of neuronal networks, to new evidence that there may be overactivity in glutamatergic systems in obsessive-compulsive disorders, and consider the implications for treatment (Rolls 2011f).

We hypothesize that cortical and related attractor networks become too stable in obsessive-compulsive disorder, so that once in an attractor state, the networks tend to remain there too long (Rolls, Loh and Deco 2008c, Rolls 2011f). The hypothesis is that the depths of the basins of attraction become deeper, and that this is what makes the attractor networks more stable. We further hypothesize that part of the mechanism for the increased depth of the basins of attraction is increased glutamatergic transmission, which increases the depth of the basins of attraction by increasing the firing rates of the neurons, and by increasing the effective value of the synaptic weights between the associatively modified synapses that define the attractor, as is made evident in Equation 2.3. The synaptic strength is effectively increased if more glutamate is released per action potential at the synapse, or if in other ways the currents injected into the neurons through the NMDA (N-methyl-d-aspartate) and/or AMPA synapses are larger. In addition, if NMDA receptor function is increased, this could also increase the stability of the system because of the temporal smoothing effect of the long time constant of the NMDA receptors (Wang 1999). This increased stability of cortical and related attractor networks, and the associated higher neuronal firing rates, could occur in different brain regions, and thereby produce different symptoms, as follows.

If these effects occurred in high order motor areas, the symptoms could include inability to move out of one motor pattern, resulting for example in re-peated movements or actions. In parts of the cingulate cortex and dorsal medial prefrontal cortex, this could result in difficulty in switching between actions or strategies (Rushworth, Behrens, Rudebeck and Walton 2007a, Rushworth et al. 2007b), as the system would be locked into one action or strategy. If an action was locked into a high order motor area due to increased stability of an attractor network, then lower order motor areas might thereby not be able to escape easily what they implement, such as a sequence of movements, so that the sequence would be repeated.

If occurring in the lateral prefrontal cortex (including the dorsolateral and ventrolateral parts), the increased stability of attractor networks could produce symptoms that include a difficulty in shifting attention, and in cognitive set shifting in which what is relevant to the task may shift from color to shape, etc. (Veale, Sahakian, Owen and Marks 1996, Watkins, Sahakian, Robertson, Veale, Rogers, Pickard, Aitken and Robbins 2005, Chamberlain, Fineberg, Blackwell, Robbins and Sahakian 2006, Chamberlain, Fineberg, Menzies, Blackwell, Bullmore, Robbins and Sahakian 2007). These are in fact important symptoms that can be found in obsessive-compulsive disorder (Menzies et al. 2008).

Planning may also be impaired in patients with OCD (Menzies et al. 2008), and this could arise because there is too much stability of attractor networks in the dorsolateral prefrontal cortex concerned with holding in mind the different short-term memory representations that encode the different steps of a plan (Rolls 2008c). Indeed, there is evidence for dorsolateral prefrontal cortex (DLPFC) dysfunction in patients with OCD, in conjunction with impairment on a version of the Tower of London task, a task often used to probe planning aspects of executive function (van den Heuvel, Veltman, Groenewegen, Cath, van Balkom, van Hartskamp, Barkhof and van Dyck 2005).

An increased firing rate of neurons in the orbitofrontal cortex, and anterior cingulate cortex, produced by hyperactivity of glutamatergic transmitter systems, would increase emotionality, which is frequently found in obsessive-compulsive disorder. Part of the increased anxiety found in obsessive-compulsive disorder could be related to an inability to complete tasks or actions in which one is locked. But part of our unifying proposal is that part of the increased emotionality in OCD may be directly related to increased firing produced by the increased glutamatergic activity in brain areas such as the orbitofrontal and anterior cingulate cortex (Rolls, Loh and Deco 2008c). The orbitofrontal cortex and anterior cingulate cortex are involved in emotion, in that they are activated by primary and secondary reinforcers that produce affective states (Rolls 2004a, Rolls 2005a, Rolls 2008c, Rolls and Grabenhorst 2008, Rolls 2009a), and in that damage to these regions alters emotional behaviour and emotional experience (Rolls, Hornak, Wade and McGrath 1994a, Hornak, Rolls and Wade 1996, Hornak, Bramham, Rolls, Morris, O'Doherty, Bullock and Polkey 2003, Hornak, O'Doherty, Bramham, Rolls, Morris, Bullock and Polkey 2004, Berlin, Rolls and Kischka 2004, Berlin, Rolls and Iversen 2005). Indeed, negative emotions as well as positive emotions activate the orbitofrontal cortex, with the emotional states produced by negative events tending to be represented in the lateral orbitofrontal cortex and dorsal part of the anterior cingulate cortex (Kringelbach and Rolls 2004, Rolls 2005a, Rolls 2008c, Rolls 2009a, Grabenhorst and Rolls 2011).

If the increased stability of attractor networks occurred in temporal lobe semantic memory networks, then this would result in a difficulty in moving from one thought to another, and possibly in stereotyped thoughts, which again may be a symptom of obsessive-compulsive disorder (Menzies et al. 2008). The obsessional states are thus proposed to arise because cortical areas concerned with cognitive functions have states that become too stable. The compulsive states are proposed to arise partly in response to the obsessional states, but also partly because cortical areas concerned with actions have states that become too stable.

The theory provides a unifying computational account of both the obsessional and compulsive symptoms, in that both arise due to increased stability of cortical attractor networks, with the different symptoms related to overstability in different cortical areas. The theory is also unifying in that a similar increase in glutamatergic activity in the orbitofrontal and anterior cingulate cortex could increase emotionality, as described earlier.

10.2.3 Glutamate and increased depth of the basins of attraction of attractor networks

We tested this approach in integrate-and-fire neuronal network simulations (Section 2.5.5), and showed that increased NMDA or AMPA excitatory receptor functionality (increased synaptic conductances) could produce overstability of noisy attractor networks, and could make them less distractible (Rolls, Loh and Deco 2008c).

We also showed that increasing the synaptic conductances activated by the GABA inhibitory neurons in the network could correct for the effect of increasing NMDA receptor activated synaptic currents on the persistent high firing rate type of test; and for the effect of increasing NMDA on the spontaneous state simulations when there was no initiating retrieval stimulus, and the network should remain in the low firing rate state until the end of the simulation run.

The implications for symptoms are that agents that increase GABA conductances might reduce and normalize the tendency to remain locked into an idea or concern or action; and would make it much less likely that the quiescent resting state would be left by jumping (because of the noisy spiking) towards a state representing a dominant idea or concern or action.

This simulation evidence, that an increase of glutamatergic synaptic efficacy can increase the stability of attractor networks and thus potentially provide an account for some of the symptoms of obsessive-compulsive disorder, is consistent with evidence that glutamatergic function may be increased in some brain systems in obsessive-compulsive disorder (Rosenberg, MacMaster, Keshavan, Fitzgerald, Stewart and Moore 2000, Rosenberg, MacMillan and Moore 2001, Rosenberg, Mirza, Russell, Tang, Smith, Banerjee, Bhandari, Rose, Ivey, Boyd and Moore 2004, Pittenger et al. 2006, Rolls 2011f) and that cerebro-spinal-fluid glutamate levels are elevated (Chakrabarty, Bhattacharyya, Christopher and Khanna 2005). Consistent with this, agents with antiglutamatergic activity such as riluzole, which can decrease glutamate transmitter release, may be efficacious in obsessive-compulsive disorder (Pittenger et al. 2006, Bhattacharyya and Chakraborty 2007).

Further evidence for a link between glutamate as a neurotransmitter and OCD comes from genetic studies. There is evidence for a significant association between the SLC1A1 glutamate transporter gene and OCD (Stewart, Fagerness, Platko, Smoller, Scharf, Illmann, Jenike, Chabane, Leboyer, Delorme, Jenike and Pauls 2007). This transporter is crucial in terminating the action of glutamate as an excitatory neurotransmitter and in maintaining extracellular glutamate concentrations within a normal range (Bhattacharyya and Chakraborty 2007). In addition, Arnold et al. postulated that N-methyl-d-aspartate (NMDA) receptors were involved in OCD, and specifically that polymorphisms in the $3'$ untranslated region of GRIN2B (glutamate receptor, ionotropic, N-methyl-d-aspartate 2B) were associated with OCD in affected families (Arnold, Rosenberg, Mundo, Tharmalingam, Kennedy and Richter 2004).

Thus we have seen how an increase in cortical excitability could increase the stability of cortical recurrent attractor networks, and that if expressed in different prefrontal systems, could produce some of the different types of symptoms found in different patients with obsessive-compulsive disorders (Rolls 2011f). Moreover, the approach links the low-level changes found in the increase in NMDA receptor efficacy in OCD to the systems-level effects such as overstability of cognitive and/or motor functioning. Implications for treatments are developed elsewhere (Rolls 2011f).

10.3 Stochastic noise, attractor dynamics, and normal aging

In Section 10.1 we saw how some cognitive symptoms such as poor short-term memory and attention could arise due to reduced depth in the basins of attraction of prefrontal cortical networks, and the effects of noise. The hypothesis is that the reduced depth in the basins of attraction would make short-term memory unstable, so that sometimes the continuing firing of neurons that implement short-term memory would cease, and the system under the influence of noise would fall back out of the short-term memory state into spontaneous firing.

Given that top-down attention requires a short-term memory to hold the object of attention in mind, and that this is the source of the top-down attentional bias that influences competition in other networks that are receiving incoming signals, then disruption of short-term memory is also predicted to impair the stability of attention (Section 2.6).

These ideas are elaborated in Section 10.1, where the reduced depth of the basins of attraction in schizophrenia is related to down-regulation of NMDA receptors, or to factors that influence NMDA receptor generated ion channel currents such as dopamine D1 receptors.

Could similar processes in which the stochasticity of the dynamics is increased because of a reduced depth in the basins of attraction contribute to the changes in short-term memory and attention that are common in aging? Reduced short-term memory and less good attention are common in aging, as are impairments in episodic memory (Grady 2008). What changes in aging might contribute to a reduced depth in the basins of attraction? I describe recent hypotheses based on our theoretical approach to understanding cortical function (Rolls and Deco 2010, Rolls, Deco and Loh 2012).

10.3.1 NMDA receptor hypofunction

One factor is that NMDA receptor functionality tends to decrease with aging (Kelly et al. 2006). This would act, as described in Section 10.1, to reduce the depth of the basins of attraction, both by reducing the firing rate of the neurons in the active attractor, and effectively by decreasing the strength of the potentiated synaptic connections that support each attractor, as the currents passing through these potentiated synapses would decrease. These two actions are clarified by considering Equation 2.3 on page 33, in which the energy E reflects the depth of the basin of attraction.

An example of the reduction of firing rate in an attractor network produced by even a small downregulation (by 4.5%) of the NMDA receptor activated ion channel conductances is shown in Fig. 10.3, based on a mean-field analysis. The effect of this reduction would be to decrease the depth of the basins of attraction, both by reducing the firing rate, and by producing an effect similar to weakened synaptic strengths, as shown in Equation 2.3.

In integrate-and-fire simulations, the effect on the firing rates of reduction in the NMDA activated channel conductances by 4.5% and in the GABA activated channel conductances by 9% are shown in Fig. 10.4.

The reduced depth in the basins of attraction could have a number of effects that are relevant to cognitive changes in aging.

First, the stability of short-term memory networks would be impaired, and it might be difficult to hold items in short-term memory for long, as the noise might push the network easily out of its shallow attractor, as described in Section 10.1.

Second, top-down attention would be impaired, in two ways. First, the short-term memory network holding the object of attention in mind would be less stable, so that the source of the top-down bias for the biased competition in other cortical areas might disappear. Second, and very interestingly, even when the short-term memory for attention is still in its persistent attractor state, it would be less effective as a source of the top-down bias, because the firing rates would be lower, as shown in Figs. 10.3 and 10.4.

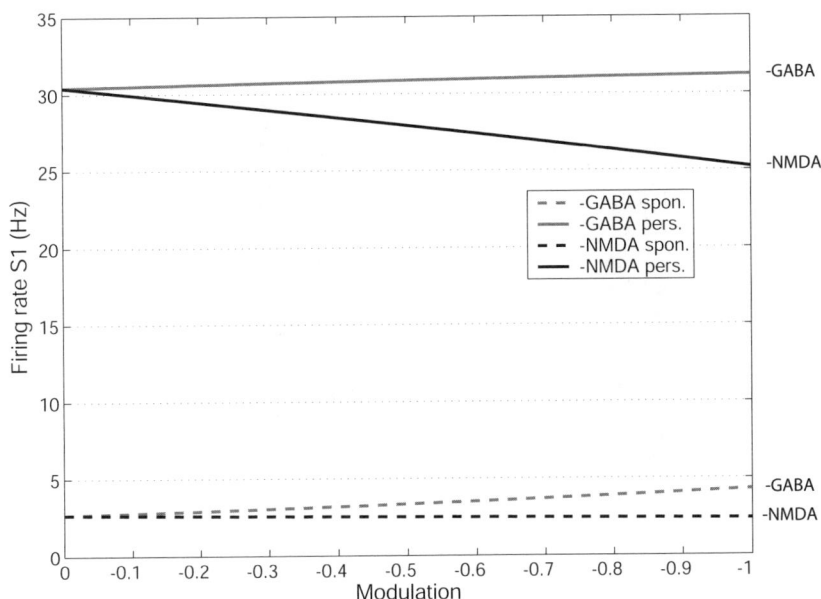

Fig. 10.3 The firing rate of an attractor network as a function of a modulation of the NMDA receptor activated ion channel conductances. On the abscissa, 0 corresponds to no reduction, and −1.0 to a reduction by 4.5%. The change in firing rate for the high firing rate short-term memory state of persistent firing, and for the spontaneous rate, are shown. The values were obtained from the mean-field analysis of the network described by Rolls and Deco (2010), and thus for the persistent state represent the firing rate when the system is in its high firing rate state. Curves are also shown for a reduction of up to 9% in the GABA receptor activated ion channel conductances. (Data of Loh, Rolls and Deco; see Rolls and Deco (2010).)

Third, the recall of information from episodic memory systems in the temporal lobe (Rolls 2008c, Dere, Easton, Nadel and Huston 2008, Rolls 2008d, Rolls 2010c) would be impaired. This would arise because the positive feedback from the recurrent collateral synapses that helps the system to fall into a basin of attraction, representing in this case the recalled memory, would be less effective, and so the recall process would be more noisy overall.

Fourth, any reduction of the firing rate of the pyramidal cells caused by NMDA receptor hypofunction (Figs. 10.3 and 10.4) would itself be likely to impair new learning involving long-term synaptic potentiation (LTP, Section 2.5).

In addition, if the NMDA receptor hypofunction were expressed not only in the prefrontal cortex where it would affect short-term memory, and in the temporal lobes where it would affect episodic memory, but also in the orbitofrontal cortex, then we would predict some reduction in emotion and motivation with aging, as these functions rely on the orbitofrontal cortex (see Rolls (2005a) and Section 10.1).

Although NMDA hypofunction may contribute to cognitive effects such

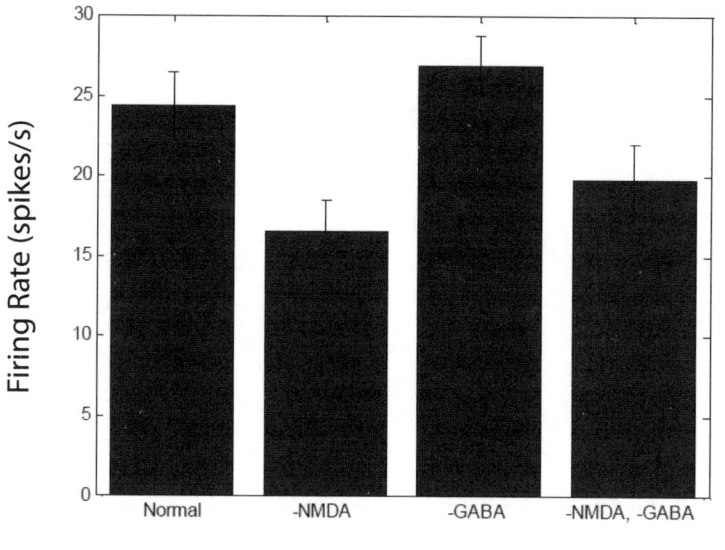

Fig. 10.4 The firing rate (mean $\pm$ sd across neurons in the attractor and simulation runs) of an attractor network in the baseline (Normal) condition, and when the NMDA receptor activated ion channel conductances in integrate-and-fire simulations are reduced by 4.5% ($-$NMDA). The firing rate for the high firing rate short-term memory state of persistent firing is shown. The values were obtained from integrate-and-fire simulations, and only for trials in which the network was in a high firing rate attractor state using the criterion that at the end of the 3 s simulation period the neurons in the attractor were firing at 10 spikes/s or higher. The firing rates are also shown for a reduction of 9% in the GABA receptor activated ion channel conductances ($-$GABA), and for a reduction in both the NMDA and GABA receptor activated ion channel conductances ($-$NMDA, $-$GABA). (Data of Loh, Rolls and Deco.)

as poor short-term memory and attention in aging and in schizophrenia, the two states are clearly very different. Part of the difference lies in the positive symptoms of schizophrenia (the psychotic symptoms, such as thought disorder, delusions, and hallucinations) which may be related to the additional downregulation of GABA in the temporal lobes, which would promote too little stability of the spontaneous firing rate state of temporal lobe attractor networks, so that the networks would have too great a tendency to enter states even in the absence of inputs, and to not be controlled normally by input signals (Loh, Rolls and Deco 2007a, Loh, Rolls and Deco 2007b, Rolls, Loh, Deco and Winterer 2008d) (see Section 10.1). However, in relation to the cognitive symptoms of schizophrenia, there has always been the fact that schizophrenia is a condition that often has its onset in the late teens or twenties, and I suggest that there could be a link here to changes in NMDA and related receptor functions that are related to aging. In particular, short-term memory is at its peak when young, and it may be the case that by the late teens or early twenties NMDA and related receptor systems (including dopamine) may be less efficacious than when younger, so that the cognitive symptoms of schizophrenia are more likely to occur at this age than earlier (Rolls and Deco 2011).

10.3.2 Dopamine

Dopamine D1 receptor blockade in the prefrontal cortex can impair short-term memory (Sawaguchi and Goldman-Rakic 1991, Sawaguchi and Goldman-Rakic 1994, Goldman-Rakic 1999, Castner, Williams and Goldman-Rakic 2000). Part of the reason for this may be that D1 receptor blockade can decrease NMDA receptor activated ion channel conductances, among other effects (Seamans and Yang 2004, Durstewitz, Kelc and Gunturkun 1999, Durstewitz, Seamans and Sejnowski 2000a, Brunel and Wang 2001, Durstewitz and Seamans 2002) (see further Section 10.1). Thus part of the role of dopamine in the prefrontal cortex in short-term memory can be accounted for by a decreased depth in the basins of attraction of prefrontal attractor networks (Loh, Rolls and Deco 2007a, Loh, Rolls and Deco 2007b, Rolls, Loh, Deco and Winterer 2008d). The decreased depth would be due to both the decreased firing rate of the neurons, and the reduced efficacy of the modified synapses as their ion channels would be less conductive (see Equation 2.3). The reduced depth of the basins of attraction can be thought of as decreasing the signal-to-noise ratio (Loh, Rolls and Deco 2007b, Rolls, Loh, Deco and Winterer 2008d). Given that dopaminergic function in the prefrontal cortex may decline with aging (Sikström 2007), and in conditions in which there are cognitive impairments such as Parkinson's disease, the decrease in dopamine could contribute to the reduced short-term memory and attention in aging.

In attention deficit hyperactivity disorder (ADHD), in which there are attentional deficits including too much distractibility, catecholamine function more generally (dopamine and noradrenaline (i.e. norepinephrine)) may be reduced (Arnsten and Li 2005), and I suggest that these reductions could produce less stability of short-term memory and thereby attentional states by reducing the depth of the basins of attraction.

10.3.3 Impaired synaptic modification

Another factor that may contribute to the cognitive changes in aging is that long-lasting associative synaptic modification as assessed by long-term potentiation (LTP) is more difficult to achieve in older animals and decays more quickly (Barnes 2003, Burke and Barnes 2006, Kelly et al. 2006). This would tend to make the synaptic strengths that would support an attractor weaker, and weaken further over the course of time, and thus directly reduce the depth of the attractor basins. This would impact episodic memory, the memory for particular past episodes, such as where one was at breakfast on a particular day, who was present, and what was eaten (Rolls 2008c, Rolls 2008d, Dere et al. 2008, Rolls 2010c). The reduction of synaptic strength over time could also affect short-term memory, which requires that the synapses that support a short-term memory attractor be modified in the first place using LTP, before the attractor is used (Kesner and Rolls 2001).

In view of these changes, boosting glutamatergic transmission is being explored as a means of enhancing cognition and minimizing its decline in aging. Several classes of AMPA receptor potentiators have been described in the last decade. These molecules bind to allosteric sites on AMPA receptors, slow desensitization, and thereby enhance signalling through the receptors. Some AMPA receptor potentiator agents have been explored in rodent models and are now entering clinical trials (Lynch and Gall 2006, O'Neill and Dix 2007). These treatments might increase the depth of the basins of attraction. Agents that activate the glycine or serine modulatory sites on the NMDA receptor (Coyle 2006) would also be predicted to be useful.

Another factor is that Ca^{2+}-dependent processes affect Ca^{2+} signalling pathways and impair synaptic function in an aging-dependent manner, consistent with the Ca^{2+} hypothesis of brain aging and dementia (Thibault, Porter, Chen, Blalock, Kaminker, Clodfelter, Brewer and Landfield 1998, Kelly et al. 2006). In particular, an increase in Ca^{2+} conductance can occur in aged neurons, and CA1 pyramidal cells in the aged hippocampus have an increased density of L-type Ca^{2+} channels that might lead to disruptions in Ca^{2+} homeostasis, contributing to the plasticity deficits that occur during aging (Burke and Barnes 2006).

10.3.4 Cholinergic function

Another factor is acetylcholine. Acetylcholine in the neocortex has its origin largely in the cholinergic neurons in the basal magnocellular forebrain nuclei of Meynert. The correlation of clinical dementia ratings with the reductions in a number of cortical cholinergic markers such as choline acetyltransferase, muscarinic and nicotinic acetylcholine receptor binding, as well as levels of acetylcholine, suggested an association of cholinergic hypofunction with cognitive deficits, which led to the formulation of the cholinergic hypothesis of memory dysfunction in senescence and in Alzheimer's disease (Bartus 2000, Schliebs and Arendt 2006). Could the cholinergic system alter the function of the cerebral cortex in ways that can be illuminated by stochastic neurodynamics?

The cells in the basal magnocellular forebrain nuclei of Meynert lie just lateral to the lateral hypothalamus in the substantia innominata, and extend forward through the preoptic area into the diagonal band of Broca (Mesulam 1990). These cells, many of which are cholinergic, project directly to the cerebral cortex (Divac 1975, Kievit and Kuypers 1975, Mesulam 1990). These cells provide the major cholinergic input to the cerebral cortex, in that if they are lesioned the cortex is depleted of acetylcholine (Mesulam 1990). Loss of these cells does occur in Alzheimer's disease, and there is consequently a reduction in cortical acetylcholine in this disease (Mesulam 1990, Schliebs and Arendt 2006). This loss of cortical acetylcholine may contribute to the memory loss in Alzheimer's disease, although it may not be the primary factor in the aetiology.

In order to investigate the role of the basal forebrain nuclei in memory, Aigner, Mitchell, Aggleton, DeLong, Struble, Price, Wenk, Pettigrew and Mishkin (1991) made neurotoxic lesions of these nuclei in monkeys. Some impairments on a simple test of recognition memory, delayed non-match-to-sample, were found. Analysis of the effects of similar lesions in rats showed that performance on memory tasks was impaired, perhaps because of failure to attend properly (Muir, Everitt and Robbins 1994). Damage to the cholinergic neurons in this region in monkeys with a selective neurotoxin was also shown to impair memory (Easton and Gaffan 2000, Easton, Ridley, Baker and Gaffan 2002).

One way in which reduced acetylcholine could impair memory is by a resulting increase in neuronal adaptation. A property of cortical neurons is that they tend to adapt with repeated input, that is, to respond less (Abbott, Varela, Sen and Nelson 1997, Fuhrmann, Markram and Tsodyks 2002). However, this adaptation is most marked in slices, in which there is no acetylcholine. One effect of acetylcholine is to reduce this adaptation (Power and Sah 2008). The mechanism is summarized elsewhere (Rolls and Deco 2010). A variety of neuromodulators, including acetylcholine (ACh), noradrenaline, and glutamate acting via G-protein-coupled receptors, reduce spike-frequency adaptation (Nicoll 1988), and would be predicted to ameliorate the cognitive symptoms of aging.

It is predicted that enhancing cholinergic function, or compensating for its effects by other treatments that increase cortical excitability, will help to reduce the instability of attractor networks involved in short-term memory and attention that may occur in aging.

Part of the interest of this stochastic neurodynamics approach to aging is that it provides a way to test combinations of pharmacological treatments, that may together help to minimize the cognitive symptoms of aging. Indeed, the approach facilitates the investigation of drug combinations that may together be effective in doses lower than when only one drug is given. Further, this approach may lead to predictions for effective treatments that need not necessarily restore the particular change in the brain that caused the symptoms, but may find alternative routes to restore the stability of the dynamics.

10.4 Conclusions

Overall in this chapter we have seen that complex psychiatric symptoms, and some of the cognitive symptoms of aging, can be approached with a neuro-approach that not only considers chemical and related changes in the brain, but now has a way of relating those changes to the symptoms by using a model that connects ion channel and transmitter changes to the operation of cognitive systems in which computations are performed, and thereby to symptoms, and to potential treatments.

In this way, a mechanistic, computational neuroscience, approach to brain function (Rolls and Treves 1998, Rolls 2008c, Rolls and Deco 2010) is leading to new concepts in treating disorders such as schizophrenia, obsessive-compulsive disorder, and in treating the cognitive symptoms of normal aging.

11 Neuroreligion

Why are religions so common in human societies? How do adaptations of the human brain make religion fit the human mind? How are we predisposed to religion? Why does religion persist (Hinde 2010)? What is the role of rationality in religion?

In this chapter, I consider these questions. I do not wish to question particular religions, or to argue against particular religions. I think that some of what is offered by many religions as ways of promoting a supporting and stable society has much to offer large numbers of individuals. My aim instead is to cast some light on the questions above, and, by understanding why many humans find religions attractive, to see how their value can be positive; and to provide rational approaches to help reduce conflict related to different religions.

11.1 Explanation and causality

11.1.1 The Hebb synapse and the detection of causality

The Hebb synapse (Section 2.5.1) is the simple way that the brain uses to relate cause to effect. It detects causality in a simple way, by associating a prior event with a subsequent event, in such a way that the first event can later be used to predict the second event. A simple example is Pavlov's bell: when the bell is rung before food is delivered, the bell (the prior event) is associated with the subsequent event (the delivery of food), and the animal learns to salivate when the bell is heard, and not just when the food is delivered.

In a Hebb synapse, which captures some of the properties of such associative learning, the synaptic connection from one neuron (which might in our example convey the conditioned stimulus, the CS, the bell) to another neuron (which might in our example respond to the taste which is an unconditioned stimulus, UCS, that elicits the unconditioned response of salivation), becomes stronger if the CS neuron firing precedes and predicts the firing of the neuron responding to the UCS (see Fig. 11.1) (Section 2.5.1). Then later when the CS is delivered, it activates the UCS (taste) neuron through the strengthened synaptic connection, and the conditioned response of salivation occurs. The synapse does not show this increase if the UCS precedes the CS. The synapse thus detects causality because the order of the two events, and the ability of the first event, the bell, to predict the second, the food, is what is learned. The mechanism implements

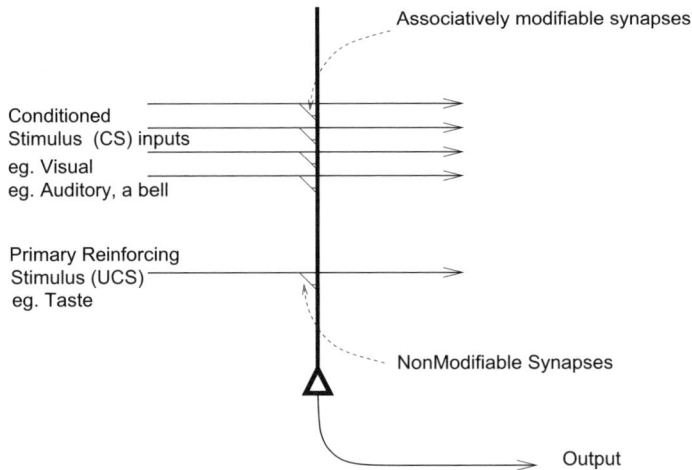

Fig. 11.1 Pattern association between a primary reinforcer, such as the taste of food (an uncond-itioned stimulus, UCS), which activates neurons through non-modifiable synapses, and a potential secondary reinforcer (a conditioned stimulus or CS), such as the sight of food or a bell that pre-cedes the taste of food, which has associatively modifiable synapses onto the same neurons. The output might represent a taste, and the taste recalled by the CS, and might lead to a response such as salivation, which could become conditioned. The triangle represents the cell body of the neuron, the thick vertical element the dendrite which receives synaptic connections from the axons of other neurons, shown as thin lines.

the form of causality summarized by *'post hoc, ergo propter hoc'* ('after that, therefore because of that').

This synaptic mechanism only works in most brain regions if there is almost no delay between the CS and the UCS (and ideally they might partly overlap in time). The biological reason for this is that if there was an incoming stream of possible conditioned stimuli, there would be no way of knowing which was the relevant stimulus as a CS, so the brain learns primarily about one stimulus that immediately precedes another stimulus.

The types of events that can be associated in this way are not at all restricted to classical conditioning, which is what has been described. Actions can also be associated with rewards that are outcomes of actions, and this is part of the basis of instrumental learning, in which actions can be learned that are instrumental in obtaining rewards (or avoiding punishers) (Appendix A).

11.1.2 Reasoning and the detection of causality

Humans (and to some extent some related animals) have a second way of detecting causality, which can bridge longer time delays and utilizes reasoning (Chapter 5). A necessary component of the neural architecture is a short-term memory that can hold the order of several items in store over periods of up to 60 s or more. I have described a way in which such short-term memories that incorporate the order of the items (in the prefrontal cortex and hippocampus) could be implemented by a series of linked attractor networks, each capable of keeping one item active (Rolls and Deco 2010) (Section 2.7). Such a short-term memory system enables series of events occurring in a particular order, typical of more complicated types of causality, to be detected. With what is described as working memory, one might be able to reorder the items in the short-term memory, and check that a different order of the same items would not produce the same effect, in a form of hypothesis-testing. Similar order-sensitive long-term memory processes could be implemented in analogous ways (Rolls and Deco 2010), and might lead to detection of causality over even longer time intervals.

11.1.3 Failure to detect causality

However, what happens if these mechanisms fail to detect causality? Say, for example, it rains, but there is no clear cause for why it rained on a particular day. Do we invent a rain-god, to provide an explanation, and a (frequently personified) agent who can be besought to produce rain, in the same way that people can be besought to provide other types of benefit? I suggest that this is one 'function' of gods, an explanatory and potentially providing function. The god arises naturally as a semi-personified agent, as people are frequently explainers and providers.

It may indeed be adaptive to come to this type of 'explanation', and to stop worrying about the cause of an effect such as rain, for then one can get on with more useful things, such as building a shelter, or gathering food, or searching for water. Of course, when the rain comes following our praying to the rain-god for rain, our beliefs in the rain-god are strengthened. Our brains may be adapted for this, in that we appear to notice and give weight to just a few suspicious coincidences (such as praying followed by rain), and to de-emphasize, or probably just not remember because it is not consolidated by the coincidence (e.g. of rain), the many instances in which praying for rain may not have been followed by rain.

A factor that increases the strength and persistence of such learning, revealed by learning theory, is that a low probability of an event being followed by a reward leads to learning that is very persistent: it shows great 'resistance to extinction' (Mazur 1998). This would tend to consolidate such beliefs about the causes of the rain.

Ascribing 'causes' or 'explanations' for effects may be adaptive in another way too. It may provide us with the feeling that we are in control, and can do something about it, for example by beseeching the 'provider'. It is well established in psychology that if rewards stop coming, and there is nothing we can do about it, people become depressed. The effect is termed 'learned helplessness' (Seligman 1978, Lee Duckworth, Steen and Seligman 2005). It is found for example in rats, which will cease trying, and 'give up', if rewards formerly under their control, are no longer provided. The rat stops trying to do anything, and becomes generally passive and does not interact with its environment. This may be adaptive, in that the rat does not waste effort and attention directed at goals that are no longer available and actions that are no longer being rewarded.

Such *'learned helplessness'* provides a useful model of reactive depression (i.e. depression produced as a reaction to a change in the environment). An example of this in humans is the depression that occurs after the death of a loved one, which may stop us trying to obtain any reinforcers. Part of an approach to treatment of this type of depression is to help the patient to learn again that actions can provide rewards (though different rewards). Helping a patient to learn how to take actions to receive reinforcers again from the world can be very beneficial in the treatment of depression.

Part of the value of an 'explanation' as provided by a religion may thus be that it reduces effects of learned helplessness, and indeed the general encouragement of positive interactions with the world is beneficial to psychological well-being (Lee Duckworth, Steen and Seligman 2005), and may be a useful spin-off of the type of 'causal explanation' that may be offered by a religion.

The situation in rats is quite analogous. If a reward is given at random times, behaviour that is described technically as 'superstitious' can result (Mazur 1998). For example, on one occasion when the rat is rotating right a reward might happen to be given at that random time. An action-outcome association might be set up, and the rat might keep rotating to the right, until when eventually a reward is given at another random time, the belief that turning right is associated with reward becomes strengthened further. The rat does not learn that not turning, or turning left, might also happen to be associated with reward, because once the behaviour of turning right is rewarded, that behaviour continues, and there is insufficient sampling of the other possibilities. In this way superstitious behaviour can be set up in rats, and it does appear to have parallels in humans.

We have seen that the human brain may be predisposed to produce 'explanations' for effects in situations in which the causes cannot easily be detected or understood. Even with many of the world's largest computers working incessantly, it is difficult to predict the weather more than 1–2 weeks in advance. When explanations are sought for issues such as 'Why are we here?', and 'What is the purpose of our existence on earth?', modern science (Dawkins 2009) is starting to provide some answers from the evidence about evolution, and from the theory of evolution, that make the types of explanation offered in Genesis centuries ago seem difficult to defend and take literally. It is interesting that some churchmen and women take those 'explanations' now more as metaphor than explanation.

My argument, from an understanding of how the brain detects causality, is thus that at least two types of brain processing are relevant to understanding why religion may be attractive because of its 'explanations'. One is the associative learning normally used to detect simple causality, described earlier. The second is the capability to reason (Chapter 5), which leads to questions being asked to which there are no clear answers, such as: 'What will happen in future?', 'What will happen after death?'. In these circumstances, a causal agent such as a god, frequently in semi-personified form which is typical of many causal agents we encounter, may reduce the feeling of uncertainty by providing an 'explanation' or prediction.

This reduction in the feeling of uncertainty may in fact be adaptive, because it helps individuals with those beliefs to focus on solving other problems, and to keep positive interactions with the world so that reinforcers are received.

As our reasoning capability is applied more and more, a lot of reasoning may lead us to understand that our brains have a tendency to find religious 'explanations' attractive in the way I have described, and to take this into account when deciding whether a rational (i.e. reasoning), non-religious, approach to these issues is more likely to lead to correct answers than a religious approach. (Where there are no clear answers, the rational approach may acknowledge this rather than supply incorrect answers.)

Such a reasoning approach is likely to be less comfortable than the 'explanations' offered by religions. But our brains are also interested in finding out the best solution to problems, which frequently involves parsimony (Section 7.11), and this may be where we will be led.

I believe that this approach to religion, as well as to many of the other issues in this book, is new in that it seeks to understand these issues by understanding some of the difficult computational problems faced by the brain, and the types of solution that the brain has found to these computational problems. These problems include detecting causality, and using gene-specified goals to direct behaviour and pleasure.

This makes the approach described in this book different from evolutionary psychology (Buss 2008, Barrett, Dunbar and Lycett 2002) (or sociobiology (Wilson 1975, Dawkins 1976)), which considers behavioural adaptations shaped by evolutionary pressure, but has not taken the brain mechanisms involved into account in a direct way. Understanding the brain mechanisms involved in these types of process helps us to understand reward processes and thereby to address aesthetics, ethics, and economics. Understanding the brain mechanisms involved in detecting causality helps one to address religion. Understanding the brain processing at the mechanistic level also provides a way to address how alterations in brain systems may lead to behavioural dysfunctions such as those described in Chapter 10, which may then, in the light of our understanding of how they are implemented in the brain, become more easily treatable.

11.2 Consciousness, the soul, everlasting life, immortality, and the fear of death

The adaptive value of a reasoning (rational) system is that it enables us to defer immediate (typically gene-specified) rewards, and to think many steps ahead to select a longer-term plan that may be more advantageous (Chapter 5). However, a side effect of being able to think ahead is that we may start to fear death, as we can see that it is irreversible. There are also uncertainties: will our stream of consciousness stop? Many religions offer hope and promises here, which are rewarding to the human brain in that they may reduce uncertainty, promise everlasting life, hold out the hope that we will see our loved ones again, etc. This is all reinforcing, and contributes to the attractiveness of some religions. In this sense, the human mind, by its ability to think ahead, and to reduce uncertainty where possible, is adapted by evolution in a way that makes religious 'explanation' and 'life after death' attractive to the human mind.

Nicholas Humphrey has suggested that the state of being conscious and having thoughts and feelings, and a sense of the continuity of the self over time, predisposes us to want that state to continue (Humphrey 2011). We value the sense of continuity, and want our thoughts and plans to continue. In a sense, this has to be a feature built into consciousness by evolution, for this then will encourage the individual to take actions to defend continuation of those feelings and of the self, and in the end that is likely to be good for the genes. However, once consciousness with these properties has evolved, the individual will not want those states to end at death[36]. This may lead to hopes and beliefs that those conscious states will continue after death, and that we may be able to continue our interactions with our loved ones after death. Humphrey (2011) uses the word 'soul' to refer to such concepts.

This could lead to rituals at death that might help survival in the following world, such as burying treasures and things often necessary during life with the dead, as in Neolithic and Egyptian burial sites. These beliefs and rituals might be pre-religious, and religions might build on these properties of the conscious mind to promise continuing mental states after death, provided perhaps that one did something of value to religious institutions, such as donating tithes (tenths of one's income) to the religious institution, or being promised salvation if one obeys the rules of the sect (as in the Ten Commandments) that might promote the survival of the sect.

---

[36]I note that once consciousness including feelings of the continuity of the self have evolved, wanting it to continue, even liking it, has to be built in to it, for otherwise it might be associated with suicide, loss of those genes from the population, and thus selection against the brain mechanisms necessary for consciousness.

Beliefs in or promises of everlasting life and immortality probably have advantages, as they may promote positive thinking and interactions rather than depression and withdrawal, and positive interactions can be advantageous for psychological well-being (Lee Duckworth, Steen and Seligman 2005).

Such beliefs may also serve the society or tribe, and help it to succeed in a possible competition against other tribes or societies (Section 4.3). Indeed, religious beliefs may help individuals to be brave in the face of death, and the brains that are adapted to support such beliefs might be selected for if such actions promote the survival of that society, and thus of the genes of the individuals with those characteristics within those societies. (The genes that promote such minds would be present in the offspring at home of those who might offer their lives in this way, and in the women who choose such brave men as protectors of their offspring.) An unfortunate consequence of such minds though is that such beliefs may lead towards not only tribalism, but to war (Section 4.3), with much suffering.

My point here is that by a rational, scientific, understanding of these adaptations of the human brain (in the area of 'neuro-religion'), we may be better able to channel the effects of such minds towards mutual support in a positive thinking society with altruism, and to guard against the dangers that could arise if tribalism leads to conflict between societies, and between religions.

It is interesting that the Greek word for soul, psyche, has been adopted for the modern science of psychology ('the study of the soul'), which in its extended form of cognitive and computational neuroscience, might now be thought of as the science of the soul. And the whole study of the mind, including consciousness, and its relation to brain function, can now perhaps be thought of as providing us with better, and quite precise, understanding of how at least a scientist might use the term soul in future: to refer to the conscious feeling of self and its continuity throughout life that are implemented by brain mechanisms that we are starting to understand.

However, the more we understand about the very close relation between the mental states that we experience, and brain function and brain computation (Rolls 2005a, Rolls 2008c, Rolls and Deco 2010), the more clear one has to be that there is no prospect of our mental states surviving the cessation of function of our brain at death. There may of course be a different type of continuity of the self after death, including the effects of one's good deeds and interactions with people, discoveries, contributions to science and art, etc. that will continue after one's death.

11.3 Power, status, and wealth of the clergy

Given that humans have been selected for competition between individuals within a society to accumulate wealth, power, and resources, as this has consequences for the survival of genes (Section 4.3), there is a vested (bevested) interest (probably usually unconscious) in the clergy in the accumulation of wealth, power, and resources, and this is a feature of many religions. For example, medieval monasteries, and the Church in England, accumulated wealth by tithes. In a sense too, the opportunity to forgive sins (or not) provides clergy with a type of power and influence. The fact that some of the clergy may be celibate does not contradict this argument: the clergy have the same genes that serve in the general population to promote wealth, power, and resources. And of course the interests of the phenotype might be relevant too (Section 5.4). An English rectory is not such a bad place in which to live.

Of course, the accumulation of wealth may also be for the general good, as when it is used to support individuals within society in behaviours that may be shaped in part by brain adaptations for reciprocal altruism (Section 4.2.2), kin altruism (Section 4.2.1), and stakeholder altruism (Section 4.2.3). Such charity may be a useful force within society (and between societies) for stability of the society. One sense in which this applies is that if the differentials of the wealth between individuals, or between societies, are too great, this may promote envy and attempts to gain more of the wealth, behaviours that are adaptive for genes and have been selected for. The implication is that some equalization of wealth across societies may reduce the tendency for strife. Religions that preach that worldly wealth should be discarded may be doing this in a context in which individual wealth and luxury can be a disruptive force in maintaining a stable and safe society. The latter is likely to be attractive to the majority because of the biological underpinnings to interest and value. In addition, preachers may understand that wealth does not always lead to happiness, in which many other factors, including good social relations in the family and at work, are frequently important (Argyle 1987), again for clear reasons related to neurosociality (Chapter 4).

The accumulation of power and wealth within a particular religion or sect might also be promoted by doctrines such as a prohibition of or limitations on birth control.

Doctrines that promote stable marriages may be seen as good for the stability of the society.

Doctrines such as care for one's children are simply consistent with gene-specified goals.

11.4 Rules of social behaviour to help a society thrive and survive

Behaviours produced by the brain adaptations that promote reciprocal altruism (Section 4.2.2), kin altruism (Section 4.2.1), and stakeholder altruism (Section 4.2.3) are frequently emphasized as important in religions. An effect of them is to help keep the religious society stable, which will be for the good of its leaders, and frequently for the reproductive success of the individuals (in fact, their genes) in that religious society. The Ten Commandments of God carried down the mountain by Moses can be seen as a set of such rules. The Freemasons emphasise altruism, which is at least partly reciprocal (within the group).

This aspect of religions and related organizations to stabilize communities by altruism including mutual help can be useful in supporting a community that helps those within it. (In the end, such behaviour will be selected for if it leads to reproductive success, to fitness of the genes in the individuals in that society.)

My suggestion as a reasoning scientist is that these altruistic forces within society that are promoted by religions can be very helpful in producing cooperation and stability, and that this aspect of religions can be very useful.

11.5 Tribalism and religion

The same forces of reciprocal altruism (Section 4.2.2), kin altruism (Section 4.2.1), and stakeholder altruism (Section 4.2.3), when coupled with the natural competition for resources, power, and wealth (Section 4.3), can lead to tribalism (Section 4.3). We can see different religions, and different sects within a religion, as tapping into these adaptations of the brain that can lead to tribalism. Where the tribalism results in mutual support for the individuals within the tribe, this can be advantageous.

The label of being within a particular religion or sect may help reciprocal altruism to extend powerfully, beyond the strictly tit-for-tat reciprocation between individuals, to support for members of the same tribe, in this case a religion or religious sect, and in this way reciprocal altruism may become a much more powerful force than it is when operating strictly between identifiable individuals (Section 4.2.2).

But where the tribalism within a society leads it into conflict with other tribes or societies, this can lead to battles, war, and suffering. Where this is coupled with religious beliefs of everlasting life or immortality if one dies in the cause, this can be a dangerous combination that can support strife or war.

Religion can be a form of tribalism: we naturally want to belong to a tribe, and exclusion may have been life-threatening, at least in the past when our genes evolved (Section 4.3).

Different religions, sects, and denominations may be seen as different attempts to make a stable grouping or tribe. The smaller sects may allow the tribalism to operate more effectively, as the individuals are more likely to become known personally to each other, which may facilitate the reciprocation, as well as catering for individual, perhaps genetically influenced, dispositions.

11.6   Fundamentalism and rationality

If one believes that one religion is true and correct, and the others not, for no rational reason (but just for a god-given 'reason', perhaps something within one's culture, perhaps a tribal belief that helps to distinguish one from other tribes), this can be a dangerous form of tribalism, and can lead to religious wars, or to wars supported by religions. Such wars can be used to support particular 'tribal' groups, which may have other vested interests. Such tribal groups may align themselves with a religion, and, as we have seen earlier, religious beliefs (of for example immortality) can help to support tribal warfare.

An issue is that with the predisposition of the human mind for religious beliefs and societies, which I have argued for above, how is one to argue rationally which, if any, is the one true religion? It may be safer to argue, with scientific understanding, that given that there are many different religions, all adapted to the ways that our minds have evolved, it may be wiser to encourage different religions to respect each other, and not try to dominate the others as a result of the human tendency for wealth, power, and resources. It may be wiser, and more rational, to promote the good that can come out of religions, and help religions to benefit and support those within the same, and in different, societies, through an understanding of the forces that evolution has incorporated into our brains that can help societies, or on the other hand can lead to dangerous competition between societies.

Although many individuals have fundamental beliefs that may not be fully rational, they may believe them earnestly, and may because of the supportive, altruistic, aspects of religions, be encouraged to believe in supporting others. I have a great respect for the good that can come of this, and indeed my father Eric Fergus Rolls was a Methodist missionary in India who strived to do his best for education in India.

11.7   Advantages and dangers of religions

We have seen that there are a number of ways in which religions fit the human brain, and have adapted themselves to it. Altruism, including reciprocal altruism, within a 'tribe' can be useful.

But fundamentalism can be dangerous. It may be wiser, and safer, to see rationally, by reasoning, that there are a number of reasons for the existence of religions; that there may be no provable evidence that any one is the true religion; and that therefore mutual respect of the different religions may be advantageous, with each one focusing on altruism within its group or tribe, and each operating within the bounds of international law, which itself is a social contract set up by those who agree to it (the international law), and will agree to defend those laws, and prosecute those who break those international laws (see Section 12.3).

We should not necessarily criticize religion, particularly the socially supportive part, but should seek to benefit from it where it can help social stability, and guard against its possible dangers where it might support tribalism that leads to conflict.

11.8   Extra-sensory perception

Some people believe in extrasensory perception. Extrasensory perception refers to reception of information not gained through the recognized senses and not inferred from experience, and includes psychic abilities such as telepathy, clairvoyance, and precognition (the feeling that one can predict in advance what will occur). Such beliefs may be supported by suspicious coincidences, in that something presaged turns out to occur. But such confirmed expectations are more memorable than a large number of expectations that are not confirmed, and because of the adaptive value of noticing possible correlations in the world, we may be insufficiently sensitive to disconfirming evidence. Bayes' theorem provides a rigorous way to deal with the probabilities in such circumstances (see Section A.2), but our brains do not naturally compute according to Bayesian principles (Chapter 2), and we are influenced much more by heuristics, useful short-cut principles, which may though not be optimal. And this appears to be the case with extrasensory perception, which has not under controlled conditions been shown to produce results more frequently than would be expected by chance (Humphrey 1995).

So our brains, because of an adaptation to detect causality by suspicious coincidences, have a tendency to believe in extrasensory perception, but no data support this ability. The Greek oracles, including the one at Delphi, lived by making predictions, but it is of interest that frequently the applicant was made to wait for days for the prophecy, during which time the soothsayers had time to evaluate the applicant, and make an informed guess about the future.

Many other supposedly paranormal phenomena do not stand up to critical examination (Humphrey 1995). Astrology is in a similar category. But our brains, because of the way they assess causality and make predictions (Section 11.1), are prone to making errors of this apparently psychic type.

11.9   Religious experiences and brain activity

During psychotic states such as schizophrenia, delusions and hallucinations are common, and are among the positive symptoms of schizophrenia (Rolls et al. 2008d, Rolls and Deco 2011, Rolls 2011f). The delusions might be that one is for example Elvis Presley, Napoleon Bonaparte, Jesus, or another religious person. The hallucinations may include hearing voices, and the voice might be thought to be the voice of God. There is also sometimes a feeling that one is being controlled by another person, perhaps by God, and one may feel that one does not have free will.

What is happening here? As described in Section 10.1, in schizophrenia there may be instability of representations in the temporal lobe of the brain, which may result in representations of for example voices becoming active when there is no recall cue for those representations, and the representations should be in the quiescent state, with only a low (spontaneous) firing rate of the neurons. This stochastic dynamical instability (Rolls and Deco 2010) is a mechanism for the generation of hallucinations, and if the voice is telling one to do something, or describing some important truths, this might be important in religious experience.

Similarly, if instead of one's decision mechanisms remaining stable until there is sufficient evidence for a decision, they were unstable, and jumped to conclusions noisily without evidence, then one might feel that one is not in control, and that the decisions are being made somehow by another system or process, or, in a personified attribution, by another person. One such 'person' might be God, telling one what to do.

Similarly, due to the instability in the temporal lobe systems, the mechanisms for which are described in Section 10.1, the thought processes may be somewhat bizarre, jumping noisily from one thought to another loosely associated thought. This could, if repeated and resulting in consolidation in memory, lead to persistent delusions.

Such thought processes might play a role in some religious experiences. It is important to realize that such thought processes are not restricted to those with schizophrenia, but that the genes that lead to schizophrenia at one end of the distribution in the population will be present to some extent in some individuals who are towards but not right at that end of the spectrum. Such genes in smaller quantities may be present in relatives of schizophrenics, and may be being selected for because in moderation they help with lateral thinking, with easily associating one idea with another, and seeing loose associations of ideas that may be beneficial in creativity, in seeing parallels, and in generating and understanding metaphors. Indeed, it is of interest that James Joyce, the author of *Finnegan's Wake*, which abounds in loose or free and very creative associations, had a daughter who was diagnosed as schizophrenic.

Thus our understanding of the brain bases of mental processes (Rolls 2005a, Rolls 2008c), and the neural mechanisms that influence their stability and also contribute to creativity (Rolls and Deco 2010, Rolls, Loh, Deco and Winterer 2008d, Rolls 2011f) (Chapters 2 and 10), may provide some insight into spiritual and religious experiences, and into religions.

12 Neuropolitics

12.1 Introduction

Can an understanding of the forces of conflict between the emotional and the rational within humans lead to useful strategies for solving political problems? The rational approach might be to understand better how humans operate, and build societies that utilize this rationality well.

12.2 The neurobiological forces and interests that influence political behaviour

12.2.1 Altruism

There are neurobiological forces that can promote altruism because altruism can be to the advantage of an individual's genes, and they may operate unconsciously. The mechanisms that promote kin altruism, reciprocal altruism, and stakeholder altruism are described in Section 4.2.

In addition, the reasoning system (Chapter 5) might produce behaviour that is altruistic with no advantage to the individual, or the individual's genes. However, even in these cases the altruistic behaviour might be shaped by neurobiological factors that encourage membership of a group and sociality. In addition, unless the altruism is anonymous, one's reputation may be enhanced through processes such as the attractiveness of wealth being made evident by conspicuous consumption.

12.2.2 Competition

In addition to the altruistic processes that promote cooperation, we have considered the neurobiological forces that promote competition within and between societies, and how together these can lead to tribalism (Section 4.3). Together, these processes provide a neurobiological foundation for helping to understand some of the important forces that influence political behaviour and politics, as described next.

12.3 A rational politics

When we see the biological origin of these forces operating within society, and how our brains are adapted to promote, probably largely unconsciously through in-built heuristics, such behaviour (though often probably using conscious strategies), we may ask what solutions might be appropriate in this situation.

As a neuroscientist, one recalls that the more recently evolved system that can help with long-term strategies to problem-solving is the reasoning or rational system, which can enable humans to defer or forgo gene-identified immediate rewards, and plan for a long-term strategy (Chapter 5). The application of this reasoning might be to lead us towards a social contract, in which the individuals within societies would agree to certain rules or laws of society (Section 9.3). The rules or laws would take into account the biological predispositions of the individuals within the society, what each individual finds important for evolutionary reasons as rewards or punishers, for then the rules or laws are more likely to be accepted by individuals to produce a stable society. The rules and laws would take into account the biological predisposition for kin and reciprocal altruism. But the rules or laws would also (given that there are biological forces promoting strong competition between individuals and between groups) include systems for regulating the competition, paying particular attention to the human heuristic of being very sensitive to loss (Section 8.3), so that actions that cause loss to individuals need to be carefully regulated. Such loss could include loss of possessions, life, land, food, water, etc. At the same time, such laws and rules could take into account that the impact of rewards tails off as their magnitude increases, so that not much more value might be placed on enormous than on great wealth (Section 8.3). The rules and laws would also benefit by taking into account reward and punishment discounting with time in the future (Section 8.4.2), noting that this differs between individuals.

Such a rationally organized social contract taking into account how humans are built neurobiologically would be in contrast to systems of absolute rights; and to god-given divine rules that may differ from society to society that rationality may not be able to justify. The Ten Commandments brought from God by Moses provided a useful prescription for rules that might help to produce a stable society, but different systems of divine rules and beliefs within different religions may be impossible to reconcile if they are held to represent fundamental god-given truths.

The rational social contract politics would be different because there would be a rational basis for the rules and laws within a society. Those within a society would operate as having a particular social contract. There could of course be 'universal', that is widely agreed internationally, rights that would be given to individuals, and to societies (Section 9.3). Rogue individuals within a society would be punished or excluded by the society, and rogue societies that did not agree to live within rationally agreed international guidelines would be punished by sanctions or other controls or restrictions. The aim would be to build a system of societies in which rational argument, rather than absolute or fundamentalist principles, would be used to agree operating principles and national and international law.

12.4 Conclusions

Ideas of a social contract have been raised previously by a number of distinguished individuals (Hobbes 1651, Locke 1689, Hume 1741, Rousseau 1762, Rawls 1999). What I add (Rolls' approach) is the argument that by a scientific understanding of the underlying human neurobiology, and the forces that this generates within society, we may be better enabled to use that other great human adaptation, rational long-term decision-making by the reasoning system (Chapter 5), to agree to rules of society, a social contract (Section 9.3), that take these neurobiologically-based predispositions into account. and that build a society that can be justified by reasoning, and shaped and selected rationally by the individuals within the society.

13 Conclusions

13.1 Introduction

I have made a number of points in this book that I believe open up new ways of thinking about human behaviour, and are of fundamental importance. I draw together here some of the points made about important issues, to make sure that the implications of the points when taken together are clear. My aim throughout has been to be descriptive of the situation in terms of the science, presenting the evidence for you, the reader, to evaluate, rather than to be prescriptive. However, there are implications of many of the points made, including for example how to keep a discussion rational without using words that may obfuscate the argument by their emotional loading.

13.2 The emotional selfish gene versus the rational selfish individual

I have presented evidence in Chapter 3 that a very efficient way for genes to influence behaviour for their own selfish advantage is to specify rewards and punishers rather than responses or actions. I also presented evidence that emotional states are states produced when these (instrumental) reinforcers (i.e. goals for action) are received. Motivational states, why we want things, can be seen as states that arise when we are trying to obtain these gene-specified reinforcers or goals for action. This provides a fundamentally Darwinian account of why we have emotions, and motivational states (Rolls 1999a, Rolls 2005a, Rolls 2011b).

I have also presented evidence in Chapter 5 that a computational ability for reasoning (i.e. for rational thought) allows such gene-specified reinforcers to be deferred or even not chosen, and longer-term goals to be chosen, with behaviour directed over the long term by planning, and by maintaining the plan to obtain these goals. There may be a conflict between the goals defined by the gene-specified rewards, example of which are shown in Table 3.1, and the rational system, because the bases for the goals may be quite different. The goals specified by the genes are typically short-term goals, something to do now, such as eat or drink, though of course these short-term goals reflect what has been found over hundreds of thousands of years to have value for the genes, that is to promote their survival into the next generation. Inherent in this process is the requirement for such genes not to over-dominate others, for always eating, and never reproducing, would not be to the advantage of any of the genes, and in this sense the genes must if not cooperate, at least tolerate each other, competing on a basis so that each wins sometimes, and not individually becoming too rewarding. Mechanisms such as sensory-specific satiety are important aspects of this process.

The rational system may indicate that deferring many of these gene-specified goals might in the long term be advantageous, at least to the individual person, if not to the survival of the genes. For example, by deferring an immediate appetite, one might be able to invest the saved money, and accrue much larger gains in say 6 months. And indeed, the interests of the individual (the phenotype, the body of the individual, the living person, the 'phene') may not be the same as those of the genes (Section 5.4 on page 205). One example is that the rational system may itself have its own goals, such as feeling pleasure when difficult problems are solved, for practice in solving difficult problems may in the end turn out to be useful for the individual. It may of course be useful for the genes too, and this is doubtless part of what has been driving the evolution of the rational system.

This point is brought home clearly by the fact that of the great apes, four of those still extant are close to extinction (two species of chimpanzee, gorillas, and orang-otangs), and one, humans, is spreading over the planet, and multiplying, in a remarkably successful way. What is the difference between these great apes? The difference, it is suggested, is the great development of the rational (i.e. reasoning) system in humans.

This encourages us to look precisely at what it is that is crucial for reasoning. It is not just the use of symbols, signs for objects and events in the world. Simple associative learning provides one important step in which arbitrary stimuli can become associated with at least reinforcers. But that is just part of the emotional system. Beyond this, chimpanzees have a good capability in the use of arbitrary symbols for objects and events which need not themselves be reinforcers (Terrace 1979, Savage-Rumbaugh, McDonald, Sevcik, Hopkins and Rupert 1986, Savage-Rumbaugh, Rumbaugh and McDonald 1985, Pinker 1994). Instead, the use of some form of syntax is the crucial component computationally. This enables symbols to be linked to form for example conditional statements (if this ..., then that ...), and crucially, for multiple step plans to be developed, and compared with each other. The syntax is essential for keeping the symbols, and how they are related to each other, to be specified, and for this specification to be kept separate for each of the steps of a multiple step plan.

Without this syntax, all the neuronal firing for each of the symbols would be active simultaneously, and there would be a tower of Babel, or babble. Indeed, babble by babies may represent a stage in language development where some use of symbols is present, but syntax is still very underdeveloped. (Exactly how syntax is implemented computationally in the brain is perhaps the great remaining mystery in computational neuroscience, but part of its solution may be the ability to have local cortical attractor networks where the anatomical locations of the local attractor established at least in part the syntactic role played by the representation in a local cortical attractor (Rolls 2008c).)

Given that humans have this reasoning system, how it interacts with the emotional system becomes crucial. A fascinating point is why the rational system does not totally dominate the emotional system. Why does the emotional system remain so potent in rational man? One reason I suggest is that the rational system may not necessarily promote what is good for the genes. One example might be that everlasting life might be attractive to the reasoning individual, who could develop artistic and scientific understanding in the same continuing self. But this would be disastrous for the genes, which compete with each other by the evolutionary process of recombination, occasional mutation, and competition, which would then fail. Our genes and emotions set us up to reproduce, and if a rational process because too strong to override this, it would become weakened in the gene population. I argue that a subtle balance in evolution between the rational and the emotional systems is likely to continue, with a compromise situation in which neither can dominate the other, in a similar way to that in which other genes specifying different goals must remain in balance with each other.

Of course in this situation, because of natural variation that is a requirement for such evolutionary processes, some individuals will be more at the rational end, and others more at the emotional end, of the spectrum. Could it be that women on average tend to be more at the emotional end of the spectrum because they must for long periods attend to the calls of their genes to reproduce with a good mate, and then to care for long periods for their young, in order to ensure survival of their genes, whereas men with occasional emotional moments might maintain their reproductive success? (I have deliberately overstated the case here, but perhaps it will help to emphasize what may be an important point in understanding emotionality in women and men, of course on average.)

In this context, it is interesting to that it is often said that women are better at multitasking than men. I suppose the converse point might also be put: that men are on average better able to focus and pay maintained attention to a task. Is there any scientific basis to this? It seems unlikely that the short-term memory and attentional systems in the prefrontal cortex are inherently different in men and women. But the task demands, the demands of the world, may be different, and women may be especially sensitive to the multiple needs of their offspring, and especially responsive to the cry of a baby, because it is important for the interests of their genes, which are to survive well into the next generation. To help that survival of their children, women have an especially strong interest and investment in their children that occupies them over the long term.

Another interesting issue to draw out here about the relation between the emotional and rational systems is that there must on many occasions be a choice between them, and that this choice will be influenced by noise (statistical fluctuations) in the brain, as well as by noise and variations in the environment. People may therefore make different decisions on different occasions, even given apparently similar evidence (Section 2.12). Because the influence of the noise can operate unconsciously, there may be no real external cause to which to attribute the basis for the decision. But the human mind, perhaps in order to maintain its sense of self, and that it is in control, may confabulate and defend a reason for a decision (see Chapter 5). This is a property of the human mind for which allowance may need to be made. For crucial decisions, monitoring, and multiple checks, with rational explanations and justifications that are carefully evaluated in the light of these points, may be necessary.

Another interesting issue is how much weight we place on the emotional and rational systems in different circumstances. As we saw in Chapter 7, Plato dismissed poetry as obfuscating, by sending the mind reeling into hypnotic trances instead of focusing on rational deductions and argument (Herwitz 2008). For rational argument and scientific work, that seems right. But that does not mean that we always have to be dominated by the rational? As we have seen, much of what activates our pleasure systems originates in gene-specified rewards, including for example bringing up children, and leaving our resources to them, as this is for the good of our genes. And listening to wonderfully emotional music such as Wagner's *Wesendonck Lieder* (or *Tristan* itself) is a reward that can be much appreciated partly because it taps into our emotional system. The same can be said of romantic poetry, including Byron, Keats, and Shelley. And here, we should remember that our rational system can enhance the pleasure that we obtain from our emotional system, by leading us towards situations that lead to high emotion, and by a top-down cognitive influence that can descend into the emotional system from even the word level to increase the magnitude of the affective representation in the first part of the human cortex in which affect is represented, the orbitofrontal cortex (De Araujo, Rolls, Velazco, Margot and Cayeux 2005, McCabe, Rolls, Bilderbeck and McGlone 2008, Grabenhorst, Rolls and Bilderbeck 2008a) (see Chapter 3). But even when doing this, we can maintain a rational monitoring of events, to ensure that the emotional system does not take decisions that might be strongly disadvantageous to the rational system[37].

Continuing with this issue, we have to understand that rational argument and discussion could be impeded by emotionally loaded words, which could by the top-down effects of cognition on emotional systems just described activate the emotional system, which might, in a way that most people may not be sensitive to, contribute in the ways described in this book to the decision that is reached on an issue. This is an issue that is being addressed in the context of animal welfare by M. Stamp Dawkins (Dawkins 2012).

As described in Chapter 9, this dominance of the rational over the emotional is an important principle of responsibility in law. The relative balance between the rational and the emotional may be influenced not only by internal noise in the brain, but also by drugs such as alcohol, and by damage to parts of the brain such as the orbitofrontal cortex. These issues need careful understanding in law. In the case of alcohol, one can argue that an individual knows in advance the possible effects of alcohol on behaviour. In the case of brain damage produced for example by a road traffic accident or tumour, the issue of culpability may be much more intricate, and very careful assessment may be needed in the light of modern neuroscience.

An underlying brain principle relevant to many of these issues is that in the evolution of the brain, the main trend seems to be to add new cortical areas, and new brain systems, on top of old ones. The multiple hierarchically organized connections between corresponding areas in the sensory areas and the prefrontal cortex (Jones and Powell 1970) can be seen as an illustration of this evolutionary history. A consequence of this is that in any situation, multiple brain systems, including not only the emotional system in the orbitofrontal cortex but also the older one in the amygdala (Rolls 2005a) are engaged, as are the newer systems in humans involved in reasoning, in rational thought. A consequence is that, given the different computational capabilities of these different systems, we should not necessarily expect a unified output of behaviour. Instead, different aspects of our behaviour may be being produced by different brain systems, and their outputs need not necessarily be expected to be fully consistent and coordinated. Moreover, in these circumstances, we may be conscious of some aspects of our decision-making and behaviour, but not of other aspects (see

---

[37]The worry is often expressed that a composer such as Wagner can be appreciated for his music, but not for everything he espoused, including some of his political views. But we should remember that every individual, or work of art, taps into a high dimensional space of what we find attractive or unpleasant, with the dimensions defined by our genes and by our rational system. So we should not be surprised when some aspects of a person, or work of art are outstandingly good, but others may be poor: these inconsistencies between different dimensions of our evaluation system are what we should expect, as they are produced by natural variation influencing whatever we are assessing, and also our own valuation system.

Chapter 5). Indeed, as argued in Chapter 6, consciousness may be especially associated with the operation of the reasoning system that can reflect on its own operations. But the overriding point is that our behaviour is the result of multiple brain systems, each designed with somewhat different evolutionary goals, and each implemented in a different way, and is therefore likely to be complex, but in ways that we are now starting to understand scientifically and computationally.

This is one reason that makes humans so interesting; and at the same time makes them so important to understand, so that we can live in a society where our different attributes can be helped to flourish, but with harmony maintained between individuals and society, each with their own driving factors.

13.3   A mechanistic approach to brain function

The theme of this book is that by understanding how the brain computes, its mechanisms, we better understand human and animal behaviour. By understanding the computations performed by neurons and neuronal networks, and the effects of noise in the brain on these, we are gaining a true understanding of the mechanisms that underlie brain function and behaviour. This understanding extends very deep, to the statistical mechanics of networks operating with noise (Rolls and Deco 2010). In this sense, we are developing a truly wonderful 'mechanics of the mind' which is enabling how our brains work to be understood from the level of molecules through neurons with ion channels, through networks of such neurons to the global properties of a system, and thus to an understanding of how processes such as memory, perception, attention, decision-making, and emotion actually are implemented in the brain (Rolls 2008c, Rolls and Deco 2010). Because the operation of the mind is being understood in terms of the computational mechanisms that implement it, the approach goes far beyond the phenomenological approaches criticized by others (Legrenzi et al. 2011).

This mechanistic approach to the mind, brain function, and behaviour can be described as reductionist. However, it in no way removes the wonder of what brain function achieves. Indeed, what brain function achieves can be complemented and made more effective by an understanding of how the brain works, as described in this book.

Now that we are making such great progress in this understanding of the brain and mind, for reasons outlined in Section 2.16, there are many implications for how we operate in our culture, in our society. Some of these implications have been brought out in this book, including for emotion, aesthetics, ethics, politics, religion, and psychiatry. We have a rich new way of understanding ourselves, and our culture: *Neuroculture*.

At the same time, there is a great area of brain function that we do not yet understand even in principle at the mechanistic level of neurons and networks: how language is implemented in the brain. That is a great and interesting problem.

Appendix 1

This Appendix provides summaries of some concepts that it is useful to have set out clearly, to help make the arguments made elsewhere in this book precise, yet without distracting the reader from the arguments in the body of the book. The Appendix will be useful to readers wishing to pursue some of the arguments and concepts further.

A.1 Rewards and punishers, and learning about rewards and punishers: instrumental learning and stimulus–reinforcer association learning

In Chapter 3, on neuroaffect, the concepts of reward, punisher, primary reinforcer, secondary reinforcers, instrumental learning, and classical conditioning are important. The present section provides a concise summary of these concepts (Rolls 2005a).

Some stimuli are innately rewarding or punishing and are called primary reinforcers (for example no learning is necessary to respond to pain as aversive), while other stimuli are learned or secondary reinforcers (for example the sight of a chocolate cake is not innately rewarding, but may become a learned reinforcer, for which we may work, by the process of association learning between the sight of the cake and its taste, where the taste is a primary reward or reinforcer). This type of learning, which is important in emotion and motivation, is called stimulus–reinforcement association learning. (A better term is stimulus–reinforcer association learning, where reinforcer is being used to mean a stimulus that might be a reward or a punisher.)

A reward is something for which an animal (including of course a human) will work. A punisher is something that an animal will work to escape or avoid (or that will decrease the probability of actions on which it is contingent). In order to exclude simple reflex-like behaviour, the concept invoked here by the term 'work' is to perform an arbitrary behaviour (and action called an operant response) in order to obtain the reward or avoid the punisher. An example of an operant response might be putting money into a vending machine to obtain food, or for a rat pressing a lever to obtain food. In these cases, the food is the reward. Another example of an operant response might be moving from one place to another in order to escape from or avoid an aversive (punishing) stimulus such as a cold draught. If the aversive stimulus starts and then the response is made, this is referred to as *escape* from the punisher. If a warning stimulus (such as a flashing light) indicates that the punisher will be delivered unless the operant response is made, then the animal may learn to perform the operant response when the warning stimulus is given in order to *avoid* the punisher.

Because the definitions of reward and punisher make it a requirement that it must be at least possible to demonstrate learning of an arbitrary operant response (made to obtain the reward or to escape from or avoid the punisher), we see that learning is implicit in the definition of reward and punisher. (Merely swimming up a chemical gradient towards a source of food as occurs in single cell organisms is called a taxis as described in Section 3.10; it does not require learning, and does not make the food qualify as a reward under the definition.) In that rewards and punishers do imply the ability to learn what to do to obtain the reward or escape from or avoid the punisher, we call rewards and punishers 'reinforcers'.

This introduction leads to the definition of **reinforcers** as stimuli that if their occurrence, termination, or omission is made contingent upon the making of a response, alter the probability of the future emission of that response (as a result of the contingency (i.e. dependency) on the response). The alteration of the probability of a response (or action) is the measure that learning has taken place. A positive reinforcer (such as food) increases the probability of emission of a response on which it is contingent; the process is termed **positive reinforcement**, and the outcome is a reward (such as food). A negative reinforcer (such as a painful stimulus) increases the probability of emission of a response which causes the negative reinforcer to be omitted (as in active avoidance) or terminated (as in escape), and the procedure is termed **negative reinforcement**. In contrast, **punishment** refers to procedures in which the probability of an action is decreased. Punishment thus describes procedures in which an action decreases in probability if it is followed by a painful stimulus, as in passive avoidance. Punishment can also be used to refer to a procedure involving the omission or termination of a reward ('extinction' and 'time out' respectively), both of which decrease the probability of responses (Gray 1975, Mackintosh 1983, Dickinson 1980, Lieberman 2000). My argument is that an affectively positive or 'appetitive' stimulus (which produces a state of pleasure) acts operationally as a **reward**, which when delivered acts instrumentally as a positive reinforcer, or when not delivered (omitted or terminated) acts to decrease the probability of responses on which it is contingent. Conversely I argue that an affectively negative or aversive stimulus (which produces an unpleasant state) acts operationally as a **punisher**, which when delivered acts instrumentally to decrease the probability of responses on which it is contingent, or when not delivered (escaped from or avoided) acts as a negative reinforcer in that it then increases the probability of the action on which its non-delivery is contingent[38].

Reinforcers, that is rewards or punishers, may be unlearned or **primary reinforcers**, or learned or secondary reinforcers. An example of a primary reinforcer is pain, which is innately a punisher. The first time a painful stimulus is ever delivered, it will be escaped from, and no learning that it is aversive is needed. Similarly, the first time a sweet taste is delivered, it can act as a positive reinforcer, so it is a primary positive reinforcer or reward. Other stimuli become reinforcing by learning, because of their association with primary reinforcers, thereby becoming '**secondary reinforcers**'. For example, a (previously neutral) sound that regularly precedes an electric shock can become a secondary reinforcer. Animals will learn operant responses reinforced by the secondary reinforcer, for example jumping to a place where the secondary reinforcer is not present or terminates. Secondary reinforcers are thus important in enabling animals to avoid primary punishers such as pain.

There is a close relation of all these processes to emotion, for as is described in Chapter 3, fear is an emotional state that might be produced by a sound that has previously been associated with an electric shock. Shock in this example is the primary punisher, and fear is the emotional state that occurs to the tone stimulus as a result of the learning of the stimulus (i.e. tone)–reinforcer (i.e. shock) association. Another example of a secondary reinforcer is a visual stimulus associated with the taste of a food. For example, the first time we see a new type of food we do not treat the sight of the new visual stimulus as reinforcing, but if the stimulus has a good taste, the sight of the object becomes a positive secondary reinforcer, and we may choose the food when we see it in future by virtue of its association with a primary reinforcer. This type of learning is thus called stimulus–reinforcer association learning. (The operation is often referred to as stimulus–reinforcement association learning.) This type of learning is very important in many emotions, because it is as a result of this type of learning that many previously neutral stimuli come to elicit emotional responses, as in the example of fear above.

Unconditioned reinforcing stimuli often elicit autonomic responses. (Autonomic responses are those mediated through the autonomic nervous system, via the vagus and sympathetic nerves, which affect smooth muscle.) Examples include alterations of heart rate and of blood pressure which might be produced by a painful stimulus; and salivation which might be produced by the taste of food. Many endocrine (hormonal) responses are also mediated through the autonomic nervous system and so are autonomic responses, for example

---

[38] Note that my definition of a punisher, which is similar to that of an aversive stimulus, is of a stimulus or event that can either decrease the probability of actions on which it is contingent, or increase the probability of actions on which its non-delivery is contingent. The term punishment is restricted to situations where the probability of an action is being decreased.

the release of adrenaline (epinephrine) from the adrenal gland during emotional excitement. Previously neutral stimuli, such as the sound in our previous example, can by pairing with unconditioned stimuli, such as shock in the previous example, come by learning the association, to produce learned autonomic responses. In the example the tone might by pairing with shock come to elicit a change in heart rate, and sweating. This type of learning is called **classical conditioning**, and also **Pavlovian conditioning** after Ivan Pavlov who performed many of the original studies of this type of learning, including learned salivation to the sound of a bell that predicted the taste of food. It is a type of learning that is very similar to stimulus–reinforcer association learning, except that in the case of classical conditioning the responses involved are autonomic and endocrine responses.

In the case of stimulus–reinforcer association learning, the effects of the learning are mediated through the skeletal motor system, in that actions are performed that are instrumental in enabling animals to obtain rewards or avoid punishers, and are described as voluntary in humans. A key difference between **instrumental learning** and classical conditioning apart from the response systems involved lies in the contingencies that operate. In classical conditioning the animal has no control over whether the unconditioned stimulus is delivered (as in the experiments of Pavlov just described). In contrast, the whole notion of instrumental learning is that what the animal does is instrumental in determining whether the reinforcer (the goal) is obtained, or escaped from or avoided. Both types of learning are important in emotions because (as described in Chapter 3) instrumental reinforcers produce emotional responses, but also typically produce autonomic responses that therefore typically occur during emotional states, and indeed mediate important effects of emotions such as preparing the body for action by increasing heart rate, etc.

A more detailed description of the nature of classical (Pavlovian) conditioning and instrumental learning, and how both are related to emotion, is provided by Rolls (2005a).

Motivation refers to the state an animal is in when it is willing to work for a reward or to escape from or avoid a punisher. So for example we say that an animal is motivated to work for the taste of food, and in this case the motivational state is called hunger. The definition of motivation thus implies the capacity to perform any, arbitrary, operant response in order to obtain the reward or escape from or avoid the punisher. By implying an operant response, we exclude simple behaviours such as reflexes and taxes (such as swimming up a chemical gradient), as described above and in Chapter 3. By implying learning of any response to obtain a reward (or avoid a punisher), motivation thus focuses on behaviours in which a goal is defined. Motivation is one of the states that are involved in the large area of brain design related to the fundamental issue of how goals for behaviour are defined, and an appropriate behaviour is selected, as described in this book and brought together into a theory in Chapter 3.

A.2 Bayes' theorem

Bayes' theorem, referred to in Section 11.8 on page 337, relates the conditional and prior probabilities of events A and B, where B has a non-vanishing probability:

$$P(A|B) = \frac{P(B|A)\,P(A)}{P(B)}. \tag{A.1}$$

Each term in Bayes' theorem has a conventional name:

$P(A)$ is the *prior probability* or marginal probability of A. It is 'prior' in the sense that it does not take into account any information about B.

$P(A|B)$ is the conditional probability of A, given B. It is also called the *posterior probability* because it is derived from or depends upon the specified value of B.

$P(B|A)$ is the conditional probability of B given A.

$P(B)$ is the prior or marginal probability of B, and acts as a normalizing constant.

References

Abbott, L. F., Rolls, E. T. and Tovee, M. J. (1996). Representational capacity of face coding in monkeys, *Cerebral Cortex* **6**: 498–505.

Abbott, L. F., Varela, J. A., Sen, K. and Nelson, S. B. (1997). Synaptic depression and cortical gain control, *Science* **275**: 220–224.

Ackley, D. H. (1987). *A Connectionist Machine for Genetic Hill-Climbing*, Kluwer Academic Publishers, Dordrecht.

Adolphs, R. (2003). Cognitive neuroscience of human social behavior, *Nature Reviews Neuroscience* **4**: 165–178.

Adolphs, R., Tranel, D. and Denburg, N. (2000). Impaired emotional declarative memory following unilateral amygdala damage, *Learning and Memory* **7**: 180–186.

Adolphs, R., Gosselin, F., Buchanan, T. W., Tranel, D., Schyns, P. and Damasio, A. R. (2005). A mechanism for impaired fear recognition after amygdala damage, *Nature* **433**: 68–72.

Adrian, E. D. (1928). *The Basis of Sensations*, Christophers, London.

Aggelopoulos, N. C., Franco, L. and Rolls, E. T. (2005). Object perception in natural scenes: encoding by inferior temporal cortex simultaneously recorded neurons, *Journal of Neurophysiology* **93**: 1342–1357.

Aigner, T. G., Mitchell, S. J., Aggleton, J. P., DeLong, M. R., Struble, R. G., Price, D. L., Wenk, G. L., Pettigrew, K. D. and Mishkin, M. (1991). Transient impairment of recognition memory following ibotenic acid lesions of the basal forebrain in macaques, *Experimental Brain Research* **86**: 18–26.

Ainslie, G. (1992). *Picoeconomics*, Cambridge University Press, Cambridge.

Albantakis, L. and Deco, G. (2009). The encoding of alternatives in multiple-choice decision making, *Proceedings of the National Academy of Sciences USA* **106**: 10308–10313.

Aleman, A. and Kahn, R. S. (2005). Strange feelings: do amygdala abnormalities dysregulate the emotional brain in schizophrenia?, *Progress in Neurobiology* **77**(5): 283–298.

Alexander, R. D. (1975). The search for a general theory of behavior, *Behavioral Sciences* **20**: 77–100.

Alexander, R. D. (1979). *Darwinism and Human Affairs*, University of Washington Press, Seattle.

Allport, A. (1988). What concept of consciousness?, *in* A. J. Marcel and E. Bisiach (eds), *Consciousness in Contemporary Science*, Oxford University Press, Oxford, pp. 159–182.

Amaral, D. G. and Witter, M. P. (1989). The three-dimensional organization of the hippocampal formation: a review of anatomical data, *Neuroscience* **31**: 571–591.

Amaral, D. G. and Witter, M. P. (1995). The hippocampal formation, *in* G. Paxinos (ed.), *The Rat Nervous System*, Academic Press, San Diego, pp. 443–493.

Amaral, D. G., Ishizuka, N. and Claiborne, B. (1990). Neurons, numbers and the hippocampal network, *Progress in Brain Research* **83**: 1–11.

Amit, D. J. (1989). *Modelling Brain Function*, Cambridge University Press, New York.

Amsel, A. (1958). The role of frustrative non-reward in non-continuous reward situations, *Psychological Bulletin* **55**: 102–119.

Amsel, A. (1962). Frustrative non-reward in partial reinforcement and discrimination learning: some recent history and a theoretical extension, *Psychological Review* **69**: 306–328.

Anderson, J. R. (1996). ACT: a simple theory of complex cognition, *American Psychologist* **51**: 355–365.

Andersson, M. (1994). *Sexual Selection*, Princeton University Press, Princeton.

Andersson, M. and Simmons, L. W. (2006). Sexual selection and mate choice, *Trends in Ecology and Evolution* **21**: 296–302.

Angeletos, G.-M., Laibson, D., Repetto, A., Tobacman, J. and Weinberg, S. (2001). The hyperbolic buffer stock model: calibration, simulation, and empirical evaluation, *Journal of Economic Perspectives* **15**: 47–68.

Antoniadis, E. A., Winslow, J. T., Davis, M. and Amaral, D. G. (2009). The nonhuman primate amygdala is necessary for the acquisition but not the retention of fear-potentiated startle, *Biological Psychiatry* **65**: 241–248.

Appleton, J. (1975). *The Experience of Landscape*, Wiley, New York.

Argyle, M. (1987). *The Psychology of Happiness*, Methuen, London.

Armstrong, D. M. and Malcolm, M. (1984). *Consciousness and Causality*, Blackwell, Oxford.

Arnold, P. D., Rosenberg, D. R., Mundo, E., Tharmalingam, S., Kennedy, J. L. and Richter, M. A. (2004). Association of a glutamate (NMDA) subunit receptor gene (GRIN2B) with obsessive-compulsive disorder: a preliminary study, *Psychopharmacology (Berl)* **174**: 530–538.

Arnsten, A. F. and Li, B. M. (2005). Neurobiology of executive functions: catecholamine influences on prefrontal cortical functions, *Biological Psychiatry* **57**: 1377–1384.

Baars, B. J. (1988). *A Cognitive Theory of Consciousness*, Cambridge University Press, New York.

Baddeley, A. (1986). *Working Memory*, Oxford University Press, New York.

Baddeley, A. D. (2002). Is working memory still working?, *European Psychologist* **7**: 85–97.

Baddeley, R. J., Abbott, L. F., Booth, M. J. A., Sengpiel, F., Freeman, T., Wakeman, E. A. and Rolls, E. T. (1997). Responses of neurons in primary and inferior temporal visual cortices to natural scenes, *Proceedings of the Royal Society B* **264**: 1775–1783.

Baker, R. R. (1996). *Sperm Wars*, Fourth Estate, London.

Baker, R. R. and Bellis, M. A. (1993). Human sperm competition: ejaculate manipulation by females and a function for the female orgasm, *Animal Behaviour* **46**: 887–909.

Baker, R. R. and Bellis, M. A. (1995). *Human Sperm Competition: Copulation, Competition and Infidelity*, Chapman and Hall, London.

Bar-On, R. (1997). *The Emotional Intelligence Inventory (EQ-i): Technical Manual*, MultiHealth Systems, Toronto.

Barkow, J. H., Cosmides, L. and Tooby, J. (eds) (1992). *The Adapted Mind*, Oxford University Press, New York.

Barlow, H. (1995). The neuron doctrine in perception, *in* M. S. Gazzaniga (ed.), *The Cognitive Neurosciences*, MIT Press, Cambridge, MA, chapter 26, pp. 415–435.

Barlow, H. (1997). Single neurons, communal goals, and consciousness, *in* M. Ito, Y. Miyashita and E. T. Rolls (eds), *Cognition, Computation, and Consciousness*, Oxford University Press, Oxford, chapter 7, pp. 121–136.

Barlow, H. B. (1972). Single units and sensation: a neuron doctrine for perceptual psychology, *Perception* **1**: 371–394.

Barnes, C. A. (2003). Long-term potentiation and the ageing brain, *Philosophical Transactions of the Royal Society of London B* **358**: 765–772.

Baron-Cohen, S. (2008). *Autism and Asperger Syndrome: the Facts*, Oxford University Press, Oxford.

Barrett, L., Dunbar, R. and Lycett, J. (2002). *Human Evolutionary Psychology*, Palgrave, Basingstoke.

Barto, A. G. (1985). Learning by statistical cooperation of self-interested neuron-like computing elements, *Human Neurobiology* **4**: 229–256.

Bartus, R. T. (2000). On neurodegenerative diseases, models, and treatment strategies: lessons learned and lessons forgotten a generation following the cholinergic hypothesis, *Experimental Neurology* **163**: 495–529.

Bateson, P. (1983). *Mate Choice*, Cambridge University Press, Cambridge.

Bateson, P. and Gluckman, P. (2011). *Plasticity, Robustness, Development and Evolution*, Cambridge University Press, Cambridge.

Battaglia, F. and Treves, A. (1998). Stable and rapid recurrent processing in realistic autoassociative memories, *Neural Computation* **10**: 431–450.

Baumgarten, A. G. (1750). *Aesthetica*.

Baxter, R. D. and Liddle, P. F. (1998). Neuropsychological deficits associated with schizophrenic syndromes, *Schizophrenia Research* **30**: 239–249.

Baylis, G. C., Rolls, E. T. and Leonard, C. M. (1987). Functional subdivisions of temporal lobe neocortex, *Journal of Neuroscience* **7**: 330–342.

Baylis, L. L., Rolls, E. T. and Baylis, G. C. (1994). Afferent connections of the orbitofrontal cortex taste area of the primate, *Neuroscience* **64**: 801–812.

Bechara, A., Damasio, A. R., Damasio, H. and Anderson, S. W. (1994). Insensitivity to future consequences following damage to human prefrontal cortex, *Cognition* **50**: 7–15.

Bechara, A., Tranel, D., Damasio, H. and Damasio, A. R. (1996). Failure to respond autonomically to anticipated future outcomes following damage to prefrontal cortex, *Cerebral Cortex* **6**: 215–225.

Bechara, A., Damasio, H., Tranel, D. and Damasio, A. R. (1997). Deciding advantageously before knowing the advantageous strategy, *Science* **275**: 1293–1295.

Bechara, A., Damasio, H., Tranel, D. and Anderson, S. W. (1998). Dissociation of working memory from decision making within the human prefrontal cortex, *Journal of Neuroscience* **18**: 428–437.

Bechara, A., Damasio, H., Tranel, D. and Damasio, A. R. (2005). The Iowa Gambling Task and the somatic marker hypothesis: some questions and answers, *Trends in Cognitive Sciences* **9**: 159–162.

Ben-Ze'ev, A. (2000). *The Subtlety of Emotions*, MIT Press, Cambridge, MA.

Benabou, R. and Pycia, M. (2002). Dynamic inconsistency and self-control: a planner–doer interpretation, *Economics Letters* **77**: 419–424.

Berger, J. (1972). *Ways of Seeing*, Penguin, Harmondsworth, Essex.

Berlin, H. and Rolls, E. T. (2004). Time perception, impulsivity, emotionality, and personality in self-harming borderline personality disorder patients, *Journal of Personality Disorders* **18**: 358–378.

Berlin, H., Rolls, E. T. and Kischka, U. (2004). Impulsivity, time perception, emotion, and reinforcement sensitivity in patients with orbitofrontal cortex lesions, *Brain* **127**: 1108–1126.

Berlin, H., Rolls, E. T. and Iversen, S. D. (2005). Borderline Personality Disorder, impulsivity, and the orbitofrontal cortex, *American Journal of Psychiatry* **58**: 234–245.

Bermond, B., Fasotti, L., Niewenhuyse, B. and Schuerman, J. (1991). Spinal cord lesions, peripheral feedback and intensities of emotional feelings, *Cognition and Emotion* **5**: 201–220.

Berridge, K. C. and Robinson, T. E. (1998). What is the role of dopamine in reward: hedonic impact, reward learning, or incentive salience?, *Brain Research Reviews* **28**: 309–369.

Berridge, K. C. and Robinson, T. E. (2003). Parsing reward, *Trends in Neurosciences* **26**: 507–513.

Bertram, B. C. R. (1975). Social factors influencing reproduction in wild lions, *Journal of Zoology* **177**: 463–482.

Bhattacharyya, S. and Chakraborty, K. (2007). Glutamatergic dysfunction–newer targets for anti-obsessional drugs, *Recent Patents CNS Drug Discovery* **2**: 47–55.

Birkhead, T. (2000). *Promiscuity*, Faber and Faber, London.

Birkhead, T. R. and Moller, A. P. (1992). *Sperm Competition in Birds*, Academic Press, London.

Birkhead, T. R. and Pizzari, T. (2002). Postcopulatory sexual selection, *Nature Reviews Genetics* **3**: 262–273.

Birkhead, T. R., Chaline, N., Biggins, J. D., Burke, T. and Pizzari, T. (2004). Nontransivity of paternity in a bird, *Evolution* **58**: 416–420.

Blackmore, S. (1999). *The Meme Machine*, Oxford University Press, Oxford.

Blair, H. T., Schafe, G. E., Bauer, E. P., Rodrigues, S. M. and LeDoux, J. E. (2001). Synaptic plasticity in the lateral amygdala: a cellular hypothesis of fear conditioning, *Learning and Memory* **8**: 229–242.

Blake, R. and Logothetis, N. K. (2002). Visual competition, *Nature Reviews Neuroscience* **3**: 13–21.

Blaney, P. H. (1986). Affect and memory: a review, *Psychological Bulletin* **99**: 229–246.

Block, N. (1995a). On a confusion about a function of consciousness, *Behavioral and Brain Sciences* **18**: 22–47.

Block, N. (1995b). Two neural correlates of consciousness, *Trends in Cognitive Sciences* **9**: 46–52.

Blood, A. J. and Zatorre, R. J. (2001). Intensely pleasureable responses to music correlate with activity of brain regions implicated in reward and emotion, *Proceedings of the National Academy of Sciences USA* **98**: 11818–11823.

Blood, A. J., Zatorre, R. J., Bermudez, P. and Evans, A. C. (1999). Emotional responses to pleasant and unpleasant music correlate with activity in paralimbic brain regions, *Nature Neuroscience* **2**: 382–387.

Boden, M. A. (ed.) (1996). *The Philosophy of Artificial Life*, Oxford University Press, Oxford.

Booth, D. A. (1985). Food-conditioned eating preferences and aversions with interoceptive elements: learned appetites and satieties, *Annals of the New York Academy of Sciences* **443**: 22–37.

Booth, M. C. A. and Rolls, E. T. (1998). View-invariant representations of familiar objects by neurons in the inferior temporal visual cortex, *Cerebral Cortex* **8**: 510–523.

Bowles, S. and Gintis, H. (2004). The evolution of strong reciprocity: cooperation in heterogeneous populations, *Theoretical Population Biology* **65**: 17–28.

Bowles, S. and Gintis, H. (2005). Prosocial emotions, *in* L. E. Blume and S. N. Durlauf (eds), *The Economy as an Evolving Complex System III*, Santa Fe Institute, Santa Fe, NM.

Boyd, R., Gintis, H., Bowles, S. and Richerson, P. J. (2003). The evolution of altruistic punishment, *Proceedings of the National Academy of Sciences USA* **100**: 3531–3535.

Brody, C., Hernandez, A., Zainos, A. and Romo, R. (2003). Timing and neural encoding of somatosensory parametric working memory in macaque prefrontal cortex, *Cerebral Cortex* **13**: 1196–1207.

Brunel, N. and Wang, X. J. (2001). Effects of neuromodulation in a cortical network model of object working memory dominated by recurrent inhibition, *Journal of Computational Neuroscience* **11**: 63–85.

Buck, L. (2000). Smell and taste: the chemical senses, *in* E. Kandel, J. H. Schwartz and T. H. Jessell (eds), *Principles of Neural Science*, 4th edn, McGraw-Hill, New York, chapter 32, pp. 625–647.

Buck, L. and Axel, R. (1991). A novel multigene family may encode odorant receptors: a molecular basis for odor recognition, *Cell* **65**: 175–187.

Buckley, M. J. and Gaffan, D. (2006). Perirhinal contributions to object perception, *Trends in Cognitive Sciences* **10**: 100–107.

Burke, S. N. and Barnes, C. A. (2006). Neural plasticity in the ageing brain, *Nature Reviews Neuroscience*

7: 30–40.

Buss, D. M. (1989). Sex differences in human mate preferences: evolutionary hypotheses tested in 37 cultures, *Behavioural and Brain Sciences* **12**: 1–14.

Buss, D. M. (1994). *The Evolution of Desire: Strategies of Human Mating*, Basic Books, New York.

Buss, D. M. (2003). *Evolution of Desire. Strategies of Human Mating*, 2nd edn, Basic Books, New York, NY.

Buss, D. M. (2006). Debating sexual selection and mating strategies, *Science* **312**: 689–697.

Buss, D. M. (2008). *Evolutionary Psychology: The New Science of the Mind*, 3rd edn, Allyn and Bacon, Boston, MA.

Buss, D. M. and Schmitt, D. P. (1993). Sexual strategies theory: an evolutionary perspective on human mating, *Psychological Review* **100**: 204–232.

Buss, D. M., Abbott, M. and Angleitner, A. (1990). International preferences in selecting mates: a study of 37 cultures, *Journal of Cross-Cultural Psychology* **21**: 5–47.

Byrne, R. W. and Whiten, A. (1988). *Machiavellian Intelligence: Social Expertise and the Evolution of Intellect in Monkeys, Apes and Humans*, Clarendon Press, Oxford.

Cacioppo, J. T., Klein, D. J., Berntson, G. C. and Hatfield, E. (1993). The psychophysiology of emotion, *in* M. Lewis and J. M. Hatfield (eds), *Handbook of Emotions*, Guildford, New York, pp. 119–145.

Camerer, C. F. and Fehr, E. (2006). When does "economic man" dominate social behavior?, *Science* **311**: 47–52.

Canli, T., Zhao, Z., Desmond, J. E., Kang, E., Gross, J. and Gabrieli, J. D. (2001). An fMRI study of personality influences on brain reactivity to emotional stimuli, *Behavioral Neuroscience* **115**: 33–42.

Canli, T., Sivers, H., Whitfield, S. L., Gotlib, I. H. and Gabrieli, J. D. (2002). Amygdala response to happy faces as a function of extraversion, *Science* **296**: 2191.

Cannon, W. B. (1927). The James–Lange theory of emotion: a critical examination and an alternative theory, *American Journal of Psychology* **39**: 106–124.

Cannon, W. B. (1929). *Bodily Changes in Pain, Hunger, Fear and Rage*, 2nd edn, Appleton, New York.

Cannon, W. B. (1931). Again the James–Lange theory of emotion: a critical examination and an alternative theory, *Psychological Review* **38**: 281–295.

Carlson, N. R. (2006). *Physiology of Behavior*, 9th edn, Pearson, Boston.

Carruthers, P. (1996). *Language, Thought and Consciousness*, Cambridge University Press, Cambridge.

Carruthers, P. (2000). *Phenomenal Consciousness*, Cambridge University Press, Cambridge.

Carter, S. C. (1998). Neuroendocrine perpectives on social attachment and love, *Psychoneuroendocrinology* **23**: 779–818.

Castner, S. A., Williams, G. V. and Goldman-Rakic, P. S. (2000). Reversal of antipsychotic-induced working memory deficits by short-term dopamine D1 receptor stimulation, *Science* **287**: 2020–2022.

Chakrabarty, K., Bhattacharyya, S., Christopher, R. and Khanna, S. (2005). Glutamatergic dysfunction in OCD, *Neuropsychopharmacology* **30**: 1735–1740.

Chalmers, D. J. (1996). *The Conscious Mind*, Oxford University Press, Oxford.

Chamberlain, S. R., Fineberg, N. A., Blackwell, A. D., Robbins, T. W. and Sahakian, B. J. (2006). Motor inhibition and cognitive flexibility in obsessive-compulsive disorder and trichotillomania, *American Journal of Psychiatry* **163**: 1282–1284.

Chamberlain, S. R., Fineberg, N. A., Menzies, L. A., Blackwell, A. D., Bullmore, E. T., Robbins, T. W. and Sahakian, B. J. (2007). Impaired cognitive flexibility and motor inhibition in unaffected first-degree relatives of patients with obsessive-compulsive disorder, *American Journal of Psychiatry* **164**: 335–338.

Cheney, D. L. and Seyfarth, R. M. (1990). *How Monkeys See the World*, University of Chicago Press, Chicago.

Chevalier-Skolnikoff, S. (1973). Facial expression of emotion in non-human primates, *in* P. Ekman (ed.), *Darwin and Facial Expression*, Academic Press, New York, pp. 11–89.

Churchland, A. K., Kiani, R. and Shadlen, M. N. (2008). Decision-making with multiple alternatives, *Nature Neuroscience* **11**: 693–702.

Clark, D. A. and Beck, A. T. (2010). Cognitive theory and therapy of anxiety and depression: convergence with neurobiological findings, *Trends in Cognitive Science* **14**: 418–424.

Clelland, C. D., Choi, M., Romberg, C., Clemenson, G. D., J., Fragniere, A., Tyers, P., Jessberger, S., Saksida, L. M., Barker, R. A., Gage, F. H. and Bussey, T. J. (2009). A functional role for adult hippocampal neurogenesis in spatial pattern separation, *Science* **325**: 210–213.

Corkin, S. (2002). What's new with the amnesic patient H.M.?, *Nature Reviews Neuroscience* **3**: 153–160.

Corrado, G. S., Sugrue, L. P., Seung, H. S. and Newsome, W. T. (2005). Linear-nonlinear-Poisson models of primate choice dynamics, *Journal of the Experimental Analysis of Behavior* **84**: 581–617.

Cosmides, I. and Tooby, J. (1999). Evolutionary psychology, *in* R. Wilson and F. Keil (eds), *MIT Encyclopedia of the Cognitive Sciences*, MIT Press, Cambridge, MA, pp. 295–298.

Coyle, J. T. (2006). Glutamate and schizophrenia: beyond the dopamine hypothesis, *Cellular and Molecular Neurobiology* **26**: 365–384.

Coyle, J. T., Tsai, G. and Goff, D. (2003). Converging evidence of NMDA receptor hypofunction in the pathophysiology of schizophrenia, *Annals of the New York Academy of Sciences* **1003**: 318–327.

Crick, F. H. C. and Koch, C. (1990). Towards a neurobiological theory of consciousness, *Seminars in the Neurosciences* **2**: 263–275.

Crick, F. H. C. and Mitchison, G. (1995). REM sleep and neural nets, *Behavioural Brain Research* **69**: 147–155.

Critchley, H. D. and Rolls, E. T. (1996a). Hunger and satiety modify the responses of olfactory and visual neurons in the primate orbitofrontal cortex, *Journal of Neurophysiology* **75**: 1673–1686.

Critchley, H. D. and Rolls, E. T. (1996b). Olfactory neuronal responses in the primate orbitofrontal cortex: analysis in an olfactory discrimination task, *Journal of Neurophysiology* **75**: 1659–1672.

Cullen, E. (1957). Adaptations in the kittiwake to cliff-nesting, *Ibis* **99**: 275–302.

Cumming, L. (2009). *A Face to the World: on Self-portraits*, Harper, London.

Daly, M. and Wilson, M. (1988). *Homicide*, Aldine De Gruyter, New York.

Damasio, A. R. (1994). *Descartes' Error: Emotion, Reason, and the Human Brain*, Grosset/Putnam, New York.

Damasio, A. R. (2003). *Looking for Spinoza*, Heinemann, London.

Darwin, C. (1871). *The Descent of Man, and Selection in Relation to Sex*, John Murray [reprinted (1981) by Princeton University Press], London.

Darwin, C. (1872). *The Expression of the Emotions in Man and Animals*, University of Chicago Press. [reprinted (1998) (3rd edn) ed. P. Ekman. Harper Collins], Glasgow.

Davis, M. (2006). Neural systems involved in fear and anxiety measured with fear-potentiated startle, *American Psychologist* **61**: 741–756.

Dawkins, M. S. (1993). *Through Our Eyes Only? The Search for Animal Consciousness*, Freeman, Oxford.

Dawkins, M. S. (1995). *Unravelling Animal Behaviour*, 2nd edn, Longman, Harlow.

Dawkins, M. S. (2012). *Why Animals Matter*, Oxford University Press, Oxford.

Dawkins, M. S. and Bonney, R. (eds) (2008). *The Future of Animal Farming: Renewing the Ancient Contract*, Blackwell, Oxford.

Dawkins, R. (1976). *The Selfish Gene*, Oxford University Press, Oxford.

Dawkins, R. (1982). *The Extended Phenotype*, Freeman, Oxford.

Dawkins, R. (1986). *The Blind Watchmaker*, Longman, Harlow.

Dawkins, R. (1989). *The Selfish Gene*, 2nd edn, Oxford University Press, Oxford.

Dawkins, R. (2009). *The Greatest Show On Earth: The Evidence For Evolution*, Bantam Press, London.

Dayan, P. and Abbott, L. F. (2001). *Theoretical Neuroscience*, MIT Press, Cambridge, MA.

De Araujo, I. E. T. and Rolls, E. T. (2004). Representation in the human brain of food texture and oral fat, *Journal of Neuroscience* **24**: 3086–3093.

De Araujo, I. E. T., Rolls, E. T., Velazco, M. I., Margot, C. and Cayeux, I. (2005). Cognitive modulation of olfactory processing, *Neuron* **46**: 671–679.

De Gelder, B., Vroomen, J., Pourtois, G. and Weiskrantz, L. (1999). Non-conscious recognition of affect in the absence of striate cortex, *NeuroReport* **10**: 3759–3763.

Debiec, J., LeDoux, J. E. and Nader, K. (2002). Cellular and systems reconsolidation in the hippocampus, *Neuron* **36**: 527–538.

Debiec, J., Doyere, V., Nader, K. and LeDoux, J. E. (2006). Directly reactivated, but not indirectly reactivated, memories undergo reconsolidation in the amygdala, *Proceedings of the National Academy of Sciences USA* **103**: 3428–3433.

Deco, G. and Rolls, E. T. (2003). Attention and working memory: a dynamical model of neuronal activity in the prefrontal cortex, *European Journal of Neuroscience* **18**: 2374–2390.

Deco, G. and Rolls, E. T. (2004). A neurodynamical cortical model of visual attention and invariant object recognition, *Vision Research* **44**: 621–644.

Deco, G. and Rolls, E. T. (2005a). Neurodynamics of biased competition and cooperation for attention: a model with spiking neurons, *Journal of Neurophysiology* **94**: 295–313.

Deco, G. and Rolls, E. T. (2005b). Sequential memory: a putative neural and synaptic dynamical mechanism, *Journal of Cognitive Neuroscience* **17**: 294–307.

Deco, G. and Rolls, E. T. (2005c). Synaptic and spiking dynamics underlying reward reversal in the orbitofrontal cortex, *Cerebral Cortex* **15**: 15–30.

Deco, G. and Rolls, E. T. (2006). A neurophysiological model of decision-making and Weber's law,

European Journal of Neuroscience **24**: 901–916.

Deco, G. and Rolls, E. T. (2011). Reconciling oscillations and firing rates.

Deco, G., Ledberg, A., Almeida, R. and Fuster, J. (2005). Neural dynamics of cross-modal and cross-temporal associations, *Experimental Brain Research* **166**: 325–336.

Deco, G., Rolls, E. T. and Romo, R. (2010). Synaptic dynamics and decision-making, *Proceedings of the National Academy of Sciences* **107**: 7545–7549.

Dehaene, S., Dehaene-Lambertz, G. and Cohen, L. (1998). Abstract representations of numbers in the animal and human brain, *Trends in Neurosciences* **21**: 355–361.

Dennett, D. C. (1991). *Consciousness Explained*, Penguin, London.

Dere, E., Easton, A., Nadel, L. and Huston, J. P. (eds) (2008). *Handbook of Episodic Memory*, Elsevier, Amsterdam.

Desimone, R. and Duncan, J. (1995). Neural mechanisms of selective visual attention, *Annual Review of Neuroscience* **18**: 193–222.

DeVries, A. C., DeVries, M. B., Taymans, S. E. and Carter, C. S. (1996). The effects of stress on social preferences are sexually dimorphic in prairie voles, *Proceedings of the National Academy of Science USA* **93**: 11980–11984.

Diamond, J. (1997). *Why is Sex Fun?*, Weidenfeld and Nicholson, London.

Dickinson, A. (1980). *Contemporary Animal Learning Theory*, Cambridge University Press, Cambridge.

Dickinson, A. (1994). Instrumental conditioning, *in* N. J. Mackintosh (ed.), *Animal Learning and Cognition*, Academic Press, San Diego, pp. 45–80.

Divac, I. (1975). Magnocellular nuclei of the basal forebrain project to neocortex, brain stem, and olfactory bulb. Review of some functional correlates, *Brain Research* **93**: 385–398.

Dixson, A. F. (1998). Sexual behaviour and evolution of the seminal vesicles in primates, *Folia Primatologica* **69**: 300–306.

Dulac, C. and Torello, A. T. (2003). Molecular detection of pheromone signals in mammals: from genes to behaviour, *Nature Reviews Neuroscience* **4**: 551–562.

Dunbar, R. (1993). Co-evolution of neocortex size, group size and language in humans, *Behavioural and Brain Sciences* **16**: 681–735.

Dunbar, R. (1996). *Grooming, Gossip, and the Evolution of Language*, Faber and Faber, London.

Duncan, J. (1996). Cooperating brain systems in selective perception and action, *in* T. Inui and J. L. McClelland (eds), *Attention and Performance XVI*, MIT Press, Cambridge, MA, pp. 549–578.

Duncan, J. and Humphreys, G. (1989). Visual search and stimulus similarity, *Psychological Review* **96**: 433–458.

Durstewitz, D. and Seamans, J. K. (2002). The computational role of dopamine D1 receptors in working memory, *Neural Networks* **15**: 561–572.

Durstewitz, D., Kelc, M. and Gunturkun, O. (1999). A neurocomputational theory of the dopaminergic modulation of working memory functions, *Journal of Neuroscience* **19**: 2807–2722.

Durstewitz, D., Seamans, J. K. and Sejnowski, T. J. (2000a). Dopamine-mediated stabilization of delay-period activity in a network model of prefrontal cortex, *Journal of Neurophysiology* **83**: 1733–1750.

Durstewitz, D., Seamans, J. K. and Sejnowski, T. J. (2000b). Neurocomputational models of working memory, *Nature Neuroscience* **3 Suppl**: 1184–1191.

Dutton, D. (2009). *The Art Instinct*, Oxford University Press, Oxford.

Easton, A. and Gaffan, D. (2000). Amygdala and the memory of reward: the importance of fibres of passage from the basal forebrain, *in* J. P. Aggleton (ed.), *The Amygdala: a Functional Analysis*, 2nd edn, Oxford University Press, Oxford, chapter 17, pp. 569–586.

Easton, A., Ridley, R. M., Baker, H. F. and Gaffan, D. (2002). Unilateral lesions of the cholinergic basal forebrain and fornix in one hemisphere and inferior temporal cortex in the opposite hemisphere produce severe learning impairments in rhesus monkeys, *Cerebral Cortex* **12**: 729–736.

Ekman, P. (1982). *Emotion in the Human Face*, 2nd edn, Cambridge University Press, Cambridge.

Ekman, P. (1992). An argument for basic emotions, *Cognition and Emotion* **6**: 169–200.

Ekman, P. (1993). Facial expression and emotion, *American Psychologist* **48**: 384–392.

Ekman, P. (1998). Introduction, *C.Darwin: The Expression of the Emotions in Man and Animals, 1872, 3rd Edition 1998*, Harper Collins, Glasgow, pp. xxi–xxxvi.

Ekman, P. (2003). *Emotions Revealed: Understanding Faces and Feelings*, Weidenfeld and Nicolson, London.

Ekman, P., Friesen, W. V. and Ellsworth, P. C. (1972). *Emotion in the Human Face: Guidelines for Research and Integration of Findings*, Pergamon Press, Oxford.

Ekman, P., Levenson, R. W. and Friesen, W. V. (1983). Autonomic nervous system activity distinguishes between the emotions, *Science* **221**: 1208–1210.

Elliffe, M. C. M., Rolls, E. T. and Stringer, S. M. (2002). Invariant recognition of feature combinations in the visual system, *Biological Cybernetics* **86**: 59–71.

Epstein, J., Stern, E. and Silbersweig, D. (1999). Mesolimbic activity associated with psychosis in schizophrenia. Symptom-specific PET studies, *Annals of the New York Academy of Sciences* **877**: 562–574.

Eysenck, H. J. and Eysenck, S. B. G. (1968). *Personality Structure and Measurement*, R. R. Knapp, San Diego.

Eysenck, H. J. and Eysenck, S. B. G. (1985). *Personality and Individual Differences: a Natural Science Approach*, Plenum, New York.

Faisal, A., Selen, L. and Wolpert, D. (2008). Noise in the nervous system, *Nature Reviews Neuroscience* **9**: 292–303.

Farah, M. J. (2000). *The Cognitive Neuroscience of Vision*, Blackwell, Oxford.

Farah, M. J. (2005). Neuroethics: the practical and the philosophical, *Trends in Cognitive Sciences* **9**: 34–40.

Farah, M. J. (2012). Neuroethics. The ethical, legal, and societal impact of neuroscience, *Annual Review of Psychology* **63**: –.

Fehr, E. (2009). Social preferences and the brain, *in* P. W. Glimcher, C. F. Camerer, E. Fehr and R. A. Poldrack (eds), *Neuroeconomics. Decision Making and the Brain*, Academic Press, London, chapter 15, pp. 215–232.

Fehr, E. and Fischbacher, U. (2003). The nature of human altruism, *Nature* **425**: 785–791.

Fehr, E. and Gächter, S. (2002). Altruistic punishment in humans, *Nature* **415**: 137–140.

Fehr, E. and Rockenbach, B. (2004). Human altruism: economic, neural, and evolutionary perspectives, *Current Opinion in Neurobiology* **14**: 784–790.

Feinstein, J. S., Adolphs, R., Damasio, A. and Tranel, D. (2011). The human amygdala and the induction and experience of fear, *Current Biology* **21**: 34–38.

Fiorillo, C. D., Tobler, P. N. and Schultz, W. (2003). Discrete coding of reward probability and uncertainty by dopamine neurons, *Science* **299**: 1898–1902.

Fisher, H. (1992). *Anatomy of Love: The Natural History of Mating, Marriage, and Why We Stray*, Fawcett Columbine, New York.

Fisher, H. (2004). *Why We Love: The Nature and Chemistry of Romantic Love*, Henry Holt, New York.

Fisher, H. (2009). *Why Him? Why Her?: Finding Real Love By Understanding Your Personality Type*, Henry Holt, New York.

Fisher, R. A. (1930). *The Genetical Theory of Natural Selection*, Clarendon Press, Oxford.

Fisher, R. A. (1958). *The Genetical Theory of Natural Selection*, 2nd edn, Dover, New York.

Fodor, J. A. (1994). *The Elm and the Expert: Mentalese and its Semantics*, MIT Press, Cambridge, MA.

Földiák, P. (1991). Learning invariance from transformation sequences, *Neural Computation* **3**: 193–199.

Fox, C. R. and Poldrack, R. A. (2009). Prospect theory and the brain, *in* P. W. Glimcher, C. F. Camerer, E. Fehr and R. A. Poldrack (eds), *Neuroeconomics. Decision Making and the Brain*, Academic Press, London, chapter 11, pp. 145–173.

Franco, L., Rolls, E. T., Aggelopoulos, N. C. and Treves, A. (2004). The use of decoding to analyze the contribution to the information of the correlations between the firing of simultaneously recorded neurons, *Experimental Brain Research* **155**: 370–384.

Franco, L., Rolls, E. T., Aggelopoulos, N. C. and Jerez, J. M. (2007). Neuronal selectivity, population sparseness, and ergodicity in the inferior temporal visual cortex, *Biological Cybernetics* **96**: 547–560.

Frederick, S., Loewenstein, T. and O'Donoghue (2002). Time discounting and time preference: a critical review, *Journal of Economic Literature* **40**: 351–401.

Freud, S. (1900). *The Interpretation of Dreams*.

Frijda, N. H. (1986). *The Emotions*, Cambridge University Press, Cambridge.

Frith, C. D. and Singer, T. (2008). The role of social cognition in decision making, *Philosophical Transactions of the Royal Society of London B Biological Sciences* **363**: 3875–3886.

Fuhrmann, G., Markram, H. and Tsodyks, M. (2002). Spike frequency adaptation and neocortical rhythms, *Journal of Neurophysiology* **88**: 761–770.

Funahashi, S., Bruce, C. and Goldman-Rakic, P. (1989). Mnemonic coding of visual space in monkey dorsolateral prefrontal cortex, *Journal of Neurophysiology* **61**: 331–349.

Furman, M. and Wang, X.-J. (2008). Similarity effect and optimal control of multiple-choice decision making, *Neuron* **60**: 1153–1168.

Fuster, J. M. (1973). Unit activity in prefrontal cortex during delayed-response performance: neuronal correlates of transient memory, *Joural of Neurophysiology* **36**: 61–78.

Fuster, J. M. (1989). *The Prefrontal Cortex*, 2nd edn, Raven Press, New York.

Fuster, J. M. (1995). *Memory in the Cerebral Cortex*, MIT Press, Cambridge, MA.

Fuster, J. M. (2000). *Memory Systems in the Brain*, Raven Press, New York.

Fuster, J. M. (2008). *The Prefrontal Cortex*, 4th edn, Academic Press, London.

Fuster, J. M., Bodner, M. and Kroger, J. K. (2000). Cross-modal and cross-temporal association in neurons of frontal cortex, *Nature* **405**: 347–351.

Gangestad, S. W. and Simpson, J. A. (2000). The evolution of human mating: trade-offs and strategic pluralism, *Behavioural and Brain Sciences* **23**: 573–644.

Gangestad, S. W. and Thornhill, R. (1999). Individual differences in developmental precision and fluctuating asymmetry: a model and its implications, *Journal of Evolutionary Biology* **12**: 402–416.

Gazzaniga, M. S. (1988). Brain modularity: towards a philosophy of conscious experience, *in* A. J. Marcel and E. Bisiach (eds), *Consciousness in Contemporary Science*, Oxford University Press, Oxford, chapter 10, pp. 218–238.

Gazzaniga, M. S. (1995). Consciousness and the cerebral hemispheres, *in* M. S. Gazzaniga (ed.), *The Cognitive Neurosciences*, MIT Press, Cambridge, MA, chapter 92, pp. 1392–1400.

Gazzaniga, M. S. and LeDoux, J. (1978). *The Integrated Mind*, Plenum, New York.

Gazzaniga, M. S. (ed.) (2009). *The Cognitive Neurosciences*, 4th edn, Bradford / MIT Press, Cambridge, MA.

Ge, T., Feng, J., Grabenhorst, F. and Rolls, E. T. (2011). Componential Granger causality, and its application to identifying the source and mechanisms of the top-down biased activation that controls attention to affective vs sensory processing, *Neuroimage* p. Electronic publication 23 August.

Gennaro, R. J. (2004). *Higher Order Theories of Consciousness*, John Benjamins, Amsterdam.

Georges-François, P., Rolls, E. T. and Robertson, R. G. (1999). Spatial view cells in the primate hippocampus: allocentric view not head direction or eye position or place, *Cerebral Cortex* **9**: 197–212.

Gerstner, W. and Kistler, W. (2002). *Spiking Neuron Models: Single Neurons, Populations and Plasticity*, Cambridge University Press, Cambridge.

Gilligan, C. (1982). *In a Different Voice*, Harvard University Press, Cambridge, MA.

Gintis, H. (2003). The hitchhiker's guide to altruism: genes, culture, and the internalization of norms, *Journal of Theoretical Biology* **220**: 407–418.

Gintis, H. (2007). A framework for the unification of the behavioral sciences, *Behavioral and Brain Sciences* **30**: 1–16.

Glickman, S. E. and Schiff, B. B. (1967). A biological theory of reinforcement, *Psychological Review* **74**: 81–109.

Glimcher, P. (2003). The neurobiology of visual-saccadic decision making, *Annual Review of Neuroscience* **26**: 133–179.

Glimcher, P. (2004). *Decisions, Uncertainty, and the Brain*, MIT Press, Cambridge, MA.

Glimcher, P. (2005). Indeterminacy in brain and behavior, *Annual Review of Psychology* **56**: 25–56.

Glimcher, P. and Rustichini, A. (2004). Neuroeconomics: the consilience of brain and decision, *Science* **306**: 447–452.

Glimcher, P. W. (2011). *Foundations of Neuroeconomic Analysis*, Oxford University Press, Oxford.

Glimcher, P. W., Camerer, C. F., Fehr, E. and Poldrack, R. A. (eds) (2009). *Neuroeconomics. Decision Making and the Brain*, Academic Press, London.

Gneezy, U. (2005). Deception: the role of consequences, *American Economic Review* **95**: 384–394.

Goff, D. C. and Coyle, J. T. (2001). The emerging role of glutamate in the pathophysiology and treatment of schizophrenia, *American Journal of Psychiatry* **158**: 1367–1377.

Gold, J. I. and Shadlen, M. N. (2007). The neural basis of decision making, *Annual Review of Neuroscience* **30**: 535–574.

Goldberg, M. E. (2000). The control of gaze, *in* E. R. Kandel, J. H. Schwartz and T. M. Jessell (eds), *Principles of Neural Science*, 4th edn, McGraw-Hill, New York, chapter 39, pp. 782–800.

Goldman-Rakic, P. (1994). Working memory dysfunction in schizophrenia, *Journal of Neuropsychology and Clinical Neuroscience* **6**: 348–357.

Goldman-Rakic, P. S. (1996). The prefrontal landscape: implications of functional architecture for understanding human mentation and the central executive, *Philosophical Transactions of the Royal Society B* **351**: 1445–1453.

Goldman-Rakic, P. S. (1999). The physiological approach: functional architecture of working memory and disordered cognition in schizophrenia, *Biological Psychiatry* **46**: 650–661.

Goleman, D. (1995). *Emotional Intelligence*, Bantam, New York.

Gombrich, E. (1977). *Art and Illusion: A Study in the Psychology of Pictorial Representation*, 5th edn, Phaidon Press, London.

Goodale, M. A. (2004). Perceiving the world and grasping it: dissociations between conscious and unconscious visual processing, *in* M. S. Gazzaniga (ed.), *The Cognitive Neurosciences III*, MIT Press, Cambridge, MA, pp. 1159–1172.

Gould, S. J. (1985). *Ontogeny and Phylogeny*, Harvard University Press, Boston.

Gould, S. J. and Lewontin, R. C. (1979). The spandrels of San Marco and the Panglossian paradigm; a critique of the adaptationist programme, *Proceedings of the Royal Society of London B* **205**: 581–598.

Grabenhorst, F. and Rolls, E. T. (2009). Different representations of relative and absolute subjective value in the human brain, *Neuroimage* **48**: 258–268.

Grabenhorst, F. and Rolls, E. T. (2010). Attentional modulation of affective vs sensory processing: functional connectivity and a top down biased activation theory of selective attention, *Journal of Neurophysiology* **104**: 1649–1660.

Grabenhorst, F. and Rolls, E. T. (2011). Value, pleasure, and choice systems in the ventral prefrontal cortex, *Trends in Cognitive Sciences* **15**: 56–67.

Grabenhorst, F., Rolls, E. T., Margot, C., da Silva, M. and Velazco, M. I. (2007). How pleasant and unpleasant stimuli combine in the brain: odor combinations, *Journal of Neuroscience* **27**: 13532–13540.

Grabenhorst, F., Rolls, E. T. and Bilderbeck, A. (2008a). How cognition modulates affective responses to taste and flavor: top-down influences on the orbitofrontal and pregenual cingulate cortices, *Cerebral Cortex* **18**: 1549–1559.

Grabenhorst, F., Rolls, E. T. and Parris, B. (2008b). Selective attention to affective value alters how the brain processes taste stimuli, *European Journal of Neuroscience* **27**: 723–729.

Grabenhorst, F., Rolls, E. T. and Parris, B. A. (2008c). From affective value to decision-making in the prefrontal cortex, *European Journal of Neuroscience* **28**: 1930–1939.

Grabenhorst, F., D'Souza, A., Parris, B. A., Rolls, E. T. and Passingham, R. E. (2010). A common neural scale for the subjective value of different primary rewards, *Neuroimage* **51**: 1265–1274.

Grabenhorst, F., Rolls, E. T. and Margot, C. (2011). A hedonically complex odor mixture captures the brain's attention, *Neuroimage* **55**: 832–845.

Grady, C. L. (2008). Cognitive neuroscience of aging, *Annals of the New York Academy of Sciences* **1124**: 127–144.

Grafen, A. (1990a). Biological signals as handicaps, *Journal of Theoretical Biology* **144**: 517–546.

Grafen, A. (1990b). Sexual selection unhandicapped by the Fisher process, *Journal of Theoretical Biology* **144**: 473–516.

Gray, J. A. (1970). The psychophysiological basis of introversion-extraversion, *Behaviour Research and Therapy* **8**: 249–266.

Gray, J. A. (1975). *Elements of a Two-Process Theory of Learning*, Academic Press, London.

Gray, J. A. (1981). Anxiety as a paradigm case of emotion, *British Medical Bulletin* **37**: 193–197.

Gray, J. A. (1987). *The Psychology of Fear and Stress*, 2nd edn, Cambridge University Press, Cambridge.

Green, M. F. (1996). What are the functional consequences of neurocognitive deficits in schizophrenia?, *American Journal of Psychiatry* **153**: 321–330.

Gregory, R. L. (1970). *The Intelligent Eye*, McGraw-Hill, New York.

Griffin, D. R. (1992). *Animal Minds*, University of Chicago Press, Chicago.

Griffin, J. (2008). *On Human Rights*, Oxford University Press, Oxford.

Gross, C. G. (2002). Genealogy of the "grandmother cell", *Neuroscientist* **8**: 512–518.

Gross, C. G., Rodman, H. R., Gochin, P. M. and Colombo, M. W. (1993). Inferior temporal cortex as a pattern recognition device, *in* E. Baum (ed.), *Computational Learning and Cognition*, Society for Industrial and Applied Mathematics, Philadelphia, pp. 44–73.

Grossman, S. P. (1967). *A Textbook of Physiological Psychology*, Wiley, New York.

Hafner, H., Maurer, K., Loffler, W., an der Heiden, W., Hambrecht, M. and Schultze-Lutter, F. (2003). Modeling the early course of schizophrenia, *Schizophrenia Bulletin* **29**: 325–340.

Hailman, J. P. (1967). How an instinct is learned, *Scientific American* **221**(6): 98–108.

Hamann, S. and Canli, T. (2004). Individual differnces in emotion processing, *Current Opinion in Neurobiology* **14**: 233–238.

Hamilton, W. D. (1964). The genetical evolution of social behaviour, *Journal of Theoretical Biology* **7**: 1–52.

Hamilton, W. D. (1996). *Narrow Roads of Gene Land*, W. H. Freeman, New York.

Hamilton, W. D. and Zuk, M. (1982). Heritable true fitness and bright birds: a role for parasites, *Science* **218**: 384–387.

Hampton, R. R. (2001). Rhesus monkeys know when they can remember, *Proceedings of the National Academy of Sciences of the USA* **98**: 5539–5362.

Harcourt, A. H., Harvey, P. H., Larson, S. G. and Short, R. V. (1981). Testis weight, body weight and breeding system in primates, *Nature* **293**: 55–57.

Harcourt, A. H., Purvis, A. and Liles, L. (1995). Sperm competition: mating system, not breeding season, affects testes size of primates, *Functional Ecology* **9**: 468–476.

Harlow, H. F. and Stagner, R. (1933). Psychology of feelings and emotion, *Psychological Review* **40**: 84–194.

Hasselmo, M. E., Rolls, E. T., Baylis, G. C. and Nalwa, V. (1989). Object-centered encoding by face-selective neurons in the cortex in the superior temporal sulcus of the monkey, *Experimental Brain Research* **75**: 417–429.

Hauser, M. D. (1996). *The Evolution of Communication*, MIT Press, Cambridge, MA.

Hebb, D. O. (1949). *The Organization of Behavior: a Neuropsychological Theory*, Wiley, New York.

Heims, H. C., Critchley, H. D., Dolan, R., Mathias, C. J. and Cipolotti, L. (2004). Social and motivational functioning is not critically dependent on feedback of autonomic responses: neuropsychological evidence from patients with pure autonomic failure, *Neuropsychologia* **42**: 1979–1988.

Heistermann, M., Ziegler, T., van Schaik, C. P., Launhardt, K., Winkler, P. and Hodges, J. K. (2001). Loss of oestrus, concealed ovulation and paternity confusion in free-ranging Hanuman langurs, *Proceedings of the Royal Society of London B* **268**: 2445–2451.

Helmholtz, H. v. (1867). *Handbuch der physiologischen Optik*, Voss, Leipzig.

Hertz, J. A., Krogh, A. and Palmer, R. G. (1991). *Introduction to the Theory of Neural Computation*, Addison-Wesley, Wokingham, UK.

Herwitz, D. (2008). *Aesthetics*, Continuum, London.

Heyes, C. (2008). Beast machines? Questions of animal consciousness, *in* L. Weiskrantz and M. Davies (eds), *Frontiers of Consciousness*, Oxford University Press, Oxford, chapter 9, pp. 259–274.

Hinde, R. A. (2010). *Why Gods Persist: A Scientific Approach to Religion*, 2nd edn, Oxford University Press, Oxford.

Hobbes, T. (1651). *Leviathan, or the Matter, Forme, and Power of a Commonwealth, Ecclesiasticall and Civil*.

Hohmann, G. W. (1966). Some effects of spinal cord lesions on experienced emotional feelings, *Psychophysiology* **3**: 143–156.

Hölscher, C., Rolls, E. T. and Xiang, J. Z. (2003). Perirhinal cortex neuronal activity related to long term familiarity memory in the macaque, *European Journal of Neuroscience* **18**: 2037–2046.

Hopfield, J. J. (1982). Neural networks and physical systems with emergent collective computational abilities, *Proceedings of the National Academy of Sciences USA* **79**: 2554–2558.

Hornak, J., Rolls, E. T. and Wade, D. (1996). Face and voice expression identification in patients with emotional and behavioural changes following ventral frontal lobe damage, *Neuropsychologia* **34**: 247–261.

Hornak, J., Bramham, J., Rolls, E. T., Morris, R. G., O'Doherty, J., Bullock, P. R. and Polkey, C. E. (2003). Changes in emotion after circumscribed surgical lesions of the orbitofrontal and cingulate cortices, *Brain* **126**: 1691–1712.

Hornak, J., O'Doherty, J., Bramham, J., Rolls, E. T., Morris, R. G., Bullock, P. R. and Polkey, C. E. (2004). Reward-related reversal learning after surgical excisions in orbitofrontal and dorsolateral prefrontal cortex in humans, *Journal of Cognitive Neuroscience* **16**: 463–478.

Horne, J. (2006). *Sleepfaring: A Journey Through the Science of Sleep*, Oxford University Press, Oxford.

Hrdy, S. B. (1996). The evolution of female orgasms: logic please but no atavism, *Animal Behaviour* **52**: 851–852.

Hrdy, S. B. (1999). *Mother Nature: Natural Selection and the Female of the Species*, Chatto and Windus, London.

Hubel, D. H. and Wiesel, T. N. (1962). Receptive fields, binocular interaction, and functional architecture in the cat's visual cortex, *Journal of Physiology* **160**: 106–154.

Hubel, D. H. and Wiesel, T. N. (1968). Receptive fields and functional architecture of monkey striate cortex, *Journal of Physiology, London* **195**: 215–243.

Hume, D. (1741). *Of the Original Contract in Essays Moral, Political, and Literary*.

Hume, D. (1757). *Four Dissertations: Of Tragedy*.

Hume, D. (1777). *Selected Essays: Of the Standard of Taste*.

Humphrey, N. (1971). Colour and brightness preferences in monkeys, *Nature* **229**: 615–617.

Humphrey, N. (2011). *Soul Dust: the Magic of Consciousness*, Princeton University Press, Princeton.

Humphrey, N. K. (1980). Nature's psychologists, *in* B. D. Josephson and V. S. Ramachandran (eds), *Consciousness and the Physical World*, Pergamon, Oxford, pp. 57–80.

Humphrey, N. K. (1986). *The Inner Eye*, Faber, London.

Humphrey, N. K. (1995). *Soul Searching*, Chatto and Windus (In the USA, published as Leaps of Faith by Basic Books), London.

Illes, J. and Sahakian, B. J. (eds) (2011). *Oxford Handbook of Neuroethics*, Oxford University Press, Oxford.

Illes, J. (ed.) (2006). *Neuroethics*, Oxford University Press, Oxford.

Insabato, A., Pannunzi, M., Rolls, E. T. and Deco, G. (2010). Confidence-related decision-making, *Journal of Neurophysiology* **104**: 539–547.

Insel, T. R. and Young, I. J. (2001). The neurobiology of attachment, *Nature Reviews Neuroscience* **2**: 129–136.

Ishizu, T. and Zeki, S. (2011). Toward a brain-based theory of beauty, *PLoS ONE* **6**: e21852.

Ishizuka, N., Weber, J. and Amaral, D. G. (1990). Organization of intrahippocampal projections originating from CA3 pyramidal cells in the rat, *Journal of Comparative Neurology* **295**: 580–623.

Itti, L. and Koch, C. (2001). Computational modelling of visual attention, *Nature Reviews Neuroscience* **2**: 194–203.

Izard, C. E. (1993). Four systems for emotion activation: cognitive and non-cognitive processes, *Psychological Review* **100**: 68–90.

Jackson, B. S. (2004). Including long-range dependence in integrate-and-fire models of the high interspike-interval variability of cortical neurons, *Neural Computation* **16**: 2125–2195.

James, W. (1884). What is an emotion?, *Mind* **9**: 188–205.

Jefferson, Y. (2004). Facial beauty–establishing a universal standard, *International Journal of Orthodontics* **15**: 9–22.

Johansen, J. P., Tarpley, J. W., LeDoux, J. E. and Blair, H. T. (2010). Neural substrates for expectation-modulated fear learning in the amygdala and periaqueductal gray, *Nature Neuroscience* **13**: 979–986.

Johnson-Laird, P. N. (1988). *The Computer and the Mind: An Introduction to Cognitive Science*, Harvard University Press, Cambridge, MA.

Johnston, V. S. and Franklin, M. (1993). Is beauty in the eyes of the beholder?, *Ethology and Sociobiology* **13**: 73–85.

Jones, E. G. and Powell, T. P. S. (1970). An anatomical study of converging sensory pathways within the cerebral cortex of the monkey, *Brain* **93**: 793–820.

Kacelnik, A. and Brito e Abreu, F. (1998). Risky choice and Weber's Law, *Journal of Theoretical Biology* **194**: 289–298.

Kadohisa, M., Rolls, E. T. and Verhagen, J. V. (2004). Orbitofrontal cortex neuronal representation of temperature and capsaicin in the mouth, *Neuroscience* **127**: 207–221.

Kadohisa, M., Rolls, E. T. and Verhagen, J. V. (2005). Neuronal representations of stimuli in the mouth: the primate insular taste cortex, orbitofrontal cortex, and amygdala, *Chemical Senses* **30**: 401–419.

Kagel, J. H., Battalio, R. C. and Green, L. (1995). *Economic Choice Theory: An Experimental Analysis of Animal Behaviour*, Cambridge University Press, Cambridge.

Kahneman, D. and Tversky, A. (1979). Prospect theory: An analysis of decision under risk, *Econometrica* **47**: 263–292.

Kahneman, D. and Tversky, A. (1984). Choices, values, and frames, *American Psychologist* **4**: 341–350.

Kandel, E. R., Schwartz, J. H., Hudspeth, A. J., Siegelbaum, S. A. and Jessell, T. H. (eds) (2012). *Principles of Neural Science*, 5th edn, McGraw-Hill, New York.

Kant, I. (1790). *Critique of Judgement*.

Kappeler, P. M. and van Schaik, C. P. (2004). Sexual selection in primates: review and selective preview, *in* P. M. Kappeler and C. P. van Schaik (eds), *Sexual Selection in Primates*, Cambridge University Press, Cambridge, chapter 1, pp. 3–23.

Karno, M., Golding, J. M., Sorenson, S. B. and Burnam, M. A. (1988). The epidemiology of obsessive-compulsive disorder in five US communities, *Archives of General Psychiatry* **45**: 1094–1099.

Katz, L. D. (2000). Emotion, representation, and consciousness, *Behavioral and Brain Sciences* **23**: 204–205.

Kelly, K. M., Nadon, N. L., Morrison, J. H., Thibault, O., Barnes, C. A. and Blalock, E. M. (2006). The neurobiology of aging, *Epilepsy Research* **68, Supplement 1**: S5–S20.

Kesner, R. P. and Rolls, E. T. (2001). Role of long term synaptic modification in short term memory, *Hippocampus* **11**: 240–250.

Kettlewell, H. B. D. (1955). Selection experiments on industrial melanism in the Lepidoptera, *Heredity* **9**: 323–335.

Keverne, E. B., Nevison, C. M. and Martel, F. L. (1997). Early learning and the social bond, *Annals of the New York Academy of Science* **807**: 329–339.

Kievit, J. and Kuypers, H. G. J. M. (1975). Subcortical afferents to the frontal lobe in the rhesus monkey studied by means of retrograde horseradish peroxidase transport, *Brain Research* **85**: 261–266.

King-Casas, B., Tomlin, D., Anen, C., Camerere, C. F., Quartz, S. R. and Montague, P. R. (2005). Getting to know you: reputation and trust in a two-person economic exchange, *Science* **308**: 78–83.

Koch, C. (1999). *Biophysics of Computation*, Oxford University Press, Oxford.

Koch, C. (2004). *The Quest for Consciousness*, Roberts, Englewood, CO.

Kohonen, T. (1977). *Associative Memory: A System Theoretical Approach*, Springer, New York.

Kohonen, T. (1989). *Self-Organization and Associative Memory*, 3rd (1984, 1st edn; 1988, 2nd edn) edn, Springer-Verlag, Berlin.

Kohonen, T., Oja, E. and Lehtio, P. (1981). Storage and processing of information in distributed memory systems, *in* G. E. Hinton and J. A. Anderson (eds), *Parallel Models of Associative Memory*, Erlbaum, Hillsdale, NJ, chapter 4, pp. 105–143.

Kosfeld, M., Heinrichs, M., Zak, P. J., Fischbacher, U. and Fehr, E. (2005). Oxytocin increases trust in humans, *Nature* **435**: 673–676.

Kral, T. V. and Rolls, B. J. (2004). Energy density and portion size: their independent and combined effects on energy intake, *Physiology and Behavior* **82**: 131–138.

Kralik, J. D. and Hauser, M. D. (2000). A taste of things to come, *Behavioral and Brain Sciences* **23**: 207–208.

Kraut, R. E. and Johnson, R. E. (1979). Social and emotional messages of smiling: an ethological approach, *Journal of Personality and Social Psychology* **37**: 1539–1553.

Krebs, J. R. and Davies, N. B. (1991). *Behavioural Ecology*, 3rd edn, Blackwell, Oxford.

Krebs, J. R. and Kacelnik, A. (1991). Decision making, *in* J. R. Krebs and N. B. Davies (eds), *Behavioural Ecology*, 3rd edn, Blackwell, Oxford, chapter 4, pp. 105–136.

Kreiman, G., Koch, C. and Freid, I. (2000). Category-specific visual responses of single neurons in the human temporal lobe, *Nature Neuroscience* **3**: 946–953.

Kringelbach, M. L. and Rolls, E. T. (2003). Neural correlates of rapid reversal learning in a simple model of human social interaction, *Neuroimage* **20**: 1371–1383.

Kringelbach, M. L. and Rolls, E. T. (2004). The functional neuroanatomy of the human orbitofrontal cortex: evidence from neuroimaging and neuropsychology, *Progress in Neurobiology* **72**: 341–372.

Kringelbach, M. L., O'Doherty, J., Rolls, E. T. and Andrews, C. (2003). Activation of the human orbitofrontal cortex to a liquid food stimulus is correlated with its subjective pleasantness, *Cerebral Cortex* **13**: 1064–1071.

Krug, R., Plihal, W., Fehm, H. L. and Born, J. (2000). Selective influence of the menstrual cycle on perception of stimuli with reproductive significance: an event-related potential study, *Psychophysiology* **37**: 111–122.

Kruger, T. H., Haake, P., Chereath, D., Knapp, W., Janssen, O. E., Exton, M. S., Schedlowski, M. and Hartmann, U. (2003). Specificity of the neuroendocrine response to orgasm during sexual arousal in men, *Journal of Endocrinology* **177**: 57–64.

Laibson, D. (1997). Golden eggs and hyperbolic discounting, *Quarterly Journal of Economics* **112**: 443–477.

Laland, K. N. and Brown, G. R. (2002). *Sense and Nonsense. Evolutionary Perspectives on Human Behaviour*, Oxford University Press, Oxford.

Lane, R. D., Sechrest, L., Reidel, R., Weldon, V., Kaszniak, A. and Schwartz, G. E. (1996). Impaired verbal and nonverbal emotion recognition in alexithymia, *Psychosomatic Medicine* **58**: 203–210.

Lane, R. D., Reiman, E., Axelrod, B., Yun, L.-S., Holmes, A. H. and Schwartz, G. (1998). Neural correlates of levels of emotional awareness. Evidence of an interaction between emotion and attention in the anterior cingulate cortex, *Journal of Cognitive Neuroscience* **10**: 525–535.

Lange, C. (1885). The emotions, *in* E. Dunlap (ed.), *The Emotions*, 1922 edn, Williams and Wilkins, Baltimore.

Langlois, J. H., Roggman, L. A., Casey, R. J., Ritter, J. M., Rieserdanner, L. A. and Jenkins, V. Y. (1987). Infant preferences for attractive faces - rudiments of a stereotype, *Developmental Psychology* **23**: 363–369.

Langlois, J. H., Ritter, J. M., Roggman, L. A. and Vaughn, L. S. (1991). Facial diversity and infant preferences for attractive faces, *Developmental Psychology* **27**: 79–84.

Langlois, J. H., Kalakanis, L., Rubenstein, A. J., Larson, A., Hallam, M. and Smoot, M. (2000). Maxims or myths of beauty? a meta-analytic and theoretical review, *Psychological Bulletin* **126**: 390–423.

Lazarus, R. S. (1991). *Emotion and Adaptation*, Oxford University Press, New York.

Leak, G. K. and Christopher, S. B. (1982). Freudian psychoanalysis and sociobiology: a synthesis, *American Psychologist* **37**: 313–322.

LeDoux, J. (2008). Emotional coloration of consciousness: how feelings come about, *in* L. Weiskrantz and M. Davies (eds), *Frontiers of Consciousness*, Oxford University Press, Oxford, pp. 69–130.

LeDoux, J. E. (1992). Emotion and the amygdala, *in* J. P. Aggleton (ed.), *The Amygdala*, Wiley-Liss, New York, chapter 12, pp. 339–351.

LeDoux, J. E. (1995). Emotion: clues from the brain, *Annual Review of Psychology* **46**: 209–235.

LeDoux, J. E. (1996). *The Emotional Brain*, Simon and Schuster, New York.

Lee Duckworth, A., Steen, T. A. and Seligman, M. E. (2005). Positive psychology in clinical practice, *Annual Review of Clinical Psychology* **1**: 629–651.

Lee, H. J., Macbeth, A. H., Pagani, J. H. and Young, W. S. (2009). Oxytocin: the great facilitator of life, *Progress in Neurobiology* **88**: 127–151.

Legrenzi, P., Umilta, C. and Anderson, F. (2011). *Neuromania: On the Limits of Brain Science*, Oxford

University Press, New York.

Levenson, R. W., Ekman, P. and Friesen, W. V. (1990). Voluntary facial action generates emotion-specific autonomic nervous system activity, *Psychophysiology* **27**: 363–384.

Levin, R. J. (2002). The physiology of sexual arousal in the human female: a recreational and procreational synthesis, *Archives of Sexual Behaviour* **31**: 405–411.

Levy, N. (2007). *Neuroethics*, Cambridge University Press, Cambridge.

Lewis, C. T. and Short, C. (1879). *A Latin Dictionary*, Oxford University Press, Oxford.

Lewis, D. A., Hashimoto, T. and Volk, D. W. (2005). Cortical inhibitory neurons and schizophrenia, *Nature Reviews Neuroscience* **6**: 312–324.

Liddell, H. G. and Scott, R. (1891). *Greek-English Lexicon*, Oxford University Press, Oxford.

Liddle, P. F. (1987). The symptoms of chronic schizophrenia: a re-examination of the positive-negative dichotomy, *British Journal of Psychiatry* **151**: 145–151.

Lieberman, D. A. (ed.) (2000). *Learning*, Wadsworth, Belmont, CA.

Lieberman, J. A., Perkins, D., Belger, A., Chakos, M., Jarskog, F., Boteva, K. and Gilmore, J. (2001). The early stages of schizophrenia: speculations on pathogenesis, pathophysiology, and therapeutic approaches, *Biological Psychiatry* **50**: 884–897.

Lim, M. M., Wang, Z., Olazabal, D. E., Ren, X., Terwilliger, E. F. and Young, L. J. (2004). Enhanced partner preference in a promiscuous species by manipulating the expression of a single gene, *Nature* **429**: 754–757.

Liu, Y. H. and Wang, X.-J. (2008). A common cortical circuit mechanism for perceptual categorical discrimination and veridical judgment, *PLoS Computational Biology* p. e1000253.

Locke, J. (1689). *The Two Treatises of Civil Government*.

Logothetis, N. K. (2008). What we can do and what we cannot do with fMRI, *Nature* **453**: 869–878.

Loh, M., Rolls, E. T. and Deco, G. (2007a). A dynamical systems hypothesis of schizophrenia, *PLoS Computational Biology* **3**: e228. doi:10.1371/journal.pcbi.0030228.

Loh, M., Rolls, E. T. and Deco, G. (2007b). Statistical fluctuations in attractor networks related to schizophrenia, *Pharmacopsychiatry* **40**: S78–84.

Lycan, W. G. (1997). Consciousness as internal monitoring, *in* N. Block, O. Flanagan and G. Guzeldere (eds), *The Nature of Consciousness: Philosophical Debates*, MIT Press, Cambridge, MA, pp. 755–771.

Lynch, G. and Gall, C. M. (2006). AMPAkines and the threefold path to cognitive enhancement, *Trends in Neuroscience* **29**: 554–562.

MacDonald, C. J. and Eichenbaum, H. (2009). Hippocampal neurons disambiguate overlapping sequences of non-spatial events, *Society for Neuroscience Abstracts* p. 101.21.

MacDonald, C. J., Lepage, K. Q., Eden, U. T. and Eichenbaum, H. (2011). Hippocampal "time cells" bridge the gap in memory for discontiguous events, *Neuron* **71**: 737–749.

Mackintosh, N. J. (1983). *Conditioning and Associative Learning*, Oxford University Press, Oxford.

Maia, T. V. and McClelland, J. L. (2004). A reexamination of the evidence for the somatic marker hypothesis: what participants really know in the Iowa gambling task, *Proceedings of the National Academy of Sciences* **101**: 16075–16080.

Maia, T. V. and McClelland, J. L. (2005). The somatic marker hypothesis: still many questions but no answers, *Trends in Cognitive Sciences* **9**: 162–164.

Maier, A., Logothetis, N. K. and Leopold, D. A. (2005). Global competition dictates local suppression in pattern rivalry, *Journal of Vision* **5**: 668–677.

Malhotra, A. K., Pinals, D. A., Weingartner, H., Sirocco, K., Missar, C. D., Pickar, D. and Breier, A. (1996). NMDA receptor function and human cognition: the effects of ketamine in healthy volunteers, *Neuropsychopharmacology* **14**: 301–307.

Malsburg, C. v. d. (1990). A neural architecture for the representation of scenes, *in* J. L. McGaugh, N. M. Weinberger and G. Lynch (eds), *Brain Organization and Memory: Cells, Systems and Circuits*, Oxford University Press, New York, chapter 19, pp. 356–372.

Marr, D. (1971). Simple memory: a theory for archicortex, *Philosophical Transactions of The Royal Society of London, Series B* **262**: 23–81.

Marti, D., Deco, G., Del Giudice, P. and Mattia, M. (2006). Reward-biased probabilistic decision-making: mean-field predictions and spiking simulations, *Neurocomputing* **39**: 1175–1178.

Martin, S. J., Grimwood, P. D. and Morris, R. G. (2000). Synaptic plasticity and memory: an evaluation of the hypothesis, *Annual Review of Neuroscience* **23**: 649–711.

Matthews, G. and Gilliland, K. (1999). The personality theories of H.J.Eysenck and J.A.Gray: a comparative review, *Personality and Individual Differences* **26**: 583–626.

Matthews, G., Zeidner, M. and Roberts, R. D. (2002). *Emotional Intelligence: Science and Myth*, MIT Press, Cambridge, MA.

Maynard Smith, J. (1982). *Evolution and the Theory of Games*, Cambridge University Press, Cambridge.

Maynard Smith, J. (1984). Game theory and the evolution of behaviour, *Behavioral and Brain Sciences* **7**: 95–125.

Maynard Smith, J. and Harper, D. (2003). *Animal Signals*, Oxford University Press, Oxford.

Mazur, J. E. (1998). *Learning and Behavior*, 4th edn, Prentice Hall, New Jersey.

McCabe, C., Rolls, E. T., Bilderbeck, A. and McGlone, F. (2008). Cognitive influences on the affective representation of touch and the sight of touch in the human brain, *Social, Cognitive and Affective Neuroscience* **3**: 97–108.

McClure, S. M., Laibson, D. I., Loewenstein, G. and Cohen, J. D. (2004). Separate neural systems value immediate and delayed monetary rewards, *Science* **306**: 503–507.

McCoy, A. N. and Platt, M. L. (2005). Expectations and outcomes: decision-making in the primate brain, *Journal of Comparative Physiology A* **191**: 201–211.

McLeod, P., Plunkett, K. and Rolls, E. T. (1998). *Introduction to Connectionist Modelling of Cognitive Processes*, Oxford University Press, Oxford.

Mead, L. S. and Arnold, S. J. (2004). Quantitative genetic models of sexual selection, *Trends in Ecology and Evolution* **19**: 264–271.

Menzies, L., Chamberlain, S. R., Laird, A. R., Thelen, S. M., Sahakian, B. J. and Bullmore, E. T. (2008). Integrating evidence from neuroimaging and neuropsychological studies of obsessive-compulsive disorder: The orbitofronto-striatal model revisited, *Neuroscience and Biobehavioral Reviews* **32**: 525–549.

Meston, C. M. and Frohlich, P. F. (2000). The neurobiology of sexual function, *Archives of General Psychiatry* **57**: 1012–1030.

Mesulam, M.-M. (1990). Human brain cholinergic pathways, *Progress in Brain Research* **84**: 231–241.

Metcalfe, J. and Mischel, W. (1999). A hot/cool-system analysis of delay of gratification: dynamics of willpower, *Psychological Review* **106**: 3–19.

Millenson, J. R. (1967). *Principles of Behavioral Analysis*, MacMillan, New York.

Miller, E. K. and Cohen, J. D. (2001). An integrative theory of prefrontal cortex function, *Annual Review of Neuroscience* **24**: 167–202.

Miller, G. A. (1956). The magic number seven, plus or minus two: some limits on our capacity for the processing of information, *Psychological Review* **63**: 81–93.

Miller, G. F. (2000). *The Mating Mind*, Heinemann, London.

Miller, G. F. (2001). Aesthetic fitness: How sexual selection shaped artistic virtuosity as a fitness indicator and aesthetic preferences as mate choice criteria, *Bulletin of Psychology and the Arts* **2**: 20–25.

Montague, P. R. and Berns, G. S. (2002). Neural economics and the biological substrates of valuation, *Neuron* **36**: 265–284.

Montague, P. R., King-Casas, B. and Cohen, J. D. (2006). Imaging valuation models in human choice, *Annual Review of Neuroscience* **29**: 417–448.

Moore, H. D. M., Martin, M. and Birkhead, T. R. (1999). No evidence for killer sperm or other selective interactions between human spermatozoa in ejaculates of different males in vitro, *Procedings of the Royal Society of London B* **266**: 2343–2350.

Moore, H. D. M., Dvorakova, K., Jenkins, N. and Breed, W. (2002). Exceptional sperm cooperation in the wood mouse, *Nature* **418**: 174–177.

Moreno-Bote, R., Rinzel, J. and Rubin, N. (2007). Noise-induced alternations in an attractor network model of perceptual bistability, *Journal of Neurophysiology* **98**: 1125–1139.

Morton, G. J., Cummings, D. E., Baskin, D. G., Barsh, G. S. and Schwartz, M. W. (2006). Central nervous system control of food intake and body weight, *Nature* **443**: 289–295.

Mueser, K. T. and McGurk, S. R. (2004). Schizophrenia, *Lancet* **363**: 2063–2072.

Muir, J. L., Everitt, B. J. and Robbins, T. W. (1994). AMPA-induced excitotoxic lesions of the basal forebrain: a significant role for the cortical cholinergic system in attentional function, *Journal of Neuroscience* **14**: 2313–2326.

Nadal, J. P., Toulouse, G., Changeux, J. P. and Dehaene, S. (1986). Networks of formal neurons and memory palimpsests, *Europhysics Letters* **1**: 535–542.

Nader, K. and Einarsson, E. O. (2010). Memory reconsolidation: an update, *Annals of the New York Academy of Sciences* **1191**: 27–41.

Nagai, Y., Critchley, H. D., Featherstone, E., Trimble, M. R. and Dolan, R. J. (2004). Activity in ventromedial prefrontal cortex covaries with sympathetic skin conductance level: a physiological account of a "default mode" of brain function, *Neuroimage* **22**: 243–251.

Nakazawa, K., Quirk, M. C., Chitwood, R. A., Watanabe, M., Yeckel, M. F., Sun, L. D., Kato, A., Carr, C. A., Johnston, D., Wilson, M. A. and Tonegawa, S. (2002). Requirement for hippocampal CA3

NMDA receptors in associative memory recall, *Science* **297**: 211–218.

Nakazawa, K., Sun, L. D., Quirk, M. C., Rondi-Reig, L., Wilson, M. A. and Tonegawa, S. (2003). Hippocampal CA3 NMDA receptors are crucial for memory acquisition of one-time experience, *Neuron* **38**: 305–315.

Nesse, R. M. (2000a). Is depression an adaptation?, *Archives of General Psychiatry* **57**: 14–20.

Nesse, R. M. (2000b). Natural selection, mental modules and intelligence, *Novartis Foundation Symposium: The Nature of Intelligence* **233**: 96–115.

Nesse, R. M. and Lloyd, A. T. (1992). The evolution of psychodynamic mechanisms, *in* J. H. Barkow, L. Cosmides and J. Tooby (eds), *The Adapted Mind*, Oxford University Press, New York, pp. 601–624.

Newcomer, J. W., Farber, N. B., Jevtovic-Todorovic, V., Selke, G., Melson, A. K., Hershey, T., Craft, S. and Olney, J. W. (1999). Ketamine-induced NMDA receptor hypofunction as a model of memory impairment and psychosis, *Neuropsychopharmacology* **20**: 106–118.

Nicoll, R. A. (1988). The coupling of neurotransmitter receptors to ion channels in the brain, *Science* **241**: 545–551.

Oatley, K. and Jenkins, J. M. (1996). *Understanding Emotions*, Blackwell, Oxford.

Oatley, K. and Johnson-Laird, P. N. (1987). Towards a cognitive theory of emotions, *Cognition and Emotion* **1**: 29–50.

O'Doherty, J., Kringelbach, M. L., Rolls, E. T., Hornak, J. and Andrews, C. (2001). Abstract reward and punishment representations in the human orbitofrontal cortex, *Nature Neuroscience* **4**: 95–102.

O'Doherty, J., Dayan, P., Friston, K. J., Critchley, H. D. and Dolan, R. J. (2003a). Temporal difference models and reward-related learning in the human brain, *Neuron* **38**: 329–337.

O'Doherty, J., Winston, J., Critchley, H. D., Perrett, D. I., Burt, D. M. and Dolan, R. J. (2003b). Beauty in a smile: the role of the medial orbitofrontal cortex in facial attractiveness, *Neuropsychologia* **41**: 147–155.

O'Doherty, J., Dayan, P., Schultz, J., Deichmann, R., Friston, K. and Dolan, R. J. (2004). Dissociable roles of ventral and dorsal striatum in instrumental conditioning, *Science* **304**: 452–454.

O'Donoghue, T. and Rabin, M. (1999). Doing it now or later, *American Economic Review* **89**: 103–124.

O'Kane, D. and Treves, A. (1992). Why the simplest notion of neocortex as an autoassociative memory would not work, *Network* **3**: 379–384.

O'Keefe, J. (1976). Place units in the hippocampus of the freely moving rat, *Experimental Neurology* **51**: 78–109.

O'Neill, M. J. and Dix, S. (2007). AMPA receptor potentiators as cognitive enhancers, *IDrugs* **10**: 185–192.

O'Rahilly, S. (2009). Human genetics illuminates the paths to metabolic disease, *Nature* **462**: 307–314.

Orians, G. H. and Heerwagen, J. H. (1992). Evolved responses to landscapes, *in* J. H. Barkow, L. Cosmides and J. Tooby (eds), *The Adapted Mind*, Oxford University Press, New York, pp. 555–579.

Osvath, M. (2009). Spontaneous planning for future stone throwing by a male chimpanzee, *Current Biology* **19**: R190–191.

Osvath, M. and Osvath, H. (2008). Chimpanzee (pan troglodytes) and orangutan (pongo abelii) forethought: self-control and pre-experience in the face of future tool use, *Animal Cognition* **11**: 661–674.

Packer, C. and Pusey, A. E. (1983). Adaptations of female lions to infanticide by incoming males, *American Naturalist* **121**: 716–728.

Panksepp, J. (1998). *Affective Neuroscience: The Foundations of Human and Animal Emotions*, Oxford University Press, New York.

Panzeri, S., Rolls, E. T., Battaglia, F. and Lavis, R. (2001). Speed of feedforward and recurrent processing in multilayer networks of integrate-and-fire neurons, *Network: Computation in Neural Systems* **12**: 423–440.

Pare, D., Quirk, G. J. and LeDoux, J. E. (2004). New vistas on amygdala networks in conditioned fear, *Journal of Neurophysiology* **92**: 1–9.

Parker, G. A., Ball, M. A., Stockley, P. and Gage, M. J. (1997). Sperm competition games: a prospective analysis of risk assessment, *Proceedings of the Royal Society of London B* **264**: 1793–1802.

Paulus, M. P. and Frank, L. R. (2006). Anterior cingulate activity modulates nonlinear decision weight function of uncertain prospects, *Neuroimage* **30**: 668–677.

Pearce, J. M. (1997). *Animal Learning and Cognition*, 2nd edn, Psychology Press, Hove, Sussex.

Peitgen, H.-O., Jürgens, H. and Saupe, D. (2004). *Chaos and Fractals: New Frontiers of Science*, Springer, New York.

Penton-Voak, I. S., Perrett, D. I., Castles, D. L., Kobayashi, T., Burt, D. M., Murray, L. K. and Minamisawa, R. (1999). Menstrual cycle alters face preference, *Nature* **399**: 741–742.

Perrett, D. I., Rolls, E. T. and Caan, W. (1982). Visual neurons responsive to faces in the monkey temporal cortex, *Experimental Brain Research* **47**: 329–342.

Petrides, M. (1996). Specialized systems for the processing of mnemonic information within the primate frontal cortex, *Philosophical Transactions of the Royal Society of London B* **351**: 1455–1462.

Phelps, E. (2004). Human emotion and memory: interactions of the amygdala and hippocampal complex, *Current Opinion in Neurobiology* **14**: 198–202.

Phillips, A. G., Mora, F. and Rolls, E. T. (1981). Intra-cerebral self-administration of amphetamine by rhesus monkeys, *Neuroscience Letters* **24**: 81–86.

Pinker, S. (1994). *The Language Instinct*, Penguin, London.

Pinker, S. and Bloom, P. (1992). Natural language and natural selection, *in* J. H. Barkow, L. Cosmides and J. Tooby (eds), *The Adapted Mind*, Oxford University Press, New York, chapter 12, pp. 451–493.

Pittenger, C., Krystal, J. H. and Coric, V. (2006). Glutamate-modulating drugs as novel pharmacotherapeutic agents in the treatment of obsessive-compulsive disorder, *NeuroRx* **3**(1): 69–81.

Pizzari, T., Cornwallis, C. K., Lovlie, H., Jakobsson, S. and Birkhead, T. R. (2003). Sophisticated sperm allocation in male fowl, *Nature* **426**: 70–74.

Platt, M. L. and Glimcher, P. W. (1999). Neural correlates of decision variables in parietal cortex, *Nature* **400**: 233–238.

Power, J. M. and Sah, P. (2008). Competition between calcium-activated K+ channels determines cholinergic action on firing properties of basolateral amygdala projection neurons, *Journal of Neuroscience* **28**: 3209–3220.

Quirk, G. J., Armony, J. L., Repa, J. C., Li, X.-F. and LeDoux, J. E. (1996). Emotional memory: a search for sites of plasticity, *Cold Spring Harbor Symposia on Quantitative Biology* **61**: 247–257.

Rachlin, H. (1989). *Judgement, Decision, and Choice: A Cognitive/Behavioural Synthesis*, Freeman, New York.

Rachlin, H. (2000). *The Science of Self-Control*, Harvard Univeristy Press, Cambridge, MA.

Rawls, J. (1971). *A Theory of Justice*, Oxford University Press, Oxford.

Rawls, J. (1999). *A Theory of Justice*, Belknap Press of Harvard University Press, Cambridge, MA.

Reisenzein, R. (1983). The Schachter theory of emotion: two decades later, *Psychological Bulletin* **94**: 239–264.

Ridley, M. (1993a). *Evolution*, Blackwell, Oxford.

Ridley, M. (1993b). *The Red Queen: Sex and the Evolution of Human Nature*, Penguin, London.

Ridley, M. (1996). *The Origins of Virtue*, Viking, London.

Ridley, M. (2003). *Nature via Nurture*, Harper, London.

Rilling, J. K., Gutman, D. A., Zeh, T. R., Pagnoni, G., Berns, G. S. and Kilts, C. D. (2002). A neural basis for social cooperation, *Neuron* **24**: 395–405.

Roberts, G. (2005). Cooperation through interdependence, *Animal Behaviour* **70**: 901–908.

Robertson, R. G., Rolls, E. T. and Georges-François, P. (1998). Spatial view cells in the primate hippocampus: Effects of removal of view details, *Journal of Neurophysiology* **79**: 1145–1156.

Robins, L. N., Helzer, J. E., Weissman, M. M., Orvaschel, H., Gruenberg, E., Burke, J. D. J. and Regier, D. A. (1984). Lifetime prevalence of specific psychiatric disorders in three sites, *Archives of General Psychiatry* **41**: 949–958.

Rodin, J. (1976). The role of perception of internal and external signals in the regulation of feeding in overweight and non-obese individuals, *Dahlem Konferenzen, Life Sciences Research Report* **2**: 265–281.

Rolls, B. J. and Rolls, E. T. (1982). *Thirst*, Cambridge University Press, Cambridge.

Rolls, E. T. (1975). *The Brain and Reward*, Pergamon Press, Oxford.

Rolls, E. T. (1981). Central nervous mechanisms related to feeding and appetite, *British Medical Bulletin* **37**: 131–134.

Rolls, E. T. (1984). Neurons in the cortex of the temporal lobe and in the amygdala of the monkey with responses selective for faces, *Human Neurobiology* **3**: 209–222.

Rolls, E. T. (1986a). Neural systems involved in emotion in primates, *in* R. Plutchik and H. Kellerman (eds), *Emotion: Theory, Research, and Experience*, Vol. 3: Biological Foundations of Emotion, Academic Press, New York, chapter 5, pp. 125–143.

Rolls, E. T. (1986b). A theory of emotion, and its application to understanding the neural basis of emotion, *in* Y. Oomura (ed.), *Emotions. Neural and Chemical Control*, Japan Scientific Societies Press; and Karger, Tokyo; and Basel, pp. 325–344.

Rolls, E. T. (1987). Information representation, processing and storage in the brain: analysis at the single neuron level, *in* J.-P. Changeux and M. Konishi (eds), *The Neural and Molecular Bases of Learning*, Wiley, Chichester, pp. 503–540.

Rolls, E. T. (1989a). Functions of neuronal networks in the hippocampus and cerebral cortex in memory, *in*

R. Cotterill (ed.), *Models of Brain Function*, Cambridge University Press, Cambridge, pp. 15–33.

Rolls, E. T. (1989b). Functions of neuronal networks in the hippocampus and neocortex in memory, *in* J. H. Byrne and W. O. Berry (eds), *Neural Models of Plasticity: Experimental and Theoretical Approaches*, Academic Press, San Diego, CA, chapter 13, pp. 240–265.

Rolls, E. T. (1989c). Information processing in the taste system of primates, *Journal of Experimental Biology* **146**: 141–164.

Rolls, E. T. (1989d). Parallel distributed processing in the brain: implications of the functional architecture of neuronal networks in the hippocampus, *in* R. G. M. Morris (ed.), *Parallel Distributed Processing: Implications for Psychology and Neurobiology*, Oxford University Press, Oxford, chapter 12, pp. 286–308.

Rolls, E. T. (1989e). The representation and storage of information in neuronal networks in the primate cerebral cortex and hippocampus, *in* R. Durbin, C. Miall and G. Mitchison (eds), *The Computing Neuron*, Addison-Wesley, Wokingham, England, chapter 8, pp. 125–159.

Rolls, E. T. (1990a). Functions of the primate hippocampus in spatial processing and memory, *in* D. S. Olton and R. P. Kesner (eds), *Neurobiology of Comparative Cognition*, L. Erlbaum, Hillsdale, NJ, chapter 12, pp. 339–362.

Rolls, E. T. (1990b). Theoretical and neurophysiological analysis of the functions of the primate hippocampus in memory, *Cold Spring Harbor Symposia in Quantitative Biology* **55**: 995–1006.

Rolls, E. T. (1990c). A theory of emotion, and its application to understanding the neural basis of emotion, *Cognition and Emotion* **4**: 161–190.

Rolls, E. T. (1992). Neurophysiological mechanisms underlying face processing within and beyond the temporal cortical visual areas, *Philosophical Transactions of the Royal Society* **335**: 11–21.

Rolls, E. T. (1994). Brain mechanisms for invariant visual recognition and learning, *Behavioural Processes* **33**: 113–138.

Rolls, E. T. (1995a). A model of the operation of the hippocampus and entorhinal cortex in memory, *International Journal of Neural Systems* **6, Supplement**: 51–70.

Rolls, E. T. (1995b). A theory of emotion and consciousness, and its application to understanding the neural basis of emotion, *in* M. S. Gazzaniga (ed.), *The Cognitive Neurosciences*, MIT Press, Cambridge, MA, chapter 72, pp. 1091–1106.

Rolls, E. T. (1996). A theory of hippocampal function in memory, *Hippocampus* **6**: 601–620.

Rolls, E. T. (1997a). Consciousness in neural networks?, *Neural Networks* **10**: 1227–1240.

Rolls, E. T. (1997b). Taste and olfactory processing in the brain and its relation to the control of eating, *Critical Reviews in Neurobiology* **11**: 263–287.

Rolls, E. T. (1999a). *The Brain and Emotion*, Oxford University Press, Oxford.

Rolls, E. T. (1999b). The functions of the orbitofrontal cortex, *Neurocase* **5**: 301–312.

Rolls, E. T. (2000a). Functions of the primate temporal lobe cortical visual areas in invariant visual object and face recognition, *Neuron* **27**: 205–218.

Rolls, E. T. (2000b). The orbitofrontal cortex and reward, *Cerebral Cortex* **10**: 284–294.

Rolls, E. T. (2000c). Précis of The Brain and Emotion, *Behavioral and Brain Sciences* **23**: 177–233.

Rolls, E. T. (2002). A theory of emotion, its functions, and its adaptive value, *in* R. Trappl, P. Petta and S. Payr (eds), *Emotions in Humans and Artifacts*, MIT Press, Cambridge, Mass, chapter 2, pp. 11–32.

Rolls, E. T. (2003). Consciousness absent and present: a neurophysiological exploration, *Progress in Brain Research* **144**: 95–106.

Rolls, E. T. (2004a). The functions of the orbitofrontal cortex, *Brain and Cognition* **55**: 11–29.

Rolls, E. T. (2004b). A higher order syntactic thought (HOST) theory of consciousness, *in* R. J. Gennaro (ed.), *Higher Order Theories of Consciousness*, John Benjamins, Amsterdam, chapter 7, pp. 137–172.

Rolls, E. T. (2005a). *Emotion Explained*, Oxford University Press, Oxford.

Rolls, E. T. (2005b). What are emotions, why do we have emotions, and what is their computational basis in the brain?, *in* J.-M. Fellous and M. A. Arbib (eds), *Who Needs Emotions? The Brain Meets the Robot*, Oxford University Press, New York, chapter 5, pp. 117–146.

Rolls, E. T. (2006a). Consciousness absent and present: a neurophysiological exploration of masking, *in* H. Ogmen and B. G. Breitmeyer (eds), *The First Half Second*, MIT Press, Cambridge, MA, chapter 6, pp. 89–108.

Rolls, E. T. (2006b). The neurophysiology and functions of the orbitofrontal cortex, *in* D. H. Zald and S. L. Rauch (eds), *The Orbitofrontal Cortex*, Oxford University Press, Oxford, chapter 5, pp. 95–124.

Rolls, E. T. (2007a). The affective neuroscience of consciousness: higher order syntactic thoughts, dual routes to emotion and action, and consciousness, *in* P. D. Zelazo, M. Moscovitch and E. Thompson (eds), *Cambridge Handbook of Consciousness*, Cambridge University Press, New York, chapter 29, pp. 831–859.

Rolls, E. T. (2007b). A computational neuroscience approach to consciousness, *Neural Networks* **20**: 962–982.

Rolls, E. T. (2007c). A neuro-biological approach to emotional intelligence, *in* G. Matthews, M. Zeidner and R. Roberts (eds), *The Science of Emotional Intelligence*, Oxford University Press, Oxford, chapter 3, pp. 72–100.

Rolls, E. T. (2007d). Sensory processing in the brain related to the control of food intake, *Proceedings of the Nutrition Society* **66**: 96–112.

Rolls, E. T. (2008a). Emotion, higher order syntactic thoughts, and consciousness, *in* L. Weiskrantz and M. Davies (eds), *Frontiers of Consciousness*, Oxford University Press, Oxford, chapter 4, pp. 131–167.

Rolls, E. T. (2008b). Functions of the orbitofrontal and pregenual cingulate cortex in taste, olfaction, appetite and emotion, *Acta Physiologica Hungarica* **95**: 131–164.

Rolls, E. T. (2008c). *Memory, Attention, and Decision-Making. A Unifying Computational Neuroscience Approach*, Oxford University Press, Oxford.

Rolls, E. T. (2008d). The primate hippocampus and episodic memory, *in* E. Dere, A. Easton, L. Nadel and J. P. Huston (eds), *Handbook of Episodic Memory*, Elsevier, Amsterdam, chapter 4.2, pp. 417–438.

Rolls, E. T. (2009a). The anterior and midcingulate cortices and reward, *in* B. Vogt (ed.), *Cingulate Neurobiology and Disease*, Oxford University Press, Oxford, chapter 8, pp. 191–206.

Rolls, E. T. (2009b). From reward value to decision-making: neuronal and computational principles, *in* J.-C. Dreher and L. Tremblay (eds), *Handbook of Reward and Decision-Making*, Academic Press, New York, chapter 5, pp. 95–130.

Rolls, E. T. (2009c). Functional neuroimaging of umami taste: what makes umami pleasant, *American Journal of Clinical Nutrition* **90**: 803S–814S.

Rolls, E. T. (2009d). The neurophysiology and computational mechanisms of object representation, *in* S. Dickinson, M. Tarr, A. Leonardis and B. Schiele (eds), *Object Categorization: Computer and Human Vision Perspectives*, Cambridge University Press, Cambridge, chapter 14, pp. 257–287.

Rolls, E. T. (2010a). The affective and cognitive processing of touch, oral texture, and temperature in the brain, *Neuroscience and Biobehavioral Reviews* **34**: 237–245.

Rolls, E. T. (2010b). Attractor networks, *WIREs Cognitive Science* **1**: 119–134.

Rolls, E. T. (2010c). A computational theory of episodic memory formation in the hippocampus, *Behavioural Brain Research* **215**: 180–196.

Rolls, E. T. (2010d). Noise in the brain, decision-making, determinism, free will, and consciousness, *in* E. Perry, D. Collerton, H. Ashton and F. Lebeau (eds), *New Horizons in the Neuroscience of Consciousness*, John Benjamins, Amsterdam, pp. 113–120.

Rolls, E. T. (2011a). Chemosensory learning in the cortex, *Frontiers in Computational Neuroscience* **5**: 78 (1–13).

Rolls, E. T. (2011b). Consciousness, decision-making, and neural computation, *in* V. Cutsuridis, A. Hussain and J. G. Taylor (eds), *Perception-Action Cycle: Models, architecture, and hardware*, Springer, Berlin, chapter 9, pp. 287–333.

Rolls, E. T. (2011c). David Marr's Vision: floreat computational neuroscience, *Brain* **134**: 913–916.

Rolls, E. T. (2011d). Face neurons, *in* A. J. Calder, G. Rhodes, M. H. Johnson and J. V. Haxby (eds), *The Oxford Handbook of Face Perception*, Oxford University Press, Oxford, chapter 4, pp. 51–75.

Rolls, E. T. (2011e). From brain mechanisms of emotion and decision-making to neuroeconomics, *in* O. Oullier, A. Kirman and J. A. S. Kelso (eds), *The State of Mind in Economics*, Cambridge University Press, Cambridge.

Rolls, E. T. (2011f). Glutamate, obsessive-compulsive disorder, schizophrenia, and the stability of cortical attractor neuronal networks, *Pharmacology, Biochemistry and Behavior* p. Electronic publication 23 June.

Rolls, E. T. (2011g). The neural representation of oral texture including fat texture, *Journal of Texture Studies* **42**: 137–156.

Rolls, E. T. (2011h). A neurobiological basis for affective feelings and aesthetics, *in* E. Schellekens and P. Goldie (eds), *The Aesthetic Mind: Philosophy and Psychology*, Oxford University Press, Oxford, chapter 8, pp. 116–165.

Rolls, E. T. (2011i). Taste, olfactory, and food texture reward processing in the brain and obesity, *International Journal of Obesity* **35**: 550–561.

Rolls, E. T. (2012a). Central neural integration of taste, smell and other sensory modalities, *in* R. L. Doty (ed.), *Handbook of Olfaction and Gustation: Modern Perspectives*, third edn, Dekker, New York, chapter 44.

Rolls, E. T. (2012b). Neural mechanisms of invariant visual object recognition, *Frontiers in Computational Neuroscience*.

Rolls, E. T. and Baylis, G. C. (1986). Size and contrast have only small effects on the responses to faces of neurons in the cortex of the superior temporal sulcus of the monkey, *Experimental Brain Research* **65**: 38–48.

Rolls, E. T. and Baylis, L. L. (1994). Gustatory, olfactory and visual convergence within the primate orbitofrontal cortex, *Journal of Neuroscience* **14**: 5437–5452.

Rolls, E. T. and Deco, G. (2002). *Computational Neuroscience of Vision*, Oxford University Press, Oxford.

Rolls, E. T. and Deco, G. (2010). *The Noisy Brain: Stochastic Dynamics as a Principle of Brain Function*, Oxford University Press, Oxford.

Rolls, E. T. and Deco, G. (2011). A computational neuroscience approach to schizophrenia and its onset, *Neuroscience and Biobehavioral Reviews* **35**: 1644–1653.

Rolls, E. T. and Grabenhorst, F. (2008). The orbitofrontal cortex and beyond: from affect to decision-making, *Progress in Neurobiology* **86**: 216–244.

Rolls, E. T. and Kesner, R. P. (2006). A theory of hippocampal function, and tests of the theory, *Progress in Neurobiology* **79**: 1–48.

Rolls, E. T. and McCabe, C. (2007). Enhanced affective brain representations of chocolate in cravers vs non-cravers, *European Journal of Neuroscience* **26**: 1067–1076.

Rolls, E. T. and Milward, T. (2000). A model of invariant object recognition in the visual system: learning rules, activation functions, lateral inhibition, and information-based performance measures, *Neural Computation* **12**: 2547–2572.

Rolls, E. T. and Rolls, B. J. (1973). Altered food preferences after lesions in the basolateral region of the amygdala in the rat, *Journal of Comparative and Physiological Psychology* **83**: 248–259.

Rolls, E. T. and Stringer, S. M. (2000). On the design of neural networks in the brain by genetic evolution, *Progress in Neurobiology* **61**: 557–579.

Rolls, E. T. and Stringer, S. M. (2001a). Invariant object recognition in the visual system with error correction and temporal difference learning, *Network: Computation in Neural Systems* **12**: 111–129.

Rolls, E. T. and Stringer, S. M. (2001b). A model of the interaction between mood and memory, *Network: Computation in Neural Systems* **12**: 89–109.

Rolls, E. T. and Stringer, S. M. (2005). Spatial view cells in the hippocampus, and their idiothetic update based on place and head direction, *Neural Networks* **18**: 1229–1241.

Rolls, E. T. and Stringer, S. M. (2006). Invariant visual object recognition: a model, with lighting invariance, *Journal of Physiology – Paris* **100**: 43–62.

Rolls, E. T. and Stringer, S. M. (2007). Invariant global motion recognition in the dorsal visual system: a unifying theory, *Neural Computation* **19**: 139–169.

Rolls, E. T. and Tovee, M. J. (1994). Processing speed in the cerebral cortex and the neurophysiology of visual masking, *Proceedings of the Royal Society, B* **257**: 9–15.

Rolls, E. T. and Tovee, M. J. (1995). Sparseness of the neuronal representation of stimuli in the primate temporal visual cortex, *Journal of Neurophysiology* **73**: 713–726.

Rolls, E. T. and Treves, A. (1998). *Neural Networks and Brain Function*, Oxford University Press, Oxford.

Rolls, E. T. and Treves, A. (2011). The neuronal encoding of information in the brain, *Progress in Neurobiology* p. Electronic publication 2 September.

Rolls, E. T. and Xiang, J.-Z. (2006). Spatial view cells in the primate hippocampus, and memory recall, *Reviews in the Neurosciences* **17**: 175–200.

Rolls, E. T., Judge, S. J. and Sanghera, M. (1977). Activity of neurones in the inferotemporal cortex of the alert monkey, *Brain Research* **130**: 229–238.

Rolls, E. T., Burton, M. J. and Mora, F. (1980). Neurophysiological analysis of brain-stimulation reward in the monkey, *Brain Research* **194**: 339–357.

Rolls, E. T., Thorpe, S. J. and Maddison, S. P. (1983). Responses of striatal neurons in the behaving monkey. 1. Head of the caudate nucleus, *Behavioural Brain Research* **7**: 179–210.

Rolls, E. T., Scott, T. R., Sienkiewicz, Z. J. and Yaxley, S. (1988). The responsiveness of neurones in the frontal opercular gustatory cortex of the macaque monkey is independent of hunger, *Journal of Physiology* **397**: 1–12.

Rolls, E. T., Sienkiewicz, Z. J. and Yaxley, S. (1989). Hunger modulates the responses to gustatory stimuli of single neurons in the caudolateral orbitofrontal cortex of the macaque monkey, *European Journal of Neuroscience* **1**: 53–60.

Rolls, E. T., Yaxley, S. and Sienkiewicz, Z. J. (1990). Gustatory responses of single neurons in the orbitofrontal cortex of the macaque monkey, *Journal of Neurophysiology* **64**: 1055–1066.

Rolls, E. T., Hornak, J., Wade, D. and McGrath, J. (1994a). Emotion-related learning in patients with social and emotional changes associated with frontal lobe damage, *Journal of Neurology, Neurosurgery and Psychiatry* **57**: 1518–1524.

Rolls, E. T., Tovee, M. J., Purcell, D. G., Stewart, A. L. and Azzopardi, P. (1994b). The responses of neurons in the temporal cortex of primates, and face identification and detection, *Experimental Brain Research* **101**: 474–484.

Rolls, E. T., Critchley, H. D. and Treves, A. (1996a). The representation of olfactory information in the primate orbitofrontal cortex, *Journal of Neurophysiology* **75**: 1982–1996.

Rolls, E. T., Critchley, H. D., Mason, R. and Wakeman, E. A. (1996b). Orbitofrontal cortex neurons: role in olfactory and visual association learning, *Journal of Neurophysiology* **75**: 1970–1981.

Rolls, E. T., Robertson, R. G. and Georges-François, P. (1997a). Spatial view cells in the primate hippocampus, *European Journal of Neuroscience* **9**: 1789–1794.

Rolls, E. T., Treves, A. and Tovee, M. J. (1997b). The representational capacity of the distributed encoding of information provided by populations of neurons in the primate temporal visual cortex, *Experimental Brain Research* **114**: 149–162.

Rolls, E. T., Treves, A., Robertson, R. G., Georges-François, P. and Panzeri, S. (1998). Information about spatial view in an ensemble of primate hippocampal cells, *Journal of Neurophysiology* **79**: 1797–1813.

Rolls, E. T., Tovee, M. J. and Panzeri, S. (1999). The neurophysiology of backward visual masking: information analysis, *Journal of Cognitive Neuroscience* **11**: 335–346.

Rolls, E. T., Stringer, S. M. and Trappenberg, T. P. (2002). A unified model of spatial and episodic memory, *Proceedings of The Royal Society B* **269**: 1087–1093.

Rolls, E. T., Aggelopoulos, N. C. and Zheng, F. (2003a). The receptive fields of inferior temporal cortex neurons in natural scenes, *Journal of Neuroscience* **23**: 339–348.

Rolls, E. T., Franco, L., Aggelopoulos, N. C. and Reece, S. (2003b). An information theoretic approach to the contributions of the firing rates and the correlations between the firing of neurons, *Journal of Neurophysiology* **89**: 2810–2822.

Rolls, E. T., Kringelbach, M. L. and De Araujo, I. E. T. (2003c). Different representations of pleasant and unpleasant odors in the human brain, *European Journal of Neuroscience* **18**: 695–703.

Rolls, E. T., Verhagen, J. V. and Kadohisa, M. (2003d). Representations of the texture of food in the primate orbitofrontal cortex: neurons responding to viscosity, grittiness, and capsaicin, *Journal of Neurophysiology* **90**: 3711–3724.

Rolls, E. T., Aggelopoulos, N. C., Franco, L. and Treves, A. (2004). Information encoding in the inferior temporal visual cortex: contributions of the firing rates and the correlations between the firing of neurons, *Biological Cybernetics* **90**: 19–32.

Rolls, E. T., Xiang, J.-Z. and Franco, L. (2005). Object, space and object-space representations in the primate hippocampus, *Journal of Neurophysiology* **94**: 833–844.

Rolls, E. T., Critchley, H. D., Browning, A. S. and Inoue, K. (2006a). Face-selective and auditory neurons in the primate orbitofrontal cortex, *Experimental Brain Research* **170**: 74–87.

Rolls, E. T., Franco, L., Aggelopoulos, N. C. and Jerez, J. M. (2006b). Information in the first spike, the order of spikes, and the number of spikes provided by neurons in the inferior temporal visual cortex, *Vision Research* **46**: 4193–4205.

Rolls, E. T., Stringer, S. M. and Elliot, T. (2006c). Entorhinal cortex grid cells can map to hippocampal place cells by competitive learning, *Network: Computation in Neural Systems* **17**: 447–465.

Rolls, E. T., Grabenhorst, F. and Parris, B. (2008a). Warm pleasant feelings in the brain, *Neuroimage* **41**: 1504–1513.

Rolls, E. T., Grabenhorst, F., Margot, C., da Silva, M. and Velazco, M. I. (2008b). Selective attention to affective value alters how the brain processes olfactory stimuli, *Journal of Cognitive Neuroscience* **20**: 1815–1826.

Rolls, E. T., Loh, M. and Deco, G. (2008c). An attractor hypothesis of obsessive-compulsive disorder, *European Journal of Neuroscience* **28**: 782–793.

Rolls, E. T., Loh, M., Deco, G. and Winterer, G. (2008d). Computational models of schizophrenia and dopamine modulation in the prefrontal cortex, *Nature Reviews Neuroscience* **9**: 696–709.

Rolls, E. T., McCabe, C. and Redoute, J. (2008e). Expected value, reward outcome, and temporal difference error representations in a probabilistic decision task, *Cerebral Cortex* **18**: 652–663.

Rolls, E. T., Grabenhorst, F. and Franco, L. (2009). Prediction of subjective affective state from brain activations, *Journal of Neurophysiology* **101**: 1294–1308.

Rolls, E. T., Critchley, H., Verhagen, J. V. and Kadohisa, M. (2010a). The representation of information about taste and odor in the primate orbitofrontal cortex, *Chemosensory Perception* **3**: 16–33.

Rolls, E. T., Grabenhorst, F. and Deco, G. (2010b). Choice, difficulty, and confidence in the brain, *Neuroimage* **53**: 694–706.

Rolls, E. T., Grabenhorst, F. and Deco, G. (2010c). Decision-making, errors, and confidence in the brain, *Journal of Neurophysiology* **104**: 2359–2374.

Rolls, E. T., Grabenhorst, F. and Parris, B. A. (2010d). Neural systems underlying decisions about affective odors, *Journal of Cognitive Neuroscience* **10**: 1068–1082.

Rolls, E. T., Deco, G. and Loh, M. (2012). A stochastic neurodynamics approach to the changes in cognition and memory in aging.

Romo, R., Hernandez, A. and Zainos, A. (2004). Neuronal correlates of a perceptual decision in ventral premotor cortex, *Neuron* **41**: 165–173.

Rosenberg, D. R., MacMaster, F. P., Keshavan, M. S., Fitzgerald, K. D., Stewart, C. M. and Moore, G. J. (2000). Decrease in caudate glutamatergic concentrations in pediatric obsessive-compulsive disorder patients taking paroxetine, *Journal of the American Academy of Child and Adolescent Psychiatry* **39**: 1096–1103.

Rosenberg, D. R., MacMillan, S. N. and Moore, G. J. (2001). Brain anatomy and chemistry may predict treatment response in paediatric obsessive–compulsive disorder, *International Journal of Neuropsychopharmacology* **4**: 179–190.

Rosenberg, D. R., Mirza, Y., Russell, A., Tang, J., Smith, J. M., Banerjee, S. P., Bhandari, R., Rose, M., Ivey, J., Boyd, C. and Moore, G. J. (2004). Reduced anterior cingulate glutamatergic concentrations in childhood OCD and major depression versus healthy controls, *Journal of the American Academy of Child and Adolescent Psychiatry* **43**: 1146–1153.

Rosenthal, D. (1990). A theory of consciousness, *ZIF Report 40/1990. Zentrum für Interdisziplinäre Forschung, Bielefeld*. Reprinted in Block, N., Flanagan, O. and Guzeldere, G. (eds.) (1997) *The Nature of Consciousness: Philosophical Debates*. MIT Press, Cambridge MA, pp. 729–853.

Rosenthal, D. M. (1986). Two concepts of consciousness, *Philosophical Studies* **49**: 329–359.

Rosenthal, D. M. (1993). Thinking that one thinks, *in* M. Davies and G. W. Humphreys (eds), *Consciousness*, Blackwell, Oxford, chapter 10, pp. 197–223.

Rosenthal, D. M. (2004). Varieties of higher order theory, *in* R. J. Gennaro (ed.), *Higher Order Theories of Consciousness*, John Benjamins, Amsterdam, pp. 17–44.

Rosenthal, D. M. (2005). *Consciousness and Mind*, Oxford University Press, Oxford.

Rousseau, J.-J. (1762). *The Social Contract, Or Principles of Political Right (Du contrat social ou Principes du droit politique)*, Paris.

Rumelhart, D. E., Hinton, G. E. and Williams, R. J. (1986). Learning internal representations by error propagation, *in* D. E. Rumelhart, J. L. McClelland and the PDP Research Group (eds), *Parallel Distributed Processing: Explorations in the Microstructure of Cognition*, Vol. 1, MIT Press, Cambridge, MA, chapter 8, pp. 318–362.

Rushworth, M. F., Noonan, M. P., Boorman, E. D., Walton, M. E. and Behrens, T. E. (2011). Frontal cortex and reward-guided learning and decision-making, *Neuron* **70**: 1054–1069.

Rushworth, M. F. S., Walton, M. E., Kennerley, S. W. and Bannerman, D. M. (2004). Action sets and decisions in the medial frontal cortex, *Trends in Cognitive Sciences* **8**: 410–417.

Rushworth, M. F. S., Behrens, T. E., Rudebeck, P. H. and Walton, M. E. (2007a). Contrasting roles for cingulate and orbitofrontal cortex in decisions and social behaviour, *Trends in Cognitive Sciences* **11**: 168–176.

Rushworth, M. F. S., Buckley, M. J., Behrens, T. E., Walton, M. E. and Bannerman, D. M. (2007b). Functional organization of the medial frontal cortex, *Current Opinion in Neurobiology* **17**: 220–227.

Rusting, C. and Larsen, R. (1998). Personality and cognitive processing of affective information, *Personality and Social Psychology Bulletin* **24**: 200–213.

Ryan, M. J. (1998). Sexual selection, receiver biases, and the evolution of sex differences, *Science* **281**: 1999–2003.

Sanfey, A. G., Rilling, J. K., Aronson, J. A., Nystrom, L. E. and Cohen, J. D. (2003). The neural basis of economic decision-making in the ultimatum game, *Science* **300**: 1755–1758.

Sanghera, M. K., Rolls, E. T. and Roper-Hall, A. (1979). Visual responses of neurons in the dorsolateral amygdala of the alert monkey, *Experimental Neurology* **63**: 610–626.

Savage-Rumbaugh, E. S., Rumbaugh, D. M. and McDonald, K. (1985). Language learning in two species of apes, *Neuroscience and Biobehavioral Reviews* **9**: 653–665.

Savage-Rumbaugh, E. S., McDonald, K., Sevcik, R. A., Hopkins, W. D. and Rupert, E. (1986). Spontaneous symbol acquisition and communicative use by pygmy chimpanzees (pan paniscus), *Journal of Experimental Psychology: General* **115**: 211–235.

Sawaguchi, T. and Goldman-Rakic, P. S. (1991). D1 dopamine receptors in prefrontal cortex: Involvement in working memory, *Science* **251**: 947–950.

Sawaguchi, T. and Goldman-Rakic, P. S. (1994). The role of D1-dopamine receptor in working memory: local injections of dopamine antagonists into the prefrontal cortex of rhesus monkeys performing an oculomotor delayed-response task, *Journal of Neurophysiology* **71**: 515–528.

Schachter, S. (1971). Importance of cognitive control in obesity, *American Psychologist* **26**: 129–144.

Schachter, S. and Singer, J. (1962). Cognitive, social and physiological determinants of emotional state, *Psychological Review* **69**: 378–399.

Scherer, K. S. (1999). Appraisal theory, *in* T. Dalgleish and M. J. Power (eds), *Handbook of Cognition and Emotion*, Wiley, New York, pp. 637–663.

Scherer, K. S. (2001). The nature and study of appraisal. A review of the issues, *in* K. S. Scherer, A. Schorr and T. Johnstone (eds), *Appraisal Processes in Emotion*, Oxford University Press, Oxford, pp. 369–391.

Scherer, K. S., Schorr, A. and Johnstone, T. (eds) (2001). *Appraisal Processes in Emotion*, Oxford University Press, Oxford.

Scheuerecker, J., Ufer, S., Zipse, M., Frodl, T., Koutsouleris, N., Zetzsche, T., Wiesmann, M., Albrecht, J., Bruckmann, H., Schmitt, G., Moller, H. J. and Meisenzahl, E. M. (2008). Cerebral changes and cognitive dysfunctions in medication-free schizophrenia – An fMRI study, *Journal of Psychiatric Research* **42**: 469–476.

Schiller, D., Monfils, M. H., Raio, C. M., Johnson, D. C., LeDoux, J. E. and Phelps, E. A. (2010). Preventing the return of fear in humans using reconsolidation update mechanisms, *Nature* **463**: 49–53.

Schirmer, A., Zysset, S., Kotz, S. A. and von Cramon, Y. D. (2004). Gender differences in the activation of inferior frontal cortex during emotional speech perception, *Neuroimage* **21**: 1114–1123.

Schliebs, R. and Arendt, T. (2006). The significance of the cholinergic system in the brain during aging and in Alzheimer's disease, *Journal of Neural Transmission* **113**: 1625–1644.

Schultz, W. (1998). Predictive reward signal of dopamine neurons, *Journal of Neurophysiology* **80**: 1–27.

Schultz, W. (2004). Neural coding of basic reward terms of animal learning theory, game theory, microeconomics and behavioural ecology, *Current Opinion in Neurobiology* **14**: 139–147.

Schultz, W. (2006). Behavioral theories and the neurophysiology of reward, *Annual Review of Psychology* **57**: 87–115.

Schultz, W., Dayan, P. and Montague, P. R. (1997). A neural substrate of prediction and reward, *Science* **275**: 1593–1599.

Scott, T. R., Yaxley, S., Sienkiewicz, Z. J. and Rolls, E. T. (1986). Gustatory responses in the frontal opercular cortex of the alert cynomolgus monkey, *Journal of Neurophysiology* **56**: 876–890.

Scoville, W. B. and Milner, B. (1957). Loss of recent memory after bilateral hippocampal lesions, *Journal of Neurology, Neurosurgery and Psychiatry* **20**: 11–21.

Seamans, J. K. and Yang, C. R. (2004). The principal features and mechanisms of dopamine modulation in the prefrontal cortex, *Progress in Neurobiology* **74**: 1–58.

Seligman, M. E. (1978). Learned helplessness as a model of depression. comment and integration, *Journal of Abnormal Psychology* **87**: 165–179.

Seymour, B., O'Doherty, J., Dayan, P., Koltzenburg, M., Jones, A. K., Dolan, R. J., Friston, K. J. and Frackowiak, R. S. (2004). Temporal difference models describe higher-order learning in humans, *Nature* **429**: 664–667.

Shackelford, T. K., Le Blanc, G. L., Weekes-Shackelford, V. A., Bleske-Rechek, A. L., Euler, H. A. and Hoier, S. (2002). Psychological adaptation to human sperm competition, *Evolution and Human Behaviour* **23**: 123–138.

Shallice, T. and Burgess, P. (1996). The domain of supervisory processes and temporal organization of behaviour, *Philosophical Transactions of the Royal Society B,* **351**: 1405–1411.

Shamay-Tsoory, S. G., Tibi-Elhanany, Y. and Aharon-Peretz, J. (2007). The green-eyed monster and malicious joy: the neuroanatomical bases of envy and gloating (schadenfreude), *Brain* **130**: 1663–1678.

Shepherd, G. M. (2004). *The Synaptic Organisation of the Brain*, 5th edn, Oxford University Press, Oxford.

Shepherd, G. M. and Grillner, S. (eds) (2010). *Handbook of Brain Microcircuits*, Oxford University Press, Oxford.

Shergill, S. S., Brammer, M. J., Williams, S. C., Murray, R. M. and McGuire, P. K. (2000). Mapping auditory hallucinations in schizophrenia using functional magnetic resonance imaging, *Archives of General Psychiatry* **57**: 1033–1038.

Short, R. V. (1998). Review of R. R. Baker and M. A. Bellis, Human Sperm Competition: Copulation, Masturbation and Infidelity, *European Sociobiological Society Newsletter* **47**: 20–23.

Sigala, N. and Logothetis, N. K. (2002). Visual categorisation shapes feature selectivity in the primate temporal cortex, *Nature* **415**: 318–320.

Sikström, S. (2007). Computational perspectives on neuromodulation of aging, *Acta Neurochirurgica, Supplement* **97**: 513–518.

Silk, J. B. (2009). Social preferences in primates, *in* P. W. Glimcher, C. F. Camerer, E. Fehr and R. A. Poldrack (eds), *Neuroeconomics. Decision Making and the Brain*, Academic Press, London, chapter 18, pp. 269–

284.

Simmen-Tulberg, B. and Moller, A. P. (1993). The relationship between concealed ovulation and mating systems in anthropoid primates: a phylogenetic analysis, *American Naturalist* **141**: 1–25.

Simmons, L. W., Firman, R. C., Rhodes, G. and Peters, M. (2004). Human sperm competition: testis size, sperm production and rate of extra-pair copulations, *Animal Behaviour* **68**: 297–302.

Singer, P. (1981). *The Expanding Circle: Ethics and Sociobiology*, Oxford University Press, Oxford.

Singer, T., Seymour, B., O'Doherty, J. P., Stephan, K. E., Dolan, R. J. and Frith, C. D. (2006). Empathic neural responses are modulated by the perceived fairness of others, *Nature* **439**: 466–469.

Singer, W. (1999). Neuronal synchrony: A versatile code for the definition of relations?, *Neuron* **24**: 49–65.

Singh, D. and Bronstad, M. P. (2001). Female body odour is a potential cue to ovulation, *Proceedings of the Royal Society of London B* **268**: 797–801.

Singh, D. and Luis, S. (1995). Ethnic and gender consensus for the effect of waist-to-hip ratio on judgements of women's attractiveness, *Human Nature* **6**: 51–65.

Singh, D., Meyer, W., Zambarano, R. J. and Hurlbert, D. F. (1998). Frequency and timing of coital orgasm in women desirous of becoming pregnant, *Archives of Sexual Behaviour* **27**: 15–29.

Skaggs, W. E., McNaughton, B. L., Gothard, K. and Markus, E. (1993). An information theoretic approach to deciphering the hippocampal code, *in* S. Hanson, J. D. Cowan and C. L. Giles (eds), *Advances in Neural Information Processing Systems*, Vol. 5, Morgan Kaufmann, San Mateo, CA, pp. 1030–1037.

Smith, R. L. (1984). Human sperm competiton, *in* R. L. Smith (ed.), *Sperm Competition and the Evolution of Animal Mating Systems*, Academic Press, London, pp. 601–660.

Soltani, A. and Wang, X.-J. (2006). A biophysically based neural model of matching law behavior: melioration by stochastic synapses, *Journal of Neuroscience* **26**: 3731–3744.

Squire, L. R. (1992). Memory and the hippocampus: A synthesis from findings with rats, monkeys and humans, *Psychological Review* **99**: 195–231.

Squire, L. R., Stark, C. E. L. and Clark, R. E. (2004). The medial temporal lobe, *Annual Review of Neuroscience* **27**: 279–306.

Stein, N. L., Trabasso, T. and Liwag, M. (1994). The Rashomon phenomenon: personal frames and future-oriented appraisals in memory for emotional events, *in* M. M. Haith, J. B. Benson, R. J. Roberts and B. F. Pennington (eds), *Future Oriented Processes*, University of Chicago Press, Chicago.

Stemmler, D. G. (1989). The autonomic differentiation of emotions revisited: convergent and discriminant validation, *Psychophysiology* **26**: 617–632.

Stewart, S. E., Fagerness, J. A., Platko, J., Smoller, J. W., Scharf, J. M., Illmann, C., Jenike, E., Chabane, N., Leboyer, M., Delorme, R., Jenike, M. A. and Pauls, D. L. (2007). Association of the SLC1A1 glutamate transporter gene and obsessive-compulsive disorder, *American Journal of Medical Genetics B: Neuropsychiatric Genetics* **144**: 1027–1033.

Stringer, S. M. and Rolls, E. T. (2000). Position invariant recognition in the visual system with cluttered environments, *Neural Networks* **13**: 305–315.

Stringer, S. M. and Rolls, E. T. (2002). Invariant object recognition in the visual system with novel views of 3D objects, *Neural Computation* **14**: 2585–2596.

Stringer, S. M., Rolls, E. T. and Trappenberg, T. P. (2005). Self-organizing continuous attractor network models of hippocampal spatial view cells, *Neurobiology of Learning and Memory* **83**: 79–92.

Strongman, K. T. (2003). *The Psychology of Emotion*, 5th edn, Wiley, New York.

Sugrue, L. P., Corrado, G. S. and Newsome, W. T. (2005). Choosing the greater of two goods: neural currencies for valuation and decision making, *Nature Reviews Neuroscience* **6**: 363–375.

Suri, R. E. and Schultz, W. (2001). Temporal difference model reproduces anticipatory neural activity, *Neural Computation* **13**: 841–862.

Sutton, R. S. and Barto, A. G. (1981). Towards a modern theory of adaptive networks: expectation and prediction, *Psychological Review* **88**: 135–170.

Sutton, R. S. and Barto, A. G. (1998). *Reinforcement Learning*, MIT Press, Cambridge, MA.

Swaddle, J. P. and Reierson, G. W. (2002). Testosterone increases perceived dominance but not attractiveness in human males, *Proceedings of the Royal Society of London B* **269**: 2285–2289.

Szabo, M., Deco, G., Fusi, S., Del Giudice, P., Mattia, M. and Stetter, M. (2006). Learning to attend: Modeling the shaping of selectivity in infero-temporal cortex in a categorization task, *Biological Cybernetics* **94**: 351–365.

Takahashi, H., Kato, M., Matsuura, M., Mobbs, D., Suhara, T. and Okubo, Y. (2009). When your gain is my pain and your pain is my gain: neural correlates of envy and schadenfreude, *Science* **323**: 937–939.

Takebayashi, M. and Funahashi, S. (2009). Monkeys exhibit preference for biologically non-signoficant stimuli, *Psychologia* **52**: 147–161.

Terrace, H. S. (1979). *Nim: A Chimpanzee who Learned Sign Language*, Knopf, New York.

Thibault, O., Porter, N. M., Chen, K. C., Blalock, E. M., Kaminker, P. G., Clodfelter, G. V., Brewer, L. D. and Landfield, P. W. (1998). Calcium dysregulation in neuronal aging and AlzheimerŠs disease: history and new directions, *Cell Calcium* **24**: 417–433.

Thornhill, R. and Gangestad, S. W. (1996). The evolution of human sexuality, *Trends in Ecology and Evolution* **11**: 98–102.

Thornhill, R. and Gangstad, S. W. (1999). The scent of symmetry: a human sex pheromone that signals fitness?, *Evolution and Human Behaviour* **20**: 175–201.

Thornhill, R. and Grammer, K. (1999). The body and face of woman: one ornament that signals quality?, *Evolution and Human Behaviour* **20**: 105–120.

Thornhill, R. and Palmer, C. T. (2000). *A Natural History of Rape*, MIT Press, Cambridge, MA.

Thornhill, R., Gangestad, S. W. and Comer, R. (1995). Human female orgasm and mate fluctuating asymmetry, *Animal Behaviour* **50**: 1601–1615.

Thorpe, S. J., Rolls, E. T. and Maddison, S. (1983). Neuronal activity in the orbitofrontal cortex of the behaving monkey, *Experimental Brain Research* **49**: 93–115.

Tinbergen, N. (1951). *The Study of Instinct*, Oxford University Press, Oxford.

Tinbergen, N. (1963). On aims and methods of ethology, *Zeitschrift fur Tierpsychologie* **20**: 410–433.

Tinbergen, N., Broekhuysen, G. J., Feekes, F., Houghton, J. C. W., Kruuk, H. and Szule, E. (1967). Egg shell removal by black-headed gull *Larus ribibundus*, *Behaviour* **19**: 74–117.

Tobler, P. N., Dickinson, A. and Schultz, W. (2003). Coding of predicted reward omission by dopamine neurons in a conditioned inhibition paradigm, *Journal of Neuroscience* **23**: 10402–10410.

Tomkins, S. S. (1995). *Exploring Affect: The Selected Writings of Sylvan S. Tomkins*, Cambridge University Press, New York.

Tovee, M. J. and Rolls, E. T. (1992). Oscillatory activity is not evident in the primate temporal visual cortex with static stimuli, *Neuroreport* **3**: 369–372.

Tovee, M. J. and Rolls, E. T. (1995). Information encoding in short firing rate epochs by single neurons in the primate temporal visual cortex, *Visual Cognition* **2**: 35–58.

Tovee, M. J., Rolls, E. T., Treves, A. and Bellis, R. P. (1993). Information encoding and the responses of single neurons in the primate temporal visual cortex, *Journal of Neurophysiology* **70**: 640–654.

Tovee, M. J., Rolls, E. T. and Azzopardi, P. (1994). Translation invariance and the responses of neurons in the temporal visual cortical areas of primates, *Journal of Neurophysiology* **72**: 1049–1060.

Tranel, D., Bechara, A. and Denburg, N. L. (2002). Asymmetric functional roles of right and left ventromedial prefrontal cortices in social conduct, decision-making and emotional processing, *Cortex* **38**: 589–612.

Treves, A. (1993). Mean-field analysis of neuronal spike dynamics, *Network* **4**: 259–284.

Treves, A. and Rolls, E. T. (1991). What determines the capacity of autoassociative memories in the brain?, *Network* **2**: 371–397.

Treves, A. and Rolls, E. T. (1992). Computational constraints suggest the need for two distinct input systems to the hippocampal CA3 network, *Hippocampus* **2**: 189–199.

Treves, A. and Rolls, E. T. (1994). A computational analysis of the role of the hippocampus in memory, *Hippocampus* **4**: 374–391.

Treves, A., Panzeri, S., Rolls, E. T., Booth, M. and Wakeman, E. A. (1999). Firing rate distributions and efficiency of information transmission of inferior temporal cortex neurons to natural visual stimuli, *Neural Computation* **11**: 601–631.

Trivers, R. (1971). The evolution of reciprocal altruism, *Quarterly Review of Biology* **46**: 35–57.

Trivers, R. (1974). Parent-offspring conflict, *American Zoologist* **14**: 249–264.

Trivers, R. L. (1976). Foreword, *The Selfish Gene by R. Dawkins*, Oxford University Press, Oxford.

Trivers, R. L. (1985). *Social Evolution*, Benjamin, Cummings, CA.

Tversky, A. and Kahneman, D. (1981). The framing of decisions and the psychology of choice, *Science* **211**: 453–458.

Tversky, A. and Kahneman, D. (1986). Rational choice and the framing of decisions, *Journal of Business* **59**: 251–278.

Tversky, A. and Kahneman, D. (1992). Advances in prospect theory – cumulative representation of uncertainty, *Journal of Risk and Uncertainty* **5**: 297–323.

Ungerleider, L. G. (1995). Functional brain imaging studies of cortical mechanisms for memory, *Science* **270**: 769–775.

Ungerleider, L. G. and Haxby, J. V. (1994). 'What' and 'Where' in the human brain, *Current Opinion in Neurobiology* **4**: 157–165.

Uvnas-Moberg, K. (1998). Oxytocin may mediate the benefits of positive social interaction and emotions, *Psychneuroendocrinology* **23**: 819–835.

van den Heuvel, O. A., Veltman, D. J., Groenewegen, H. J., Cath, D. C., van Balkom, A. J., van Hartskamp,

J., Barkhof, F. and van Dyck, R. (2005). Frontal-striatal dysfunction during planning in obsessive-compulsive disorder, *Archives of General Psychiatry* **62**: 301–309.

vandenBerghe, P. L. and Frost, P. (1986). Skin colour preferences, sexual dimorphism and sexual selection: a case for gene culture evolution, *Ethnic and Racial Studies* **9**: 87–113.

Veale, D. M., Sahakian, B. J., Owen, A. M. and Marks, I. M. (1996). Specific cognitive deficits in tests sensitive to frontal lobe dysfunction in obsessive-compulsive disorder, *Psychological Medicine* **26**: 1261–1269.

Veblen, T. (1899). *The Theory of the Leisure Class*, Macmillan, New York.

Verhagen, J. V., Rolls, E. T. and Kadohisa, M. (2003). Neurons in the primate orbitofrontal cortex respond to fat texture independently of viscosity, *Journal of Neurophysiology* **90**: 1514–1525.

Verhagen, J. V., Kadohisa, M. and Rolls, E. T. (2004). The primate insular taste cortex: neuronal representations of the viscosity, fat texture, grittiness, and the taste of foods in the mouth, *Journal of Neurophysiology* **92**: 1685–1699.

Vickers, D. (1979). *Decision Processes in Visual Perception*, Academic Press, New York.

Vickers, D. and Packer, J. (1982). Effects of alternating set for speed or accuracy on response time, accuracy and confidence in a unidimensional discrimination task, *Acta Psychologica* **50**: 179–197.

Voellm, B. A., De Araujo, I. E. T., Cowen, P. J., Rolls, E. T., Kringelbach, M. L., Smith, K. A., Jezzard, P., Heal, R. J. and Matthews, P. M. (2004). Methamphetamine activates reward circuitry in drug naive human subjects, *Neuropsychopharmacology* **29**: 1715–1722.

Waelti, P., Dickinson, A. and Schultz, W. (2001). Dopamine responses comply with basic assumptions of formal learning theory, *Nature* **412**: 43–48.

Wagner, H. (1989). The peripheral physiological differentiation of emotions, *in* H. Wagner and A. Manstead (eds), *Handbook of Social Psychophysiology*, Wiley, Chichester, pp. 77–98.

Walker, M. P. and Stickgold, R. (2006). Sleep, memory, and plasticity, *Annual Review of Psychology* **57**: 139–166.

Wallis, G. and Rolls, E. T. (1997). Invariant face and object recognition in the visual system, *Progress in Neurobiology* **51**: 167–194.

Wang, X. J. (1999). Synaptic basis of cortical persistent activity: the importance of NMDA receptors to working memory, *Journal of Neuroscience* **19**: 9587–9603.

Wang, X. J. (2001). Synaptic reverberation underlying mnemonic persistent activity, *Trends in Neurosciences* **24**: 455–463.

Wang, X. J. (2002). Probabilistic decision making by slow reverberation in cortical circuits, *Neuron* **36**: 955–968.

Wang, X.-J. (2008). Decision making in recurrent neuronal circuits, *Neuron* **60**: 215–234.

Wang, X. J., Tegner, J., Constantinidis, C. and Goldman-Rakic, P. S. (2004). Division of labor among distinct subtypes of inhibitory neurons in a cortical microcircuit of working memory, *Proceedings of the National Academy of Sciences USA* **101**: 1368–1373.

Watkins, L. H., Sahakian, B. J., Robertson, M. M., Veale, D. M., Rogers, R. D., Pickard, K. M., Aitken, M. R. and Robbins, T. W. (2005). Executive function in Tourette's syndrome and obsessive-compulsive disorder, *Psychological Medicine* **35**: 571–582.

Watson, J. B. (1929). *Psychology: From the Standpoint of a Behaviorist*, 3rd edn, Lippincott, Philadelphia.

Watson, J. B. (1930). *Behaviorism: Revised Edition*, University of Chicago Press, Chicago.

Wedell, N., Gage, M. J. and Parker, G. (2002). Sperm competition, male prudence and sperm limited females, *Proceedings of the Royal Society of London B* **260**: 245–249.

Weiskrantz, L. (1968). Emotion, *in* L. Weiskrantz (ed.), *Analysis of Behavioural Change*, Harper and Row, New York, pp. 50–90.

Weiskrantz, L. (1997). *Consciousness Lost and Found*, Oxford University Press, Oxford.

Weiskrantz, L. (1998). *Blindsight*, 2nd edn, Oxford University Press, Oxford.

Weiss, A. P. and Heckers, S. (1999). Neuroimaging of hallucinations: a review of the literature, *Psychiatry Research* **92**: 61–74.

Weissman, M. M., Bland, R. C., Canino, G. J., Greenwald, S., Hwu, H. G., Lee, C. K., Newman, S. C., Oakley-Browne, M. A., Rubio-Stipec, M., Wickramaratne, P. J. et al. (1994). The cross national epidemiology of obsessive compulsive disorder. The Cross National Collaborative Group, *Journal of Clinical Psychiatry* **55 Suppl**: 5–10.

Whiten, A. and Byrne, R. W. (1997). *Machiavellian Intelligence II: Extensions and Evaluations*, Cambridge University Press, Cambridge.

Williams, G. C. (1966). *Adaptation and Natural Selection*, Princeton University Press, Princeton.

Wills, T. J., Lever, C., Cacucci, F., Burgess, N. and O'Keefe, J. (2005). Attractor dynamics in the hippocampal representation of the local environment, *Science* **308**: 873–876.

Wilson, E. O. (1975). *Sociobiology: The New Synthesis*, Harvard University Press, Cambridge, MA.

Wilson, F. A. W. and Rolls, E. T. (1990a). Learning and memory are reflected in the responses of reinforcement-related neurons in the primate basal forebrain, *Journal of Neuroscience* **10**: 1254–1267.

Wilson, F. A. W. and Rolls, E. T. (1990b). Neuronal responses related to reinforcement in the primate basal forebrain, *Brain Research* **509**: 213–231.

Wilson, F. A. W. and Rolls, E. T. (1990c). Neuronal responses related to the novelty and familiarity of visual stimuli in the substantia innominata, diagonal band of broca and periventricular region of the primate, *Experimental Brain Research* **80**: 104–120.

Wilson, H. R. (1999). *Spikes, Decisions and Actions: Dynamical Foundations of Neuroscience*, Oxford University Press, Oxford.

Wilson, M. A. (2002). Hippocampal memory formation, plasticity, and the role of sleep, *Neurobiology of Learning and Memory* **78**: 565–569.

Wilson, M. A. and McNaughton, B. L. (1994). Reactivation of hippocampal ensemble memories during sleep, *Science* **265**: 603–606.

Winkielman, P. and Berridge, K. C. (2003). What is an unconscious emotion?, *Cognition and Emotion* **17**: 181–211.

Winkielman, P. and Berridge, K. C. (2005). Unconscious affective reactions to masked happy versus angry faces influence consumption behavior and judgments of value, *Personality and Social Psychology Bulletin* **31**: 111–135.

Winslow, J. T. and Insel, T. R. (2004). Neuroendrocrine basis of social recognition, *Current Opinion in Neurobiology* **14**: 248–253.

Witter, M. P. (1993). Organization of the entorhinal–hippocampal system: a review of current anatomical data, *Hippocampus* **3**: 33–44.

Wolkin, A., Sanfilipo, M., Wolf, A. P., Angrist, B., Brodie, J. D. and Rotrosen, J. (1992). Negative symptoms and hypofrontality in chronic schizophrenia, *Archives of General Psychiatry* **49**: 959–965.

Woolf, V. (1928). *A Room of One's Own*.

Wrangham, R. W. (1993). The evolution of sexuality in chimpanzees and bonobos, *Human Nature* **4**: 47–49.

Wright, R. (1994). *The Moral Animal: Why We Are the Way We Are: The New Science of Evolutionary Psychology*, Vintage Books, New York.

Wynne-Edwards, K. E. (2001). Hormonal changes in mammalian fathers, *Hormones and Behaviour* **40**: 139–145.

Yanal, R. J. (1991). Hume and others on the paradox of tragedy, *The Journal of Aesthetics and Art Criticism* **49**: 75–76.

Yaxley, S., Rolls, E. T. and Sienkiewicz, Z. J. (1988). The responsiveness of neurones in the insular gustatory cortex of the macaque monkey is independent of hunger, *Physiology and Behavior* **42**: 223–229.

Young, L. J. (2008). Molecular neurobiology of the social brain, *Hormones and Social Behavior* **198**: 57–64.

Zahavi, A. (1975). Mate selection: a selection for a handicap, *Journal of Theoretical Biology* **53**: 205–214.

Zahavi, A. (1978). Decorative patterns and the evolution of art, *New Scientist* **19**: 182–184.

Zeller, A. C. (1987). Communication by sight and smell, *in* B. S. Smuts, D. L. Cheney, R. M. Seyfarth, R. W. Wrangham and T. T. Stuhsaker (eds), *Primate Societies*, University of Chicago Press, London, pp. 433–439.

Zhang, X. and Firestein, S. (2002). The olfactory receptor gene superfamily of the mouse, *Nature Neuroscience* **5**: 124–133.

Zhao, G. Q., Zhang, Y., Hoon, M. A., Chandrashekar, J., Erlenbach, I., Ryba, N. J. and Zucker, C. S. (2003). The receptors for mammalian sweet and umami taste, *Cell* **115**: 255–266.

Zink, C. F., Pagnoni, G., Martin, M. E., Dhamala, M. and Berns, G. S. (2003). Human striatal responses to salient nonrewarding stimuli, *Journal of Neuroscience* **23**: 8092–8097.

Zink, C. F., Pagnoni, G., Martin-Skurski, M. E., Chappelow, J. C. and Berns, G. S. (2004). Human striatal responses to monetary reward depend on saliency, *Neuron* **42**: 509–517.

Index

Appendix 2 Colour Plates

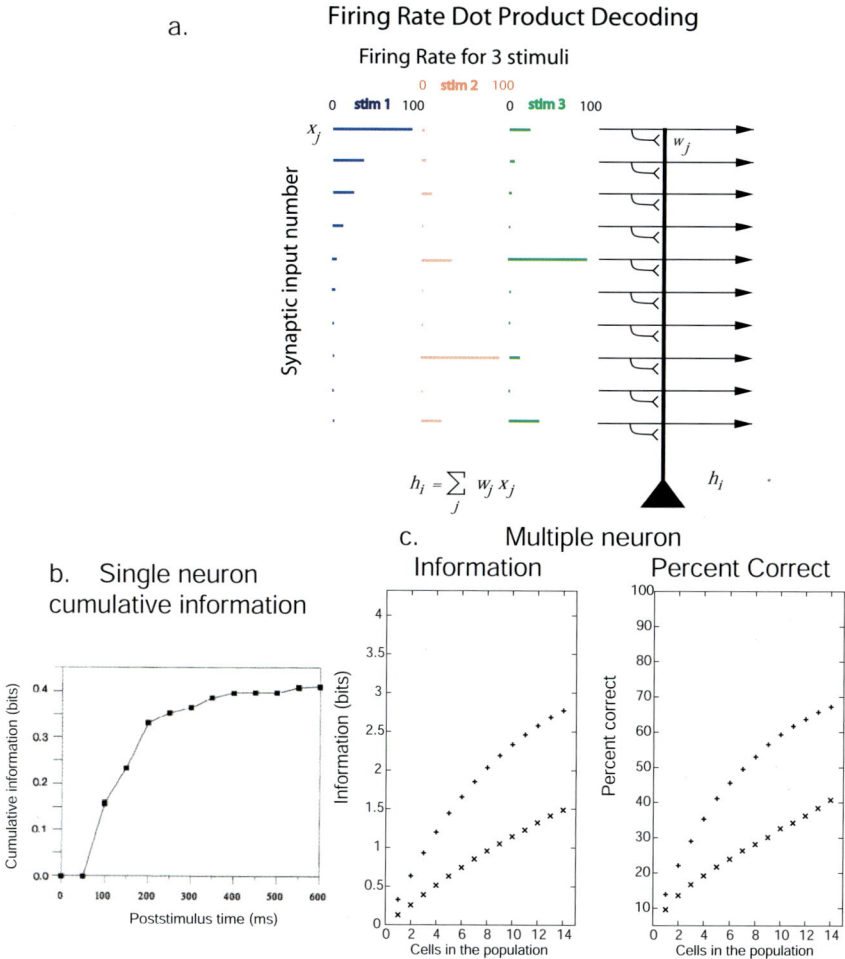

Firing Rate Dot Product Decoding

Firing Rate for 3 stimuli

$$h_i = \sum_j w_j x_j$$

Fig. 2.5 Information encoding by firing rates. (a). Each stimulus is encoded by an approximately exponential firing rate distribution of a population of neurons. The distribution is ordered to show this for stimulus 1, and other stimuli are represented by similar distributions with each neuron tuned independently of the others. The code can be read by a dot (inner) product decoding performed by any receiving neuron of the firing rates with the synaptic weights, and it is the almost linear increase in information with the number of neurons illustrated in (c) which shows that the tuning profiles to the set of stimuli are almost independent. (b). Much information is available from single neurons, and from populations of neurons, using the number of spikes in short time windows of e.g. 50 ms. This is shown by the cumulative information from the firing rates for different time periods from single neurons in the cortex in the superior temporal sulcus to a set of 10 face and non-face stimuli, and is remarkable in that the neurons do not start to respond to the visual stimuli until 80–90 ms, so that the data point at 100 ms shows the information from 20 ms or less of firing. (c) The information available about which of 20 faces had been seen that is available from the responses measured by the firing rates in a time period of 500 ms (+) or a shorter time period of 50 ms (x) of different numbers of temporal cortex cells. The corresponding percentage correct from different numbers of cells is also shown. Decoding with dot product decoding reveals similar principles.

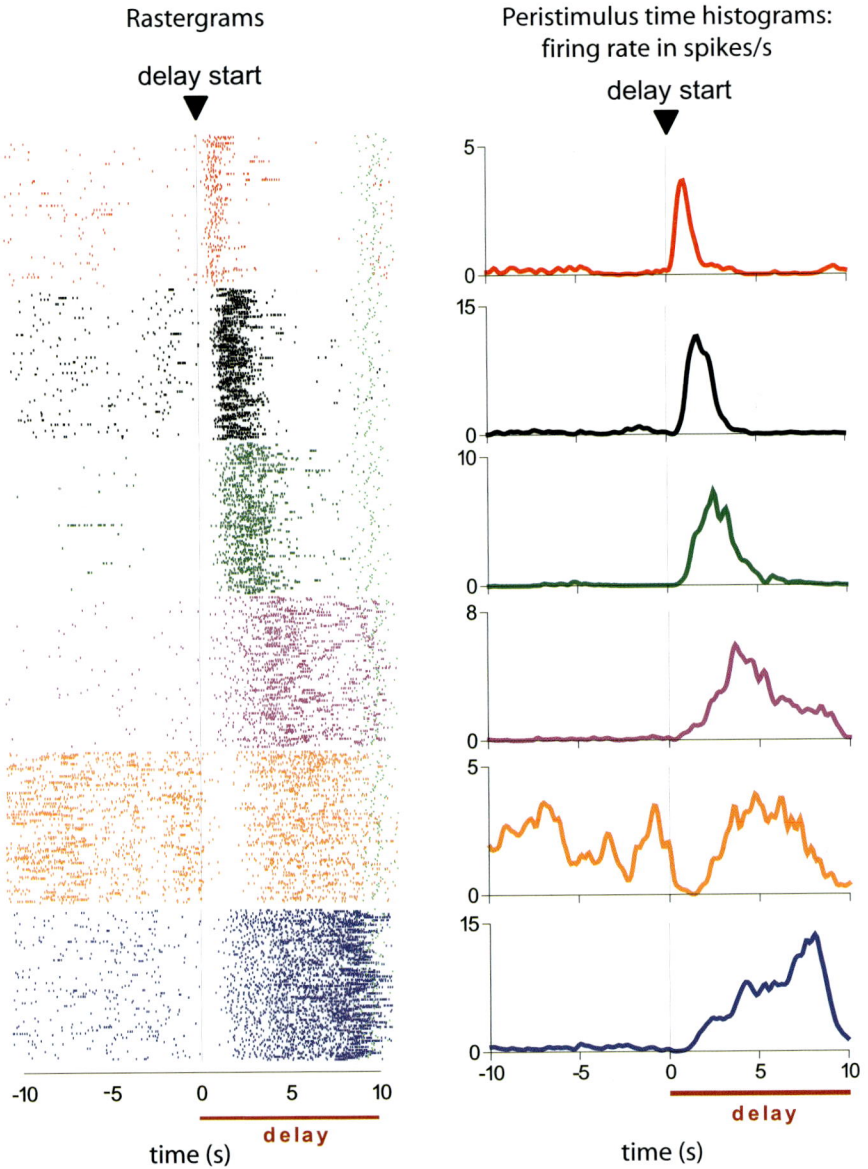

Fig. 2.12 Time encoding neurons. The activity of 6 simultaneously recorded hippocampal neurons each of which fires with a different time course during the 10 s delay in a visual object-delay-odor paired associate task. Each peristimulus time histogram (right, showing the average firing rate in spikes/s as a function of time) and set of rastergrams (left) is for a different neuron. In the rastergrams, each row is a separate trial, and each dot shows the spike time of a single neuron. The onset of the delay is shown. The visual stimulus was shown before the delay period; and the odor stimulus was delivered after the delay period. (After MacDonald and Eichenbaum 2009 with permission.)

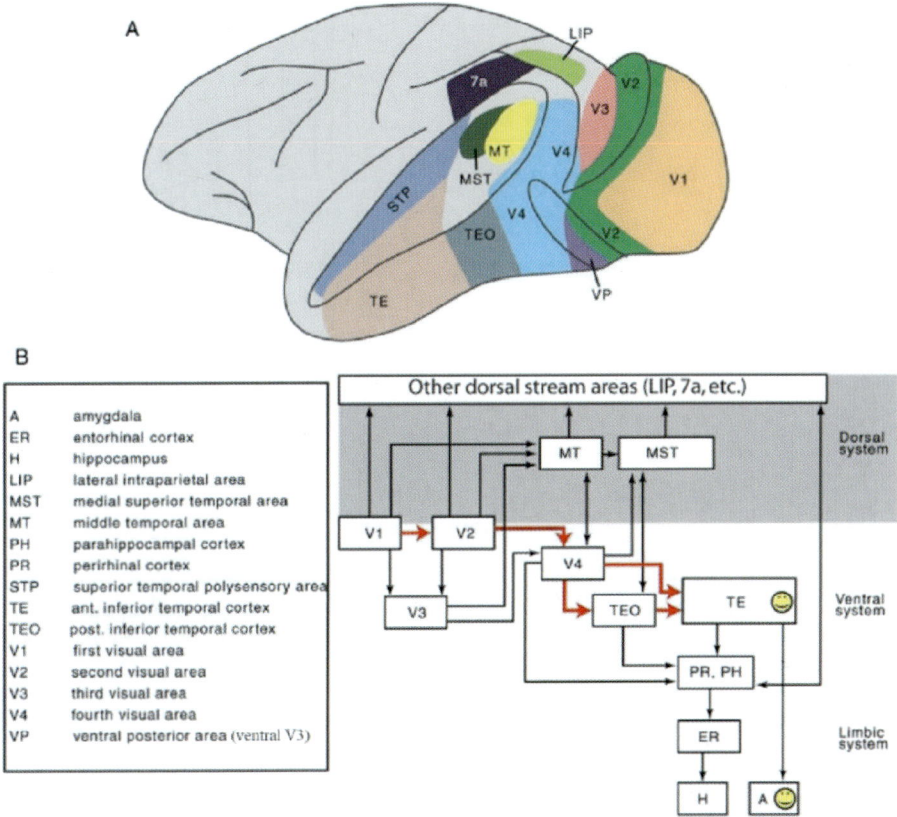

Fig. 2.20 (A). A lateral view of the monkey brain illustrating the multiplicity of functional areas within both processing streams. (B). Some of the pertinent connections of the inferior temporal cortex with other cortical areas and medial temporal lobe structures. Red lines indicate the main afferent pathway to area TE, which includes areas V1, V2, V4, and TEO. For simplicity, only projections from lower order to higher order areas are shown, but each of these feed-forward projections is reciprocated by a feedback projection. Faces indicate areas in which neurons selectively responsive to faces have been found. (Adapted from Gross et al. 1993.)

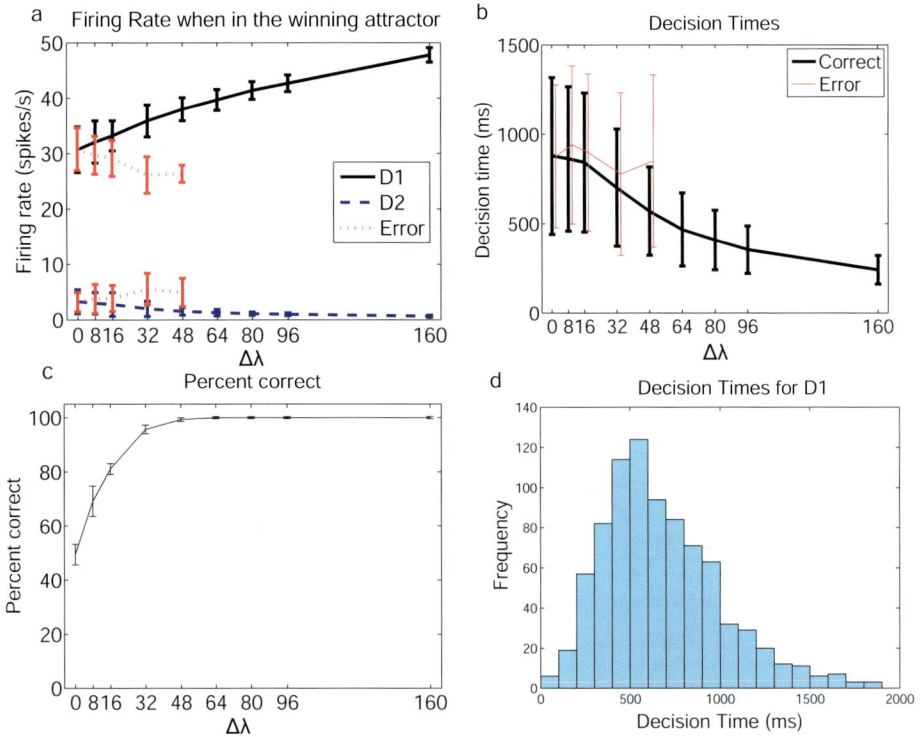

Fig. 2.27 (a) Firing rates (mean ± sd) on correct trials when in the D1 attractor as a function of ΔI. $\Delta\lambda=0$ corresponds to difficult, and $\Delta\lambda=160$ spikes/s corresponds to easy. The firing rates on correct trials for the winning population D1 are shown by solid lines, and for the losing population D2 by dashed lines. All the results are for 1000 simulation trials for each parameter value, and all the results shown are statistically highly significant. The results on error trials are shown by the dotted lines, and in this case the D2 attractor wins, and the D1 attractor loses the competition. (b) Decision or reaction times (mean ± sd) for the D1 population to win on correct trials as a function of the difference in inputs $\Delta\lambda$ to D1 and D2. (c) Per cent correct performance, i.e. the percentage of trials on which the D1 population won, as a function of the difference in inputs $\Delta\lambda$ to D1 and D2. (d) The distribution of decision times for the model for $\Delta\lambda=32$ illustrating the long tail of slow responses. Decision times are shown for 837 correct trials, the level of performance was 95.7% correct, and the mean decision time was 701 ms. (After Rolls, Grabenhorst and Deco, 2010c.)

Fig. 2.30 The Rubin vase is on the right. Noise in the brain influences the transitions from seeing this as a vase, or as two faces in profile looking at each other.

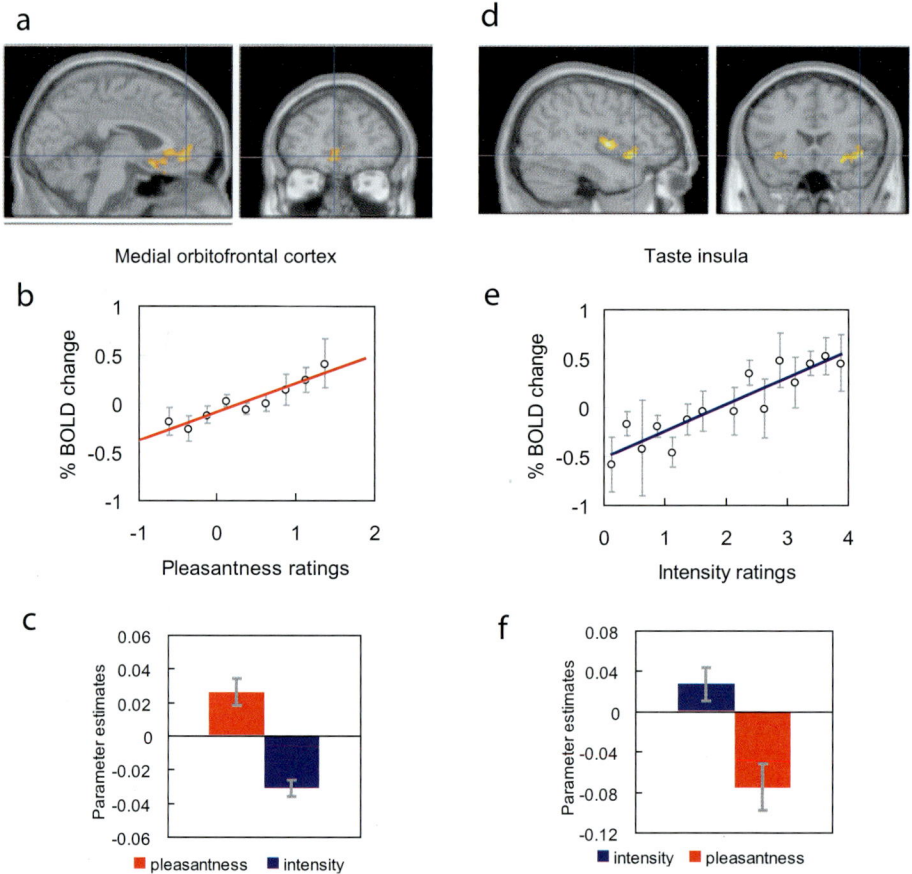

Fig. 3.5 (a-c). Human subjective pleasure related to activations in the medial orbitofrontal cortex and pregenual cingulate cortex. (a). Brain scans with a midline view (left) and coronal (cross-section) view on the right showing the activations in colour. The cursor is located in the pregenual cingulate cortex, and in the midline view the activation can be seen extending back to the medial orbitofrontal cortex. (b). The fMRI BOLD signal is directly proportional in both these brain regions to the subjective pleasantness ratings of the taste. (c). When the humans pay attention to the pleasantness of the taste, there are larger activations in these regions (reflected in the parameter estimates) than when the participants pay attention to the intensity of the identical taste. (d-f). Human subjective intensity ratings are related to activations in Tier 1 structures such as the primary taste cortex in the insula. (d). Brain scans with a parasaggittal (parallel to the midline) view (left) and coronal (cross-section) view on the right showing the activations in colour. The cursor is located in the primary taste cortex in the anterior insula. (e). The fMRI BOLD signal is directly proportional in the taste insula to the subjective intensity ratings of the taste. (f). When the humans pay attention to the intensity of the taste, there are larger activations in the primary taste cortex (reflected in the parameter estimates) than when the participants pay attention to the pleasantness of the identical taste. (After Grabenhorst, Rolls and Parris 2008b.)

Fig. 7.1 Two Forms (January) 1967. Barbara Hepworth.

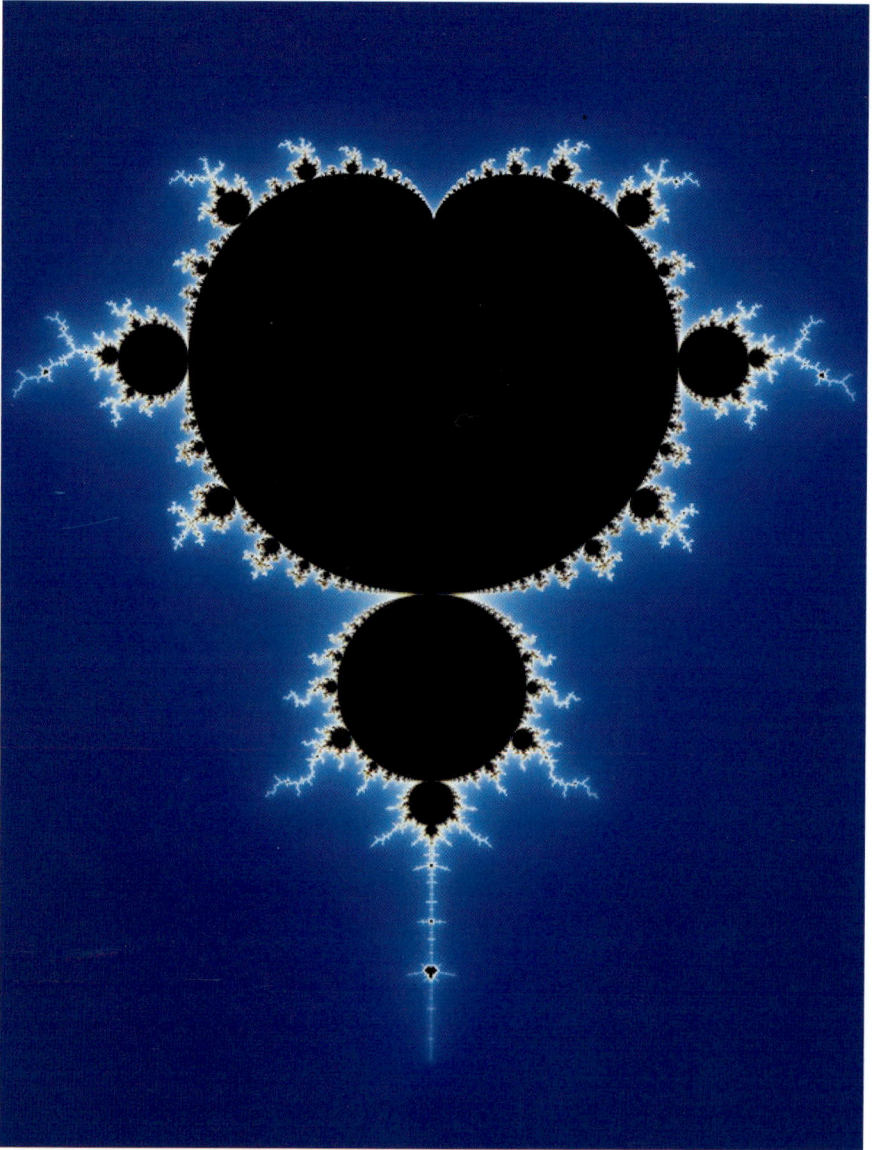

Fig. 7.2 Fractal image. A fractal is a rough or fragmented geometric shape that can be split into parts, each of which is (at least approximately) a reduced-size copy of the whole, a property called self-similarity. This example is of the Mandelbrot set. Fractal images can be produced by mathematical algorithms.